Temperate Woodland Conservation and Management

David Lindenmayer, Andrew Bennett and Richard Hobbs

CSIRO
PUBLISHING

National Library of Australia Cataloguing-in-Publication entry

Temperate woodland conservation and management/edited by David Lindenmayer, Andrew Bennett and Richard Hobbs.

9780643100374 (pbk.)

Forest and forestry – Australia.
Forest management – Australia.
Forest conservation – Australia.

Lindenmayer, David.
Bennett, Andrew (Andrew F.)
Hobbs, R. J. (Richard J.)

333.75170994

Published by
CSIRO PUBLISHING
36 Gardiner Road, Clayton VIC 3168
Private Bag 10, Clayton South VIC 3169
Australia

Telephone: [+613] 9545 8555
Local call: 1300 788 000 (Australia only)
Fax: +61 3 9662 7555
Email: csiropublishing@csiro.au
Web site: www.publishing.csiro.au

Front cover: photo by Esther Beaton
Back cover: photo by Jaime Plaza

Set in 10/12 Adobe Minion Pro and ITC Stone Sans
Edited by Janet Walker
Cover and text design by James Kelly
Typeset by Desktop Concepts Pty Ltd, Melbourne
Index by Russell Brooks
Printed by Ingram Lightning Source

CSIRO PUBLISHING publishes and distributes scientific, technical and health science books and journals from Australia to a worldwide audience and conducts these activities autonomously from the research activities of the Commonwealth Scientific and Industrial Research Organisation (CSIRO).

Feb26_RP_ILS

Contents

List of contributors

Joel R. Allan
School of Earth and Environmental Sciences, University of Adelaide, Adelaide, SA 5005. Email: joel.allan@adelaide.edu.au

Simon Attwood
Biodiversity Conservation Branch, Department of Environment, Water, Heritage and the Arts, GPO Box 787, Canberra, ACT 2601. Email: simon.attwood@environment.gov.au

Andrew F. Bennett
School of Life and Environmental Sciences, Deakin University, 221 Burwood Highway, Burwood, VIC 3125. Email: andrew.bennett@deakin.edu.au

John Benson
Royal Botanic Gardens and Domain Trust, Sydney, NSW 2000. Email: john.benson@rbgsyd.nsw.gov.au

Michiala E. Bowen
Landscape Ecology and Conservation Group, Centre for Spatial Environmental Research, School of Geography, Planning and Environmental Management, University of Queensland, St Lucia, QLD 4072. Email: michiala.bowen@uq.edu.au

Keith Bradby
Director, Gondwana Link Ltd, PO Box 5276, Albany, WA 6332. Email: bradby@gondwanalink.org

Geoff W. Brown
Arthur Rylah Institute for Environmental Research, Department of Sustainability and Environment, 123 Brown Street, Heidelberg, VIC 3084. Email: geoff.brown@dse.vic.gov.au

Emma Burns
Biodiversity Conservation Branch, Department of Environment, Water, Heritage and the Arts, GPO Box 787, Canberra, ACT 2601. Email: emma.burns@environment.gov.au

David Carr
Greening Australia Ltd, PO Box 74, Yarralumla, ACT 2600. Email: dbcarr@bigpond.com

Michael F. Clarke
Department of Zoology, La Trobe University, Bundoora, VIC 3086. Email: m.clarke@latrobe.edu.au

Mason Crane
Fenner School of Environment and Society, Australian National University, Canberra, ACT 0200. Email: mason.crane@anu.edu.au

William E. Davis, Jr.
Boston University, c/23 Knollwood Drive, East Falmouth, MA 02536, USA

Jim Donaldson
Formerly of Land & Water Australia. Email: donaldson@netspeed.com.au

Josh Dorrough
Visiting Research Scientist, CSIRO Sustainable Ecosystems, Canberra. Email: josh.dorrough@csiro.au

Don A. Driscoll
Fenner School of Environment and Society, Australian National University, Canberra, ACT 0200. Email: don.driscoll@anu.edu.au

Teresa J. Eyre
Biodiversity and Ecological Sciences Unit, Queensland Department of Environment and Resource Management, Forestry Building, 80 Meiers Road, Indooroopilly, QLD 4068. Email: teresa.eyre@derm.qld.gov.au

Graham Fifield
Greening Australia Capital Region, PO Box 538, Jamison Centre, ACT 2615. Email: graham.fifield@act.greeningaustralia.org.au

Joern Fischer
Fenner School of Environment and Society, Australian National University, Canberra, ACT 0200. Email: joern.fischer@anu.edu.au

Hugh A. Ford
Department of Zoology, University of New England, Armidale, NSW 2351. Email: hford@une.edu.au

David Freudenberger
Greening Australia Ltd, PO Box 74, Yarralumla, ACT 2600. Email: dfreudenberger@greeningaustralia.org.au

Philip Gibbons
Fenner School of Environment and Society, Australian National University, Canberra, ACT 0200. Email: philip.gibbons@anu.edu.au

Louise Gilfedder
Resource Management & Conservation Division, Department of Primary Industries, Parks, Water and Environment, PO Box 44, Hobart, TAS 7001. Email: louise.gilfedder@dpipwe.tas.gov.au

Merilyn J. Grey
Department of Zoology, La Trobe University, Bundoora, VIC 3086. Email: m.grey@latrobe.edu.au

Richard Hobbs
School of Plant Biology, University of Western Australia, 35 Stirling Highway, Crawley, WA 6009. Email: richard.hobbs@uwa.edu.au

Jamie B. Kirkpatrick
School of Geography and Environmental Studies, University of Tasmania, Private Bag 78, GPO Hobart, TAS 7001. Email: j.kirkpatrick@utas.edu.au

Ted Lefroy
Centre for Environment, University of Tasmania, Private Bag 55, Hobart, TAS 7001. Email: ted.lefroy@utas.edu.au

David Lindenmayer
Fenner School of Environment and Society, Australian National University, Canberra, ACT 0200. Email: david.lindenmayer@anu.edu.au

Linda F. Lumsden
Arthur Rylah Institute for Environmental Research, Department of Sustainability and Environment, PO Box 137, Heidelberg, VIC 3084; School of Life and Environmental Sciences, Deakin University, 221 Burwood Highway, Burwood, VIC 3125. Email: lindy.lumsden@dse.vic.gov.au

Ian D. Lunt
Institute for Land, Water & Society, Charles Sturt University, PO Box 789, Albury, NSW 2640. Email: ilunt@csu.edu.au

Anita Lyons
Centre for Ecosystem Diversity and Dynamics in the Department of Environmental Biology, Curtin University of Technology, PO Box U1987, Perth, WA 6845; Australian National Insect Collection, CSIRO Entomology, GPO Box 1700, Canberra, ACT 2601. Email: anita.lyons@csiro.au

Christopher MacGregor
Fenner School of Environment and Society, Australian National University, Canberra, ACT 0200. Email: christopher.macgregor@environment.gov.au

Ralph Mac Nally
Australian Centre for Biodiversity, School of Biological Sciences, Monash University, Melbourne, VIC 3800. Email: ralph.macnally@sci.monash.edu.au

Jonathan D. Majer
Centre for Ecosystem Diversity and Dynamics in the Department of Environmental Biology, Curtin University of Technology, PO Box U1987, Perth, WA 6845. Email: j.majer@curtin.edu.au

Richard E. Major
Terrestrial Ecology, Australian Museum, 6 College St, Sydney, NSW 2010. Email: richard.major@austmus.gov.au

Adrian D. Manning
Fenner School of Environment and Society, Australian National University, Canberra, ACT 0200. Email: adrian.manning@anu.edu.au

Martine Maron
Landscape Ecology and Conservation Group, Centre for Spatial Environmental Research, School of Geography, Planning and Environmental Management, University of Queensland, St Lucia, QLD 4072. Email: m.maron@uq.edu.au

Tara G. Martin
CSIRO Sustainable Ecosystems, 306 Carmody Rd, St Lucia, QLD 4067. Email: tara.martin@csiro.au

Clive A. McAlpine
Landscape Ecology and Conservation Group, Centre for Spatial Environmental Research, School of Geography, Planning and Environmental Management, University of Queensland, St Lucia, QLD 4072. Email: c.mcalpine@uq.edu.au

Sue McIntyre
CSIRO Sustainable Ecosystems, PO Box 284, Canberra, ACT 2601. Email: sue.mcintyre@csiro.au

Damian Michael
Fenner School of Environment and Society, Australian National University, Canberra, ACT 0200. Email: michaeldamian@hotmail.com

Rebecca Montague-Drake
Fenner School of Environment and Society, Australian National University, Canberra, ACT 0200. Email: becmd@westnet.com.au

Geoff Park
North Central Catchment Management Authority, PO Box 18, Huntly, VIC 3551. Email: geoff.park@nccma.vic.gov.au

David C. Paton
School of Earth and Environmental Sciences, University of Adelaide, Adelaide, SA 5005. Email: david.paton@adelaide.edu.au

Stuart Pearson
Australian Defence Force Academy, School of Physical, Environmental and Mathematical Sciences, University of New South Wales, Canberra, ACT 2600. Email: stuart.pearson@adfa.edu.au

Melinda Pickup
Greening Australia Capital Region, PO Box 538, Jamison Centre, ACT 2615. Email: mpickup@act.greeningaustralia.org.au

Suzanne M. Prober
CSIRO Sustainable Ecosystems, Wembley, WA. Email: suzanne.prober@csiro.au

James Q. Radford
School of Life and Environmental Sciences, Deakin University, 221 Burwood Highway, Burwood, VIC 3125; Bush Heritage Australia, PO Box 329, Flinders Lane, Melbourne, VIC 8009. Email: jradford@bushheritage.org.au

Harry F. Recher
School of Natural Sciences, Edith Cowan University, Joondalup, WA 6029; The Australian Museum, 6-8 College Street, Sydney, NSW 2000. Email: hjrecher@bigpond.com

Jonathan R. Rhodes
Landscape Ecology and Conservation Group, Centre for Spatial Environmental Research, School of Geography, Planning and Environmental Management, and The Ecology Centre, University of Queensland, St Lucia, QLD 4072. Email: j.rhodes@uq.edu.au

Doug Robinson
Trust for Nature, 2/385 Little Lonsdale Street, Melbourne, VIC 3000. Email: dougr@tfn.org.au

Jim Robinson
Greening Australia Victoria, PO Box 525, Heidelberg, VIC 3084. Email: jrobinson@gavic.org.au

Daniel J. Rogers
Department for Environment and Heritage, South Australian Government, GPO Box 1047, Adelaide, SA 5001. Email: daniel.rogers@sa.gov.au

Peter G. Spooner
Institute for Land, Water & Society, Charles Sturt University, Albury, NSW 2640. Email: pspooner@csu.edu.au

Sue Streatfield
Greening Australia Capital Region, PO Box 538, Jamison Centre, ACT 2615. Email: sue.streatfield@act.greeningaustralia.org.au

Richard Thackway
Land and Forest Sciences Program, Bureau of Rural Sciences, GPO Box 858, Canberra, ACT 2601. Email: richard.thackway@brs.gov.au

Kevin R. Thiele
Western Australian Herbarium, Department of Environment & Conservation, Kensington, WA 6983. Email: kevin.thiele@dec.wa.gov.au

Rodney van der Ree
Australian Research Centre for Urban Ecology, Royal Botanic Gardens, Melbourne, C/- School of Botany, University of Melbourne, VIC 3010. Email: rvdr@unimelb.edu.au

Matthew J. Ward
Department for Environment and Heritage, South Australian Government, GPO Box 1047, Adelaide, SA 5001. Email: matthew.ward@sa.gov.au

David M. Watson
Institute for Land, Water and Society, Charles Sturt University, PO Box 789, Albury, NSW 2640. Email: dwatson@csu.edu.au

Andrew West
Adelaide & Mt Lofty Ranges NRM Board, 205 Greenhill Road, Eastwood, SA 5063. Email: andrew.west@adelaide.nrm.sa.gov.au

Jann Williams
NRM Insights Pty Ltd and University of Tasmania, PO Box 3263 UMDC, Ulverstone, TAS 7315. Email: jannw@bigpond.net.au

Nigel Willoughby
Department for Environment and Heritage, South Australian Government, GPO Box 1047, Adelaide, SA 5001. Email: nigel.willoughby@sa.gov.au

Charlie Zammit
Biodiversity Conservation Branch, Department of Environment, Water, Heritage and the Arts, GPO Box 787, Canberra, ACT 2601. Email: charlie.zammit@environment.gov.au

1

A DECADE OF RESEARCH AND MANAGEMENT INSIGHTS IN AUSTRALIA'S TEMPERATE WOODLANDS

David Lindenmayer, Andrew Bennett and Richard Hobbs

Introduction

For much of the past 50 years, tall wet eucalypt forests and rainforests have been the focus of many of Australia's ecological researchers and environmental policy makers. Over the past 10–20 years, however, there has been a shift in emphasis away from wet forests toward the ecology, management and conservation of Australia's temperate woodland environments and the biota they support. Concurrently there has been a body of work about ecologically sustainable agriculture and its integration with other environmental values like the conservation of woodland biota and the preservation of native grassland ecosystems. Arising from this change in focus were seminal books on landscape change and habitat fragmentation (e.g. Saunders *et al.* 1987; Saunders and Hobbs 1991) and a decade ago a book was published including specifically targeted research and management of temperate woodlands (Hobbs and Yates 2000). Since these earlier contributions, there has been considerable new research on temperate woodlands as well as major changes in natural resource management practice and policy. How far have we come in that time? What can we say that is new and important? What are the key knowledge gaps that need to be closed?

This book is, in part, an attempt to answer these (and other) questions by presenting some of the latest thinking in temperate woodland ecology, woodland management, and woodland policy. It is a compilation of opinion pieces by leading thinkers in temperate woodland ecology, management practice and policy making. Some have worked a long time in temperate woodlands whereas others have more recently joined the woodland scientific or policy fraternities. There are authors from universities, State and Federal Government agencies, regional catchment management bodies, non-government organisations and private landholders. The scope of the presentations is as broad as the range of organisations represented by the authors. Indeed, the chapters in the book cover work completed or ongoing in all of the nation's major jurisdictions, except the Northern Territory (where there is no temperate woodland). The map on the next page shows the location of the work that features in different chapters. There are also national overview chapters from a number of authors. The book includes contributions from plant ecologists, animal ecologists, natural resource managers and policy makers. Some chapter authors have chosen a single species focus whereas others concentrate on particular

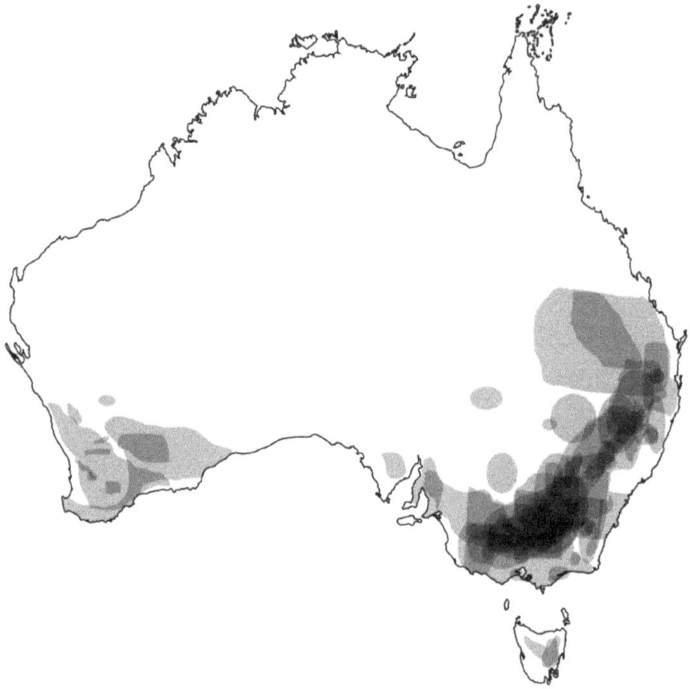

assemblages (e.g. plants, invertebrates or birds), or particular ecological processes in woodland ecosystems (e.g. the regeneration of native overstorey eucalypt trees).

The structure of the chapters

Each chapter in this book has been written in a way that makes new ecological insights, suggestions for altered management practices, and recommendations for new policy directives readily accessible to a wide audience with interests in sustainable agriculture, woodland conservation and management and the conservation of biodiversity. To do this, we asked the author/s of each chapter to outline their key perspectives on woodland research, woodland management or woodland policy. Each chapter had been written according to a broadly similar structure that encompassed:

- a box summarising each of the key lessons;
- an introductory paragraph outlining the background history and broad location to the work on which the lessons was based;
- a paragraph or two describing each lesson, the reasons why that lesson is important, and the underlying rationale for the lesson;
- a concluding paragraph (or two) including where there are key knowledge gaps that need to be filled; and
- a small set of references to direct readers to additional scientific or policy-related material.

The specific instructions to authors were to keep chapters as short as possible (less than 3500 words) and written in a way that was accessible to a wide audience. The resulting chapters vary in style – and we have elected not to edit them heavily but rather leave them as opinion pieces with strong points of view – and sometimes there is general agreement among chapters and at other times not.

We were initially concerned about a risk of substantial overlap and redundancy between chapters. This concern proved to be unfounded. Rather, the breadth of material and new insights is striking and makes for truly fascinating reading. The contributions are not entirely idiosyncratic, however, and there are some important common themes that emerge from the body of work that is presented. We discuss these in some detail in the synthesis chapter at the end of this book. We leave it to our readers to judge whether our approach has been successful or not.

Chapter arrangement

We have arranged the chapters in five parts. The primary focus of the first is on national level woodland research. The second part is largely concerned with research at the State level. The order of the chapters runs from west to east of Australia; that is, chapters from Western Australia appear first, then South Australia, Victoria, Tasmania, the Australian Capital Territory, New South Wales, Queensland. The third part of the book is about national level management and policy making. It is followed by a section (Part D) which focuses on state-level management and policy making. The order of the chapters in that section is broadly the same as in the second part; viz: west to east. We readily admit that our distinction in the separate parts between research and management/policy chapters is somewhat arbitrary. This is because of large number of authors stressed the importance of the interface of science, management and policy with a focus on the challenge of making change happen. Such discussions in some chapters include not only how things get adopted but also an assessment of the characteristics which unify those changes that are actually adopted on the ground or in policy. The fifth and final part has just one chapter. This is our synthesis which draws together some overview perspectives based on the collective material in the first two sections of the book.

Finally, this book eventuated because of a fellowship awarded by the now sadly defunct organisation, Land & Water Australia. The demise of that body is a much-regretted loss – for woodland conservation and management, for biodiversity conservation, for ecologically sustainable land management, and for many other reasons. Indeed, it is notable that a significant number of the chapters in this book are based on work that was either wholly or in part, supported by Land and Water Australia. It is our hope that an informed future government might create a new and even better organisation to promote the conservation and management of this nation's vegetation, its biodiversity and its precious environmental resources.

References

Hobbs, R.J. and Yates, C.J., eds. (2000). *Temperate Eucalypt Woodlands in Australia: Biology, Conservation, Management and Restoration*. Surrey Beatty and Sons: Chipping Norton.

Saunders, D.A., Arnold, G.W., Burbidge, A.A. and Hopkins, A.J., eds. (1987). *Nature Conservation: The Role of Remnants of Native Vegetation*. Surrey Beatty and Sons: Chipping Norton.

Saunders, D.A. and Hobbs, R.J., eds. (1991). *Nature Conservation 2: The Role of Corridors*. Surrey Beatty and Sons: Chipping Norton.

PART A
Research: national

2

WOODLAND RESTORATION

David Carr, Jim Robinson and David Freudenberger

1. Build soil moisture before planting.
2. Eliminate weed competition for at least two years after planting.
3. Plan seed supply well in advance.
4. Select the right species and provenance.
5. Only use high quality seed.
6. Plant in the correct season to minimise heat stress and water deficit.
7. Only use healthy seedlings that meet specifications.
8. Do not plant if ground preparation is inadequate.
9. Protect seedlings from browsing and grazing until they are tall enough to resist damage.
10. Soil nutrients, particularly nitrogen and phosphorous, should be depleted before planting to resist re-invasion by exotic weeds.
11. Continue to monitor the site and adjust management according to what you see.
12. Minimise disturbance in native vegetation in good condition.
13. We still have a lot to learn.

Introduction

In fragmented landscapes, native vegetation is reduced to small isolated patches, often with reduced species and structural diversity. These patches can be managed to improve their diversity and to stop further degradation through a range of techniques including grazing regulation and weed, fire and feral animal control. Healthy and sustainable animal populations, however, depend on the ability to move through the landscape to seek food, mates and shelter and to escape predators and catastrophic events. Even plants 'move' through the landscape as pollen and seeds disperse. Depending on how a species 'perceives' the landscape (Manning *et al.* 2004), fragmentation can hinder or prevent this movement.

The conservation of ecosystems and their functions depends on three broad actions:

1. Protect – we should firstly secure and protect what we have.
2. Repair – improve the condition of existing patches.
3. Restore – enlarge existing patches, increase connectivity between patches and create new patches.

Restoration of native vegetation is a method used to increase the functional connectivity of the landscape and to increase the habitat available to wildlife. Restoration can also be used to buffer existing plant communities against weather, feral animals and weed invasion. Restoration is usually achieved through seedling transplant, direct seeding or assisted natural regeneration particularly on sites where the existing vegetation has been largely removed and the soil seedbank has been depleted. More detailed information on these techniques can be found in Corr (2003) or on the Florabank website (Carr *et al.* 2008).

This chapter discusses lessons about successful restoration works at the patch scale. It does not discuss the important issue of designing restoration projects, nor where restoration should occur in the landscape. These lessons for restoration are a checklist for planning on-ground works, as they apply to all woodland restoration projects, wherever they occur.

These lessons emerge from the collective field experience of Greening Australia over 25 years and from observations of the restoration successes and failures of the many landholders and organisations we have worked with in this time. Restoration research is uncommon in Australia, but has provided some valuable information to improve restoration practice. This experience has been gained from revegetation in all areas of Australia where woodlands occur.

Lessons

1. Build soil moisture before planting

Access to moisture throughout the soil profile is essential to the survival and growth of newly germinated seeds or newly transplanted seedlings. Soil moisture must be built up in the 12–24 months prior to planting by removing plants that will use this moisture and by minimising soil disturbance that will increase evaporation. The usual method for achieving this is to use herbicides to maintain a weed-free soil surface after initial soil preparation. If the upper soil horizon is not moist at the time of planting, seedlings should be watered.

2. Eliminate weed competition for at least two years after planting

Weeds compete with newly established plants for moisture. Many weeds are much more efficient at drawing moisture from the soil than new seedlings, so reduced growth or plant death usually results. Plants which struggle in the first few years of life never reach their full potential or growth rate. Weeds should be controlled in an area one metre in diameter around new plants for at least two years to ensure the plants are vigorous enough to out-compete weeds.

3. Plan seed supply well in advance

In many regions throughout Australia, seed is not readily available for the full suite of species suitable for woodland restoration. The poor financial return to seed collectors for difficult-to-collect species, sporadic seeding in some species, and a small and inconsistent market for seed is responsible for the poor diversity of seed available. Most of the species in the groundcover layer of grassy woodlands are unlikely to be readily available from seed suppliers or nurseries. Allow plenty of time for these species to be collected, germinated and grown to a size suitable for transplanting into the field. Be prepared to engage seed collectors at least two years before the plants are required to allow sufficient time for locating populations, collecting seed and growing in the nursery.

In most regions where woodlands occur, the skills to collect, germinate and grow woodland grasses and herbs are rare or do not exist, so a program of training and innovation will be needed for existing seed collectors and nurseries. Landcare groups or regional NRM bodies will need to identify priority species for restoration projects, then identify, and if necessary secure, high quality populations of these species for continuing seed supply.

4. Select the right species and provenance

Choose species that occur locally or in similar soil-landscape environments. Where possible, match the relative abundance of species found in native vegetation on similar sites. It is not realistic to restore all species that occur on a site because of limitations of seed collection, propagation and establishment. Look for those species which are structurally or functionally important. These are not always trees. For example, they may be species that provide food for pollinators throughout the year or that help reduce the invasion by weeds.

Seed for the plants should come from large populations growing on similar sites within the same bioregion (Commonwealth of Australia 2005), so they are adapted to the environment of the planting site (Broadhurst *et al.* 2009). Rainfall, soil, altitude, aspect and slope position are important environmental factors to guide the sourcing of seed.

An exception to this principle is where the environment has changed to the extent that some local species can no longer survive; for example, from secondary salinity, altered soil structure, frost, increased aridity or waterlogging. In these instances, a 'nursery' planting may first be required to reduce wind exposure, water logging, salinity or other factors hostile to restoration of locally native species.

5. Only use high quality seed

Using seed of poor physical or genetic quality will result in poor germination, poor growth, lower resistance to diseases and pests, an inability to adapt to climate change and environmental shocks, an inability to regenerate and reduced genetic diversity (Broadhurst 2007). Quality seed will greatly increase the chances that long-term restoration goals can be met, as well as short-term budget and plant survival goals. Seed supply should be planned well in advance of planting to ensure sufficient quantity, diversity and quality can be gathered. This applies equally to seed for both direct seeding and for nursery propagation. High temperatures and humidity, and pests and diseases are the main reasons seeds rapidly lose viability making proper storage practices critical (Ellis and Roberts 1980).

6. Plant in the correct season to minimise heat stress and water deficit

Planting season varies across Australia; however, planting season will depend on a balance between maximising stored moisture, minimum germination temperatures, post-planting air temperature, rainfall and humidity. Project planners should have a clear idea of the optimal and acceptable planting windows and the periods of the year when planting should not be undertaken.

7. Only use healthy seedlings that meet specifications

The size and root structure of seedlings is critical to their survival after transplanting and to their long-term survival. Poor root development in the nursery can result in early death, slow growth, instability in windy conditions and long-term self-strangulation (Thomas *et al.* 2008). Seedlings that are not hardened properly will be set back or killed when transplanted into hot

Figure 2.1 Choose healthy seedlings for woodland restoration projects. (Photo: Dave Carr)

or frosty conditions. Nursery containers should be chosen to produce healthy roots with strong growing tips and a low shoot-to-root ratio (Figure 2.1). Nursery systems should allow for maximum air pruning of roots. Seedlings should also be ordered or germinated with enough growing time to meet specifications. In particular, root development should be sufficient so that potting media does not collapse on the removal of a seedling from its container.

8. Do not plant if ground preparation is inadequate

Ground preparation techniques are used to soften the soil and increase water infiltration, root area and soil aeration. Ground preparation will vary according to soil type, site history and establishment technique being used. Many plantings will fail if the recommended ground preparation is not followed.

Ripping on contours when the subsoil is dry is suggested for compacted soils including cropping pans and to allow for easy use of manual planting tools. It is not recommended for deep sands or cracking clay soils. Mounding of non-sandy soils may be useful to maximise establishment and growth rates in waterlogged, heavy or saline soils.

Soil treatments such as water crystals, gypsum, compost, coconut fibre can be added to improve the structure and water-holding capacity of the soil, but they add considerable cost to planting projects. A thin narrow strip of coarse sand has been used successfully as mulch for surface-sown seeds. Inoculants of soil biota can be added to both direct seeding and seedlings to increase the nitrogen-fixing ability of *Acacia* and some pea species. Wattle-Grow is a commercial *Bradyrhizobium* inoculant developed by CSIRO that is available as a powder.

9. Protect seedlings from browsing and grazing until they are tall enough to resist damage

Young seedlings are palatable to a wide range of mammal, bird and insect predators. Browsing or grazing can kill seedlings or delay their growth substantially. Continued defoliation is likely to affect growth permanently. Potential grazers and browsers should be identified before planting and controlled or managed through guarding, fencing, shooting, poisoning, warren ripping or other appropriate measures.

10. Soil nutrients, particularly nitrogen and phosphorous, should be depleted before planting to resist re-invasion by exotic weeds

Many weeds are favoured by high soil fertility and will continue to reinvade restoration sites unless nutrient levels are reduced (Duncan *et al.* 2008; Smallbone *et al.* 2008; Prober *et al.* 2009). Techniques to remove nutrients include weed harvest, scalping, sugar (or organic carbon) application (Prober *et al.* 2009), and mining nutrients using unfertilised crops. The subsequent establishment of dense *Themeda* swards will continue to lock up nutrients, making them unavailable to weeds.

11. Continue to monitor the site and adjust management according to what you see

Monitoring of sites and adaptive or responsive management is critical to success. Weed growth, pests and diseases, seed germination and plant growth and survival should be monitored for the first few years after planting. After the initial establishment phase, competition between species, plant structure and use of the site by wildlife can be monitored. The most important part of monitoring is to have clear goals for the project at the start and to have decided what constitutes success. The success (or otherwise) of the project can then be assessed against these goals. Use the results of monitoring to make management decisions to ensure plant survival and growth.

12. Minimise disturbance of native vegetation in good condition

Native vegetation in good condition should not be disturbed. This lesson should be applied in all sites where revegetation is conducted on degraded lands. The exception is where revegetation is used to enhance the condition of existing native vegetation. On many sites, the original woodland trees have been cleared leaving a 'derived grassland' with a high diversity of native species. Alternatively, the trees have been retained, but the ground and understorey strata have been substantially altered through grazing, nutrient enrichment and cultivation. In these environments, high levels of disturbance for revegetation can damage or destroy the existing native vegetation and encourage weeds at the expense of existing plants (Environment ACT 2005).

On sites where there is a reasonably intact ground layer, disturbance for weed control and ground preparation should be kept to a minimum. Sites should only be ripped (where necessary), but not cultivated or mounded. Mechanical or hand augers can create planting spots with minimal disturbance to other vegetation. Weed control along planting lines should be around planting spots only. Any weeds that germinate from the soil seedbank should be selectively controlled.

Where retained trees have a highly degraded understorey, restoration work should not damage trees. Ripping and cultivation should not occur within the drip-lines of existing trees. In these areas use only those herbicides which are active on the weeds being controlled and avoid spray drift.

13. There is still a lot to learn

In recent years there have been significant improvements in techniques used for restoring woodlands. These include the use of 'moisture-banking', nutrient depletion and weed seedbank depletion techniques referred to earlier. The Grassy Groundcover project jointly developed by Greening Australia and the University of Melbourne has developed techniques to successfully re-establish diverse, resilient grassland communities (Gibson-Roy *et al.* 2007a, 2007b). This method reverses the typical forestry-derived tree-planting technique by using nutrient depletion and ground-layer competition rather than tree density to control weed competition. The technique requires further development to be carried out at large scales, but is likely to become widely used in woodland restoration.

There are some major challenges to be overcome to successfully restore woodlands at large scales. Direct seeding could be substantially improved to increase its reliability and decrease the amount of seed used (Carr *et al.* 2009). Cheap efficient methods of depleting soil nutrients and soil-weed seedbanks need to be developed. We need to understand much more about the interactions between plants and the soil biota and to develop methods and propagules for the establishment of mycorrhizae, *Frankia* and *Bradyrhizobium*. There are many species still excluded from restoration programs because we do not know enough about their seed physiology and ecology including dormancy, storage behaviour, germination and field establishment.

Conclusions

The 13 Lessons outlined here are designed to help restoration practitioners increase establishment and survival success when restoring woodland vegetation. Successful restoration of woodlands is expensive and time-consuming, so managing to ensure maximum success is important. The importance of controlling weeds to reduce competition for moisture with young seedlings is probably the most important lesson learned by successful restoration practitioners. Each of the lessons highlighted in this chapter emphasises the importance of planning to achieve successful outcomes. At least two years preparation in advance of planting and two years follow-up maintenance is required.

Our challenge is to continue to move from revegetation to restoration. Can we recreate resilient, self-sustaining woodlands with increasing diversity above and below the ground, at the scale required to address the historical (and continuing) decline in woodlands and the fauna they support?

Bios

David Carr is currently the National Technical Capacity Manager for Greening Australia. Based in Armidale, he is responsible for technical leadership and guidance for Greening Australia as it moves to landscape-scale ecological restoration. He managed Greening Australia's national projects, *Exchange* and *Florabank*. David has been with Greening Australia for 15 years, including time in Armidale working with farmers on biodiversity conservation, farm forestry and revegetation. He has worked on species selection research, seed supply and management, revegetation techniques and extension methods throughout Australia. David has published several books on plant identification and farm forestry for northern NSW.

Jim Robinson has worked for Greening Australia in Victoria since 1990 and is one of Australia's pre-eminent authorities on revegetation. He started his career as a farmer, but sold his business and moved into revegetation activities at the time that Prime Minister Hawke announced the national planting of one billion trees. Jim's experience covers working with Aboriginal people, plant identification, seed collection, plant propagation, bush foods, revegetation, farm forestry and property planning. He is active throughout the year training people in revegetation techniques, developing new techniques, collecting seed and direct seeding.

David Freudenberger is Greening Australia's Director of Science and Major Projects. He has 25 years of ecological research experience, the most recent 15 years with CSIRO. David is an author of over 130 research publications including 49 papers in peer-reviewed scientific journals. He has had research leadership roles in CSIRO and is a graduate of the Australian Rural Leadership Program. His role in Greening Australia is to foster transformative landscape scale initiatives to conserve biodiversity, improve water quality in our rivers, contribute to sustainable agriculture and enhance community well being.

References

Broadhurst, L.M. (2007). Managing genetic diversity in remnant vegetation: implications for local provenance seed collection and landscape restoration. Land & Water Australia Technical Note01/2007.

Broadhurst, L., Lowe, A., Coates, D., Cunningham, S., McDonald, M., Vesk, P. and Yates, C. (2009). Seed supply for broadscale restoration: maximising evolutionary potential. *Evolutionary Applications* **1**, 587–597.

Carr, D., Bonney, N., Huxtable, D. and Bartle, J. (2009). Improving direct seeding for woody crops in temperate Australia: a review. Joint Venture Agroforestry Program Publication N° 09/047. Rural Industries Research and Development Corporation: Canberra.

Carr, D., Rawlings, K. and Atkinson, P. (2008). Native Vegetation Management Tool: guidelines for designing, implementing and monitoring revegetation for biodiversity conservation. URL http://www.florabank.org.au/default.asp?V_DOC_ID=954 (Cited 17 November, 2009)

Commonwealth of Australia (2005). Interim Biogeographic Regionalisation for Australia, V 6.1. Department of the Environment and Heritage: Canberra.

Corr, K. (2003). Revegetation techniques: a guide for establishing native vegetation in Victoria. Greening Australia Victoria: Heidelberg.

Duncan, D.H., Dorrough, J., White, M. and Moxham, C. (2008). Blowing in the wind? Nutrient enrichment of remnant woodlands in an agricultural landscape. *Landscape Ecology* **23**, 107–119.

Ellis, R.H. and Roberts, E.H. (1980). Improved equations for the prediction of seed longevity. *Annals of Botany* **45**, 13–30.

Environment ACT (2005). National Recovery Plan for Natural Temperate Grassland of the Southern Tablelands (NSW and ACT): an endangered ecological community. Environment ACT: Canberra.

Gibson-Roy, P., Delpratt, D. and Moore, G. (2007a). Restoring Western (Basalt) Plains grassland. 2. Field emergence, establishment and recruitment following direct seeding. *Ecological Management & Restoration* **8**, 123–132.

Gibson-Roy, P., Delpratt, J. and Moore, G. (2007b). Restoring the Victorian Western (Basalt) Plains grassland. 1. Laboratory trials of viability and germination, and the implications for direct seeding. *Ecological Management & Restoration* **8**, 114–122.

Manning, A.D., Lindenmayer, D.B. and Nix, H.A. (2004). Continua and Umwelt: novel perspectives on viewing landscapes. *Oikos* **104**, 621–628.

Prober, S.M. and Lunt, I.D. (2009). Restoration of *Themeda australis* swards suppresses soil nitrate and enhances ecological resistance to invasion by exotic annuals. *Biological Invasions* **11**, 171–181.
Smallbone, L.T., Lunt, I.D. and Prober, S.M. (2008). Soil nitrate promotes growth of an exotic grass more than native forbs. *Ecological Management & Restoration* **9**, 60–63.
Thomas, D.S., Heagney, G.H. and Harper, P. (2008). Nursery transplant practices determine seedling root quality of two subtropical eucalypts. *New Forests* **30**, 125–134.

3

PRIORITISING CONSERVATION IN TEMPERATE WOODLANDS

Philip Gibbons

1. Conservation priorities for woodlands should be set according to the difference between: (i) the likely improvement over time with conservation actions; and (ii) the potential loss over the same time without additional conservation actions.
2. Lesson #1 means that setting priorities for woodland conservation requires a better understanding of the gains likely to be achieved in woodlands with different conservation actions.
3. Lesson #1 also means that setting priorities for woodland conservation requires a better understanding of the trajectory of woodlands under the status quo.
4. High quality, or relatively unmodified, woodlands are not necessarily the highest priority for conservation actions.
5. Planting trees in cleared areas should not be a priority for woodland conservation.
6. Mitigating threats posed by human population growth and agricultural intensification represent higher priorities for woodland conservation than mitigating the effects of climate change.

Introduction

Given that the resources for woodland conservation are limited, where should our conservation priorities lie? A focus of my work has been to develop tools to assess proposals for developing woodlands and prioritise conservation investment in woodlands. My field work has focused on various woodland communities (Yellow Box *Eucalyptus melliodora*, Blakely's Red Gum *E. blakelyi*, Grey Box *E. microcarpa*, Poplar Box *E. populnea*, Coolibah *E. coolabah* and Mugga Ironbark *E. sideroxylon*) throughout the wheat–sheep belt of New South Wales (see map on next page). My findings and decision support tools, however, are applied across temperate woodlands and other ecosystems in several states of Australia.

Lessons

1. Conservation priorities for woodlands should be set according to the difference between: (i) the likely improvement over time with conservation actions; and (ii) the potential loss over the same time without additional conservation actions

In 1976 Jared Diamond wrote, 'the question is not which refuge system contains more total species, but which contains more species doomed to extinction in the absence of refuges' (Diamond 1976). Diamond was telling us that our conservation priorities are not necessarily the sites with the highest biodiversity values, but those sites likely to lose the greatest biodiversity values without intervention – a message that is prominent in the systematic conservation planning literature (Pressey and Bottrill 2008). If we accept that conservation priorities in our temperate woodlands should be set according to the difference between (i) the likely improvement over time with conservation actions and (ii) the potential loss over the same time without additional conservation actions (see Figure 3.1a), then this changes the types of temperate woodlands that we should conserve urgently and the data we need to make these decisions.

2. Setting priorities for woodland conservation requires a better understanding of the trajectory of woodlands with different conservation actions

The model for prioritising conservation actions in Figure 3.1a requires an understanding of the amount of improvement in biodiversity we are likely to get when employing conservation actions in temperate woodlands. In a workshop I recently attended, however, there was little consensus among field staff on appropriate conservation actions for the woodland sites we visited. This is not surprising because the efficacy of management actions can vary from site to site. McIntyre and Lavorel (2007) identified several barriers that can inhibit restoration in

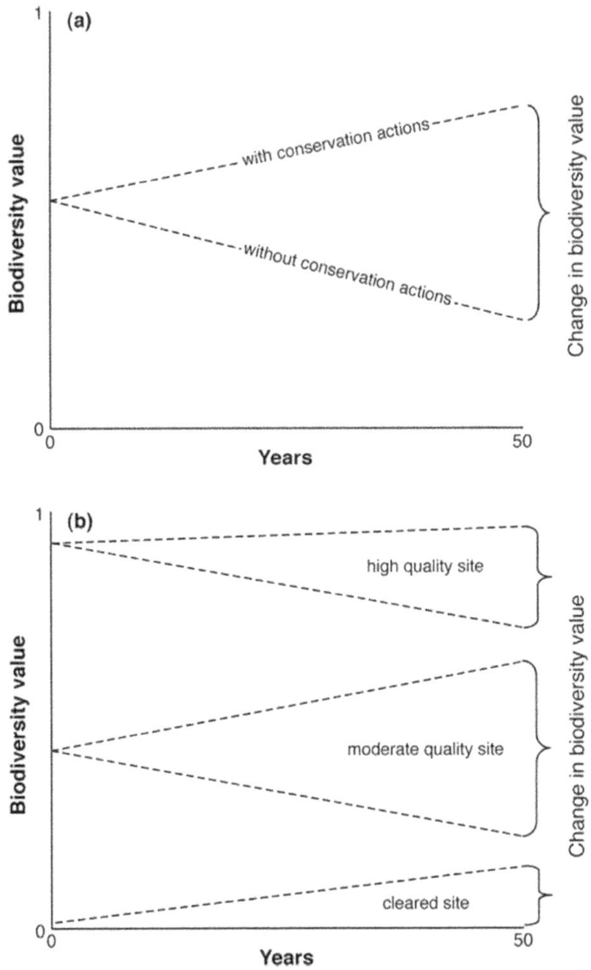

Figure 3.1 Conservation priorities for woodlands should be based on the difference between biodiversity values with and without conservation actions over time (a), which has implications for the types of sites that become priorities for conservation (b).

woodland ecosystems depending on land-use history. This model was supported by Briggs *et al.* (2008) who found little effect of two commonly funded conservation actions (fencing and grazing management) on exotic cover and the regeneration of some tree species.

Where there is uncertainty about the efficacy of management actions in temperate woodlands, then investors should monitor all funded activities and apply the principles of adaptive management to them. There have been efforts to improve monitoring after conservation actions in temperate woodlands (e.g. the Federal Government's Environmental Stewardship Program); however, monitoring programs cannot provide us with information about the efficacy of conservation actions unless there is consistent recording of management inputs (e.g. days grazed) as well as biodiversity outcomes (e.g. plant species richness) (Zerger *et al.* 2009). This is a key omission in many monitoring programs. To apply the principles of adaptive management effectively also means that a degree of flexibility must be maintained throughout any conservation program. That is, management actions should not be set in concrete from the outset so that modifications and adjustments can occur if monitoring indicates that the objectives are not being met.

3. Setting priorities for woodland conservation requires a better understanding of the trajectory of woodlands under the status quo

The model for prioritising conservation actions in Figure 3.1a also requires an understanding of likely changes in the extent and condition of temperate woodlands on a site without additional investment. Several threats (e.g. invasion of exotic plants, lack of regeneration, salinity) and permitted activities under law (e.g. firewood collection, grazing) can potentially erode woodland condition and extent over time. Conversely, stronger native vegetation laws afford increased protection to remnant vegetation.

One of the reasons that we don't know how damaging or effective these threats and laws are is because woodlands are not monitored effectively across broad scales. This is because synoptic satellite sensors such as Landsat (which is the most common satellite used for mapping land cover change in Australia [e.g. National Carbon Accounting System]) do not have the spatial resolution to detect significant areas of temperate woodlands. Classifications based on Landsat imagery can only typically detect trees where the canopy is relatively dense (see Figure 3.2). Several intact temperate woodland stands, however, including those dominated by White Box, Poplar Box and Coolibah have, on average, a sparse canopy cover (<20%) (Gibbons *et al.* 2008). Further, vegetation layers produced from synoptic satellite sensors such as Landsat do not consistently detect small or narrow, linear remnants (Figure 3.2). This is a serious issue because scattered trees and small remnants represent a large proportion of extant temperate woodlands (Gibbons and Boak 2002) and are often under greater threat from land clearing (Gibbons *et al.* 2009), tree mortality (Gibbons *et al.* 2008), a lack of natural tree regeneration (Weinberg *et al.* in review) and salinity (Seddon *et al.* 2007) than larger remnants. Thus, existing mapping is unlikely to be sensitive to changes in the extent of temperate woodlands brought about by these threats. There are existing satellite sensors that can be used to effectively map temperate woodlands (Levin *et al.* 2009), but these are still too expensive to purchase at a scale and frequency that is required to monitor temperate woodlands at regular intervals across large areas.

4. High quality, or relatively unmodified, woodlands are not necessarily the highest priority for conservation actions

If we accept the logic in Figure 3.1, then high quality remnants are not necessarily a high priority for conservation. This is for two principal reasons. First, high quality remnants have little capacity for improvement with management relative to remnants of moderate to poor quality that have not been fertilised (McIntyre and Lavorel 2007) (see Figure 3.1b). Second, high quality remnants are not necessarily under a high degree of threat relative to moderate to poor quality remnants. High quality woodland remnants can no longer be cleared for agricultural production under revised land clearing laws in several states of Australia, although this is not necessarily the case in peri-urban areas. Many intact remnants are grazed intermittently and overstorey regeneration is typically sufficient where such conditions exist. This contrasts with moderate to poor quality woodland remnants which represent the majority of our remaining woodlands. Clearing and agricultural intensification are not necessarily excluded in moderate to poor quality remnants under existing legislation, grazing by livestock is often continuous and regeneration is often absent; that is, poor to moderate quality remnants are often at a higher risk of decline and ultimate loss under the status quo than high quality remnants. For example, in central-western NSW, Weinberg *et al.* (in review) found that around half of all remnants – principally small, modified remnants – contained no

Figure 3.2 Broadscale mapping of tree cover from Landsat imagery (grey) overlaid with very high-resolution satellite imagery (Ikonos) illustrating that temperate woodland vegetation that occurs with a sparse canopy cover, in small patches, or in a narrow linear configuration is poorly mapped and therefore poorly monitored across broad scales in Australia. (Includes material © 2002 Space Imaging LLC, distributed by Raytheon Australia)

regeneration. Investing in high quality remnants would do nothing to address the regeneration crisis in this landscape. Investing public funds in high quality remnants can also undermine the existing legal requirement, or duty of care, in many areas that requires land managers to mitigate threats to remnant vegetation.

5. Planting trees in cleared areas should not be a priority for woodland conservation

If we accept the logic in Figure 3.1b, then plantings – and planting corridors in particular – should not be a focus of woodland conservation. This is for two reasons. The first is that plantings that are established on totally cleared sites can only achieve a modest biodiversity gain relative to effective interventions on sites that are in moderate condition and under considerable threat. Plantings can form part of successful interventions on modified sites, however. Second, a planting that is in a narrow corridor does not afford additional benefits because studies consistently indicate that the amount and quality of habitat is more important to biota than its configuration (e.g. Radford *et al.* 2005; Turner 2005). Conservation efforts in temperate woodlands should focus on stabilising and increasing the extent and condition of our existing temperate woodlands rather than planting new vegetation on highly modified sites for relatively modest biodiversity benefit.

6. Mitigating threats posed by human population growth and agricultural intensification represent higher priorities for woodland conservation than mitigating the effects of climate change

While climate change is an obvious emerging threat, I argue that there are two more immediate threats to temperate woodlands. Elevated residential property prices and continuing strong population growth in Australia (projected 55% increase by 2050, Population Reference Bureau 2009) are creating strong demand for urban growth in cities and regional centres throughout parts of Australia. The urban footprint of several expanding population centres (e.g. Melbourne, Albury, Canberra and Sydney) is encroaching upon woodland communities listed as threatened. An emerging shortage of global food production is spawning intensification of agricultural production which can be detrimental to native woodlands. Intensive grazing by livestock, broadscale use of fertiliser and cultivation are particularly damaging to temperate woodlands. In central-western NSW, Ozolins *et al.* (2001) observed a 25% increase in cultivation from the 1960s to the 1990s. We must address the issues of human population growth and agricultural intensification to protect remaining temperate woodlands effectively from these emerging threats and ensure our remaining temperate woodlands are functional, resilient ecosystems. This also happens to be the best way to prepare our temperate woodlands for the impacts of climate change (Steffen *et al.* 2009).

Conclusions

I began this chapter with the question: Where should our conservation priorities lie in temperate woodlands? The answer is in places where we stand to achieve greatest gains in biodiversity values. These will be areas that contain important values that are under threat and where these threats can be mitigated confidently with management. These are not necessarily sites in the best condition. To support these decisions, we need a greater understanding of the efficacy of different management actions for conserving temperate woodlands. We also need a greater understanding of the trajectory of our temperate woodlands under the existing suite of threats (which we know to include exotic plant invasion, grazing, fertiliser application, the lack of regeneration and tree mortality) and regulations that aim to mitigate some of these. We also need to consider emerging threats, particularly clearing for urban expansion and agricultural intensification.

Acknowledgements

Comments from David Lindenmayer and Andrew Bennett improved an earlier version of this chapter. This work was completed with funding from the Applied Environmental Decision Analysis hub of the Australian Government's Commonwealth Environmental Research Facility.

Bio

Phil Gibbons began his career in the late 1980s working in the wood production forests of East Gippsland, Victoria and southern NSW. He completed a PhD on hollow-dependent fauna in 1999 and has subsequently focused on applied research that has included carbon storage in Australian vegetation, the conservation value of paddock trees and methods for rapid biodiversity assessment. He has published two books and approximately 50 peer reviewed articles. His work informs codes of forest practice, land clearing regulations and incentive programs across Australia. He is currently a Senior Research Fellow at the Australian National University.

References

Briggs, S.V., Taws, N.M., Seddon, J.A. and Vanzella, B. (2008). Condition of fenced and unfenced remnant vegetation in inland catchments in south-eastern Australia. *Australian Journal of Botany* **56**, 590–599.

Cunningham, R.B., Lindenmayer, D.B., Crane, M., Michael, D., MacGregor, C., Montague-Drake, R. and Fischer, J. (2008). The combined effects of remnant vegetation and tree planting on farmland birds. *Conservation Biology* **22**, 742–752.

Diamond, J.M. (1976). Island biogeography and conservation: strategy and limitations. *Science* **193**, 1027–1029.

Gibbons, P. and Boak, M. (2002). The value of paddock trees for regional conservation in an agricultural landscape. *Ecological Management and Restoration* **3**, 205–210.

Gibbons, P., Briggs, S.V., Ayers, D.A., Doyle, S., Seddon, J., McElhinny, C., Jones, N., Sims, R. and Doody, S.J. (2008). Rapidly quantifying reference conditions in modified landscapes. *Biological Conservation* **141**, 2483–2493.

Gibbons, P., Briggs, S.V., Ayers, D.A., Seddon, J.A., Doyle, S.J., Cosier, P., McElhinny, C., Pelly, V. and Roberts, K. (2009). An operational method to assess impacts of land clearing on terrestrial biodiversity. *Ecological Indicators* **9**, 26–40.

Gibbons, P., Lindenmayer, D.B., Fischer, J., Manning, A.D., Weinberg, A., Seddon, J., Ryan, P. and Barrett, G. (2008). The future of scattered trees in agricultural landscapes. *Conservation Biology* **22**, 1309–1319.

Levin, N., McAlpine, C., Phinn, S., Price, B., Pullar, D., Kavanagh, R.P. and Law, B.S. (2009). Mapping forest patches and scattered trees from SPOT images and testing their ecological importance for woodland birds in a fragmented agricultural landscape. *International Journal of Remote Sensing* **30**, 3147–3169.

McIntyre, S. and Lavorel, S. (2007). A conceptual model of land use effects on the structure and function of herbaceous vegetation. *Agriculture Ecosystems & Environment* **119**, 11–21.

Ozolins, A., Brack, C. and Freudenberger, D. (2001). Abundance and decline of isolated trees in the agricultural landscapes of Central New South Wales. *Pacific Conservation Biology* **7**, 195–203.

Population Reference Bureau (2009). 'World Population Data Sheet 2009.' Population Reference Bureau: Washington, D.C.

Pressey, R.L. and Bottrill, M. (2008). Opportunism, threats and the evolution of systematic conservation planning. *Conservation Biology* **22**, 1340–1345.

Radford, J.Q., Bennett, A.F. and Cheers, G.J. (2005). Landscape-level thresholds of habitat cover for woodland-dependent birds. *Biological Conservation* **124**, 317–337.

Seddon, J.A., Zerger, A., Doyle, S.J. and Briggs, S.V. (2007). The extent of dryland salinity in remnant woodland and forest within an agricultural landscape. *Australian Journal of Botany* **55**, 533–540.

Steffen, W., Burbidge, A., Hughes, L., Kitching, R., Lindenmayer, D., Musgrave, W., Stafford Smith, M. and Werner, P. (2009). *Australia's Biodiversity and Climate Change*. CSIRO Publishing: Melbourne.

Turner, I.M. (2005). Landscape ecology: what is the state of the science? *Annual Reviews of Ecology, Evolution and Systematics* **36**, 319–344.

Weinberg, A., Gibbons, P., Briggs, S.V. and Bonser, S. (in review) Broadscale failure of tree regeneration in an agricultural landscape.

Zerger, A., Freudenberger, D., Thackway, R., Wall, D. and Cawsey, M. (2009). *VegTrack*: a structured vegetation restoration activity database. *Ecological Management and Restoration* **10**, 136–144.

4

WHOLE OF PADDOCK REHABILITATION (WOPR): A PRACTICAL APPROACH TO RESTORING GRASSY BOX WOODLANDS

Sue Streatfield, Graham Fifield and Melinda Pickup

1. Woodland restoration needs to consider the production needs of farming communities.
2. Whole of Paddock Rehabilitation (WOPR) is achieving large-scale revegetation by directing funds toward stewardship payments rather than fencing activities.
3. Native trees and shrubs can be integrated into grazing and cropping systems for multiple on and off-farm benefits.
4. On-ground initiatives need to be supported by rural communities to be delivered at the scale required to make landscape-scale changes.

Introduction

Greening Australia Capital Region (GA CR) has been working with land managers in south-eastern New South Wales since 1982. During this time, we have worked with over 1500 farmers to improve the management of native vegetation on private land (see map on next page). In rural areas, this has taken the form of fencing bushland blocks, fencing and revegetating riparian corridors and planting linear shelterbelts. These activities have long formed the cornerstone of conservation on farms throughout Australia (Bennett *et al.* 2000).

Lessons

1. Woodland restoration needs to consider the production needs of the farming communities

Vast areas of NSW that once supported Grassy Box woodlands are now used for agriculture. Restoration of highly degraded woodland areas requires consideration of the production and economic needs of the farming community. Conservation and production have traditionally been viewed as opposing or separate land uses in Australian agricultural systems. Traditional conservation activities, such as fencing and revegetation, have often focused on hilltop

remnants and the lower riparian areas which result in narrow linear vegetation corridors criss-crossing the more productive landscapes (e.g. Freudenberger *et al.* 2004). For many of these activities, a large proportion of the cost is associated with fencing materials and labour. Furthermore, the high cost of fencing small blocks and linear strips often represents a relatively low return on investment (~$1500/ha) for environmental outcomes due to the high perimeter to area ratio.

To ensure successful production and environmental outcomes, larger scale, more cost-effective and lower input solutions are required to restore degraded woodlands in agricultural systems. Although there is increasing understanding of the problems and issues facing degraded woodlands in agricultural landscapes (e.g. Fischer *et al.* 2009; Gibbons *et al.* 2008), the greatest environmental challenge is finding practical and cost-effective solutions for rehabilitating these areas (Manning and Lindenmayer 2009). Greening Australia Capital Region believes that integrating conservation with production, by returning native trees and shrubs back into grazing systems, is one solution to woodland restoration.

2. Whole of Paddock Rehabilitation (WOPR) is achieving large scale revegetation by directing funds toward stewardship payments, rather than fencing activities

Greening Australia's Whole of Paddock Rehabilitation (WOPR) represents a novel, cost-effective approach to revegetation for multiple environmental and production benefits. WOPR involves revegetating entire paddocks larger than 10 ha with widely spaced (40–50 m) vegetation belts. Each belt consists of four rows of vegetation established using direct seeding along the contour. Up to 30 local native tree and shrub species are selected for the seed mix and WOPR can increase the vegetation cover in the paddock to around 30% (see Figure 4.2).

As part of the WOPR program, stewardship payments of $50/ha/yr are provided to land managers to rest paddocks from grazing for a five-year period while the trees and shrubs

Case Study 1 – A paddock in Binalong, NSW

Like many paddocks in Binalong, this 30 ha block was heavily cleared during the 1950s and later sown to clover, ryegrass and phalaris to increase carrying capacity. Contour banks were installed to halt increasing erosion, but later failed. Bushfires in 1989 killed many of the remaining scattered trees and contributed to a rapidly rising water table and dryland salinity. By 1994, the salt levels across the paddock had reached between 2 and 3 ds/m with salt scalds, salt-tolerant grasses and rushes replacing the once desirable pastures.

Declining paddock productivity prompted a revegetation revolution in the form of 40 km of direct-seeded native trees and shrubs. Livestock were re-introduced to the paddock after a five-year grazing spell (see Figure 4.1). By 1996, salt levels had started to decline and by 2001, the problem had all but disappeared. Final soil tests showed salt levels of only 0.05–0.07 ds/m or less than 5% of the original levels (Hufton 2002). Informal bird surveys undertaken over the years recorded more than 20 species using the site and pasture surveys in 2008 show a re-colonisation of native pasture species (Greening Australia 2009).

Figure 4.1 Merino sheep grazing in a 12-year old WOPR site.

establish. Given the reduced fencing costs associated with WOPR, funding can be used to provide payments which help offset the loss of agricultural production during the initial establishment period. After five years, when the vegetation belts can tolerate rotational grazing, grazing is re-introduced under a 10-year management agreement. The on-ground establishment costs of WOPR ($550/ha) are around a third of the cost of traditional conservation initiatives. In this way, WOPR represents a cost-effective and efficient means of integrating woodland restoration into production systems.

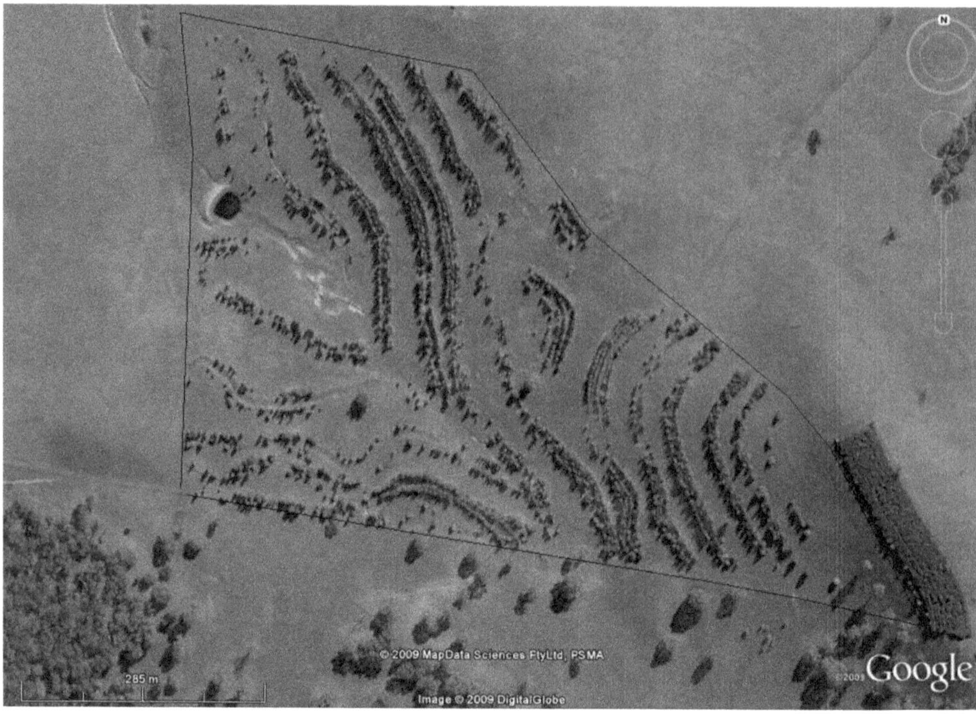

Figure 4.2 An aerial view of a WOPR paddock nine years after seeding.

3. Native trees and shrubs can be integrated into grazing and cropping systems for multiple on and off-farm benefits

The key to improving environmental quality in agricultural landscapes is to provide conservation initiatives that are compatible with, and increase the value of, production systems. The larger scale of WOPR (>10 ha paddocks), combined with rotational grazing after establishment, means it can potentially provide a range of benefits for both production and conservation. The potential environmental benefits of WOPR include reduced dryland salinity (Hufton 2002), an increase in farm wildlife (Barrett *et al.* 2008; Taws 2007), improved soil health and function, better paddock tree health, and regeneration of native grasses and trees (Greening Australia 2009). For some faunal groups (e.g. birds; Kavanagh *et al.* 2007), the biodiversity benefits of revegetation may increase proportionally with the size of the revegetation planting (Munro *et al.* 2007). This suggests that WOPR paddocks may play an important role in improving biodiversity outcomes in agricultural areas and provide critical habitat links across the landscape. Given the importance of revegetation for the long-term future of paddock trees and associated species (e.g. the Superb Parrot *Polytelis swainsonii*; Manning and Lindenmayer 2009), WOPR could assist in establishing the next generation of paddock trees in agricultural landscapes.

Pasture surveys on a WOPR site established in 1994 in Binalong in NSW, show a change in paddock composition from exotic pastures to native grasses. Furthermore, the tree and mid-storey shrubs in the revegetation belts have also thinned (Greening Australia 2009). This example suggests that, over time, the species and vegetation structure within the WOPR paddock may begin to resemble an open woodland dominated by native perennial grasses and

Case Study 2 – Wallendbeen, NSW

Nestled within the mixed farming districts around Wallendbeen, this property contains the listed White Box (*Eucalyptus albens*), Yellow Box (*Eucalyptus melliodora*), Blakely's Red Gum (*Eucalyptus blakelyi*) woodland community. A 30 ha paddock was identified for WOPR as it was unsuited to cropping due to its rocky nature. In 2008, the paddock was direct seeded with 20 local tree and shrub species. The paddock has been incorporated into a farm plan and the revegetation will provide benefits for the landholder's production system, including shade and shelter for livestock. There are also strong aesthetic motivations for the work.

The five-year rest from grazing should provide multiple production and conservation benefits. The paddock contains some native grasses, annual weeds and a mix of naturalised annual grasses and perennials in the drainage lines. The five-year rest from grazing will allow the native grasses Kangaroo Grass (*Themeda australis*), Redleg Grass (*Bothriochloa macra*) and Wallaby Grass (*Austrodanthonia spp.*) to bulk up, set seed and multiply. There are eight remnant paddock trees, with varying levels of senescence and dieback. Given the understorey composition (e.g. some native grasses), the grazing spell should allow the paddock trees to regenerate, providing a small part of the solution to a landscape-scale problem of paddock tree dieback (Fischer *et al.* 2009).

scattered woodland trees. Active thinning of revegetation stands could also be used to 'choose' the large, well established eucalypts that will become the scattered trees of the future (Manning and Lindenmayer 2009).

The potential production benefits of WOPR include increased shade and shelter for livestock (Gregory 1995), improved native pastures that are more drought-tolerant (Wilson 1996), the provision of a supplementary (browse) feed source for livestock and a potential reduction in livestock parasite load through access to *Acacia* browse (Fifield 2006). The anti-parasitic effects of *Acacia* browse and browse extracts for sheep have been documented in studies undertaken in Africa and South America (e.g. Cenci *et al.* 2007; Max *et al.* 2007), while anecdotal evidence from a recent study in south-eastern NSW also supports these findings (Fifield 2006). Although further research in Australian agricultural systems is required, *Acacia* species in WOPR paddocks have the potential to provide an anti-parasitic supplementary feed source for livestock. Previous research has also found that pasture production is greatest when tree cover across the farm is ~30% (Walpole 1999). This suggests that increasing tree cover in paddocks through WOPR may have benefits for pasture productivity.

4. On-ground initiatives need to be supported by rural communities to be delivered at the scale required to make landscape-scale changes

The real test for any new program or innovation is the willingness of farmers to integrate it into their farming practice. In 2008/9, Greening Australia Capital Region, with support from the Lachlan Catchment Management Authority and the NSW Department of Climate Change, Environment and Water, established a series of WOPR pilot sites across the Southern Tablelands and South-west Slopes of NSW. Fifteen properties were involved in the WOPR pilot program which involved resting and revegetating a total of 374 ha of paddock. The establishment of a 65 ha block highlights the willingness of farmers to rehabilitate large areas under this scheme. Even without an official promotional campaign, this pilot has attracted a waiting list

Case Study 3 – The cropper WOPR, Young, NSW

Amid the cropping country of Young lies the 'Cropper WOPR', an example of native revegetation in a woodland cropping system. For the landholder, WOPR offers an opportunity to address an exposed and highly vulnerable area of the farm, to create much-needed shade and shelter for lambs and off-shears sheep, to temporarily rest an eroded gully system from grazing, to reinstate the next generation of paddock trees and to increase birdlife.

In Figure 4.3 below, the bright green areas represents areas which have been direct seeded with native trees and shrubs; the paler green, riparian areas which are being rested from grazing. Unlike the grazing scenarios presented in case studies one and two where production is forfeited for the first five years, the landholder in this example can continue to crop the areas between the tree belts (10 ha, represented in blue). The stewardship payments are therefore paid only on the area under rest and revegetation (27 ha). This case study highlights the flexible nature of WOPR and that we are only beginning to understand the potential variations on the WOPR model required for large-scale woodland restoration.

Figure 4.3 A diagrammatic representation of the cropper WOPR at Young, NSW.

of nearly 40 farmers interested in future participation. Greening Australia Capital Region is also looking to purchase grass seeding machinery to re-establish native grasses in WOPR paddocks to complement the woody revegetation. This takes greater advantage of the grazing spell to re-establish a broader range of layers of grassy woodland vegetation.

To quantify the environmental and production outcomes of WOPR, Greening Australia Capital region has also implemented a long-term monitoring and evaluation program across several paddocks in the trial program. This monitoring program includes treatment (WOPR paddocks) and control (no revegetation) sites and is collecting baseline and subsequent data.

Conclusions

Whole of Paddock Rehabilitation is a grassroots solution to a grassroots problem. It has important implications for the future of woodland communities in agricultural systems and for reducing the landscape-scale loss of biodiversity. It addresses some of the key issues affecting Box Gum woodland communities across NSW, namely paddock tree dieback (Fischer *et al.* 2009), habitat simplification and the loss of native grasses and other elements of the understorey (Prober and Thiele 2005). By integrating conservation with production, the program is well suited to the more productive parts of the landscape which typically supported Box Gum woodlands communities in NSW. The larger spatial extent and diversity of species established within WOPR provides a number of production benefits including shade, shelter and fodder for pasture and livestock. In addition, WOPR paddocks provide critical woodland habitat, especially for migratory birds, and can improve the connectivity and critical patch size of habitat in highly cleared and fragmented landscapes.

The challenges associated with managing Grassy Box woodlands in agricultural landscapes are great and require initiatives that address the needs of both production and conservation. WOPR not only presents a solution for the next generation of paddock trees, but also represents a new generation of on-ground cost effective, farmer friendly initiatives that are needed to transform these productive and highly modified parts of the landscape. The task now is to gather the financial support required to deliver this highly successful restoration model across multiple landscapes and regions.

Acknowledgements

Greening Australia Capital Region would like to specially thank Leon and Bubbles Garry and John Weatherstone for their ingenuity, innovation and commitment in pioneering Whole of Paddock Rehabilitation. Thanks also to all the many land managers who have established WOPR paddocks as part of the pilot program. GA CR would also like to thank the many stakeholders and partners who are helping to trial and promote WOPR including the Lachlan Catchment Management Authority, NSW Department of Environment and Climate Change and the ANU Fenner School of Environment and Society. Thanks also to David Freudenberger from Greening Australia Ltd for prompting the inclusion of this chapter and for comments on an earlier draft.

Bios

Sue Streatfield graduated with First class Honours from the School of Resource and Environmental Management at the Australian National University in 1994. She has spent the last 15 years creating and overseeing innovative on-ground incentive and research projects with Greening Australia Capital Region, including the development of the WOPR program.

Graham Fifield graduated with First class Honours from the Fenner School of Environment and Society at the Australian National University in 2006. His work at Greening

Australia has involved delivering incentives to small farms through the region and more recently delivering the innovative WOPR program.

Melinda Pickup graduated with a PhD from the Department of Botany and Zoology at the Australian National University and CSIRO Plant Industry in 2008. Since this time she has worked as a field scientist and science communicator with Greening Australia.

References

Barrett, G.W., Freudenberger, D., Drew, A., Stol, J., Nicholls, A.O. and Cawsey, E.M. (2008). Colonisation of native tree and shrub plantings by woodland birds in an agricultural landscape. *Wildlife Research* **35**, 19–32.

Bennett, A.F., Kimber, S. and Ryan, P. (2000). 'Revegetation and wildlife: a guide to enhancing revegetation habitats for wildlife conservation in rural environments'. Bushcare National and Research and Development Program, Research report 2/00. Environment Australia, Canberra.

Cenci, F.B., Louvandini, H., McManus, C.M., DellPorto, A., Costa, D.M., Araujo, S.C., Minho, A.P. and Abdalla, A.L. (2007). Effects of condensed tannin from *Acacia mearnsii* on sheep infected naturally with gastrointestinal helminthes. *Veterinary Parasitology* **144** (1–2), 132–137.

Fifield, G. (2006). Acacia browse as seasonal forage for fine wool Merino sheep on the Southern Tablelands and Southwest Slopes of NSW. Honours thesis, The Australian National University.

Fischer, J., Stott, J., Zerger, A., Warren, G., Sherren, K. and Forrester, R.I. (2009). Reversing a tree regeneration crisis in an endangered ecoregion. *Proceedings of the National Academy of Sciences of the United States of America* **106**, 10386–10391.

Freudenberger, D., Harvey, J. and Drew, A. (2004). Predicting the biodiversity benefits of the Saltshaker Project, Boorowa, NSW. *Ecological Management & Restoration* **5**, 5–14.

Gibbons, P., Lindenmayer, D.B., Fischer, J., Manning, A.D., Weinberg, A., Seddon, J., Ryan, P. and Barrett, G. (2008). The future of scattered trees in agricultural landscapes. *Conservation Biology* **22**, 1309–1319.

Greening Australia (2009). Introducing Whole of Paddock Rehabilitation (WOPR): a new approach to regreening the farm. http://www.greeningaustralia.org.au/uploads//Our%20 Resources%20-20pdfs/ACT_WOPR 09.pdf

Gregory, N.G. (1995). The role of shelterbelts in protecting livestock: a review. *New Zealand Journal of Agricultural Research* **38**, 423–450.

Hufton, L. (2002). 'Quantifying dryland salinity control'. A report prepared for the Binalong Landcare Group.

Kavanagh, R.P., Stanton, M.A. and Herring, M.W. (2007). Eucalypt plantings on farms benefit woodland birds in south-eastern Australia. *Austral Ecology* **32**, 635–650.

Manning, A.D. and Lindenmayer, D.B. (2009). Paddock trees, parrots and agricultural production: An urgent need for large-scale, long-term restoration in south-eastern Australia. *Ecological Management & Restoration* **10**, 126–135.

Max, R.A., Kimambo, A.E., Kassuku, A.A., Mtenga, L.A. and Buttery, P.J. (2007). Effect of tanniniferous browse meal on nematode faecal egg counts and internal parasite burdens in sheep and goats. *South African Journal of Animal Science* **37**, 97–106.

Munro, N.T., Lindenmayer, D.B. and Fischer, J. (2007). Faunal response to revegetation in agricultural areas of Australia: a review. *Ecological Management & Restoration* **8**, 199–207.

Prober, S.M. and Thiele, K.R. (2005). Restoring Australia's temperate grasslands and grassy woodlands: integrating function and diversity. *Ecological Management & Restoration* **6**, 16–27.

Taws, N. (2007). *Bringing Birds Back: A Glovebox Guide to Bird Identification and Habitat Restoration*. Greening Australia: Canberra.

Walpole, S.C. (1999). Assessment of the economic and ecological impacts of remnant vegetation on pasture productivity. *Pacific Conservation Biology* **5**, 28–35.

Wilson, A.D. (1996). Native and low-input grasses for pastoral and marginal cropping land. *New Zealand Journal of Agricultural Research* **39**, 465–469.

5

SCATTERED PADDOCK TREES: THE LIVING DEAD OR LIFELINE TO THE FUTURE?

Adrian D. Manning and Joern Fischer

'There are those that are standing, living and breathing, but as dead as is the litter, since they have no reproductive future. They are the living dead.' (Janzen 1986, p. 307)

1. Much remnant tree cover in productive areas occurs as scattered paddock trees.
2. Scattered trees are rapidly disappearing.
3. False assumptions are commonly made about scattered trees, and these underpin their neglect.
4. Scattered trees provide valuable ecosystem functions at the local scale.
5. Scattered trees provide valuable ecosystem functions at the landscape scale.
6. Scattered trees provide valuable ecological continuity through time.
7. We need to review existing policy settings critically to reflect current science.
8. We need to improve conditions for natural tree regeneration.
9. We need to plant and direct-seed scattered trees where natural regeneration is unlikely.

Introduction

The aim of this chapter is to examine the value of scattered paddock trees that often remain as a result of the land conversion process. Our central argument is that it is important to focus research, management, and policy initiatives not only on relatively intact woodland patches, but also on scattered paddock trees. We summarise our reasons for this argument in nine lessons. The first three lessons address practical and conceptual issues concerning the plight of scattered paddock trees. The next three summarise key ecological functions provided by scattered trees. The last three lessons point towards important management and policy priorities for the future.

The lessons we summarise in this chapter arise from our collective experience working in woodland ecosystems over the last 10 years (see map on next page). We have worked on a wide range of issues during that time, including paddock trees, their regeneration, and their habitat value for native reptiles, birds, mammals and ground-dwelling invertebrates.

The management of scattered trees is affected by a number of practical and conceptual issues. These include: that scattered trees occur on land that is preferred for agriculture; that they are being rapidly lost from the landscape due to a range of threatening processes; and that a series of false assumptions underpin their neglect.

Lessons

1. Much remnant tree cover in productive areas occurs as scattered paddock trees

Prior to European settlement, temperate woodlands were widespread throughout Australia's current agricultural zone. Since then, woodland communities have been disproportionately affected by land clearing because they often coincided with lowland parts of the landscape that were most suitable for agricultural production. This bias in past land clearing means that much remnant woodland cover persists as small patches and scattered paddock trees. For example, in the Upper Lachlan River catchment in New South Wales, more than half of the remaining cover of Yellow Box (*Eucalyptus melliodora*) has been recorded in patches smaller than one hectare (Gibbons and Boak 2002). In the same region, approximately one-third of all remnant tree cover, irrespective of topographic position, occurs as scattered paddock trees (Fischer *et al.* 2010). Scattered paddock trees are particularly important in lowland environments, where they are often the only remnants of the original woodland communities and can account for a substantial amount of remnant woodland cover. Focussing only on large patches of trees, by contrast, is likely to introduce a bias towards upland environments, which are typically dominated by dry sclerophyll forest communities rather than woodlands.

2. Scattered paddock trees are rapidly disappearing

Scattered paddock trees face a range of threats that ultimately result in low tree recruitment and accelerated mortality (Table 5.1). In south-eastern Australia, regeneration of scattered

Table 5.1 Examples of threats to scattered trees and their consequences (see Manning and Lindenmayer 2009 for references within).

Threats	Clearing, vegetation modification and fragmentation, altered grazing regimes, removal of fallen timber and dead trees (often for firewood), physical disturbance (e.g. 'tidying up', cultivation, development near trees), herbicide drift
Consequences	Changes to soil health and nutrient cycles (including disruption of mycorrhizal fungi), lack of tree regeneration, weed invasion, declining tree health, invasive exotic pest animals, changed hydrology, elevated risk of salinity

trees has ceased at a landscape scale, and regional-scale declines have been predicted under status quo management (Fischer *et al.* 2009). Failure to address critical threats means that scattered paddock trees are being lost, and along with them the many ecological values and services that they provide (see below). Scattered trees are lost at a faster rate in cropping areas because they are considered incompatible with large machinery (Ozolins *et al.* 2001; Maron and Fitzsimmons 2007). Scattered trees have been poorly protected under legislation (see Lessons #7 to #9). Yet every scattered tree counts, especially considering: (1) the many decades it takes for scattered trees to mature; (2) the cessation of recruitment at a landscape scale; (3) ongoing and accelerating mortality of scattered trees; and (4) the disproportionate role of scattered trees (particularly old trees) in maintaining important ecosystem functions (see Lessons #4 to #6).

3. False assumptions are commonly made about scattered trees, and these underpin their neglect

Several false assumptions about scattered trees underpin their continued decline in Australian landscapes (Table 5.2). Addressing these false assumptions is critical to ensuring their future survival.

Scattered paddock trees are recognised globally as 'keystone structures' because they have a disproportionately positive effect on ecosystems relative to the small area they occupy

Table 5.2 Some traditional assumptions about scattered paddock trees and evidence that disproves their validity.

Assumption	Why this is not always true
Only large patches and corridors are worthwhile, because animals do not perceive scattered trees as habitat	Large patches and corridors (ecological networks) are important for conservation, but so are small patches and scattered trees. Different organisms perceive landscapes differently to each other and humans (Manning *et al.* 2004). Some species can live in and use the landscape matrix. For example, the Superb Parrot (*Polytelis swainsonii*) uses scattered tree landscapes in the south-west slopes of New South Wales as continuous habitat (Manning *et al.* 2006b).
Scattered trees will not regenerate because their seed is not viable	It had been thought that scattered trees produce seed that is less viable (Burrows 2000). In some situations, however, scattered trees do produce ample viable seed (Ottewell *et al.* 2009).
The value of remnant patches is independent of scattered trees in the surrounding landscape	The landscape context within which patches occur has a profound influence on the ecological communities within (Donald and Evans 2006). Scattered trees have been shown to have positive effects on bird diversity in woodland patches that they surround (Cunningham *et al.* 2008).

(Manning *et al.* 2006a). The values that contribute to their keystone status fall into three categories: the local scale; the landscape scale; and continuity through time.

4. Scattered trees provide valuable ecosystem functions at the local scale

Several ecological functions are tied to the presence of scattered trees, and a loss of scattered trees leads to significant declines in these functions (Figure 5.1). Scattered trees provide three types of benefits at the local scale: abiotic benefits; benefits to plants; and benefits to animals (reviewed in Manning *et al.* 2006a).

Trees influence their abiotic environment both above and below ground. For example, trees alter the microclimate underneath them by providing shade. Water infiltration is also higher underneath trees, because tree roots increase soil porosity. Finally, soil nutrient levels change around scattered trees as a result of the accumulation of leaf litter, fallen timber and animal dung. Larger, older scattered trees provide disproportionate inputs into woodland ecosystems. For example, Killey *et al.* (in press) found that Yellow Box trees that were 100 cm in diameter were 10 times more likely to produce coarse woody debris than a tree with half that diameter, and they produced 10 times greater volume of coarse woody debris. Similarly, McElhinny *et al.* (2010) found that a tree with a diameter of 120 cm produced 70 times more litter than a tree with a diameter of 20 cm.

The changes in abiotic conditions around scattered trees provide suitable habitat conditions for a different set of plant species than those inhabiting open paddocks, thus contributing to plant species diversity. Scattered woodland trees in Australia provide habitat for a wide range of animals, including many canopy-dwelling invertebrates, birds, arboreal marsupials and bats. In addition, some reptiles and invertebrates extensively use habitat features derived from scattered trees, such as fallen timber or leaf litter (for references, see review by Manning *et al.* 2006a).

5. Scattered trees provide valuable ecosystem functions at the landscape scale

Scattered trees provide several important ecological functions at the landscape scale. For example, they enhance connectivity by providing continuous although sparse cover, thereby allowing for multi-directional movements of species around the landscape (Figure 5.1). Scattered trees can complement reserves and ecological networks by replacing abrupt boundaries with gradual ones. They also increase landscape heterogeneity and can make an otherwise unusable landscape useable by providing 'stepping stones', foraging habitat, nest sites and shelter to a variety of animal species (Manning *et al.* 2009).

6. Scattered trees provide valuable ecological continuity through time

Scattered trees not only provide connectivity through space, but also through time. Trees are long-lived organisms, and can act as 'biological legacies' (*sensu* Franklin *et al.* 2000). They provide habitat for organisms from the original woodland and they can act as a focus for tree regeneration, potentially allowing tree-associated organisms to persist until larger areas of woodland are restored. This is illustrated in Figure 5.1, which shows a tree in the past which was located in a structurally complex woodland. Then, in the present, the tree persists but in a highly simplified paddock environment grazed by livestock. In the future, the same tree could potentially provide seedlings and habitat structure in a new, more complex self-perpetuating woodland ecosystem that allows for well-planned livestock grazing (e.g. rotational grazing; see below).

Scattered trees could also have an important role in facilitating adaptive responses to climate change. For woodland organisms, the habitat now is much more fragmented than the

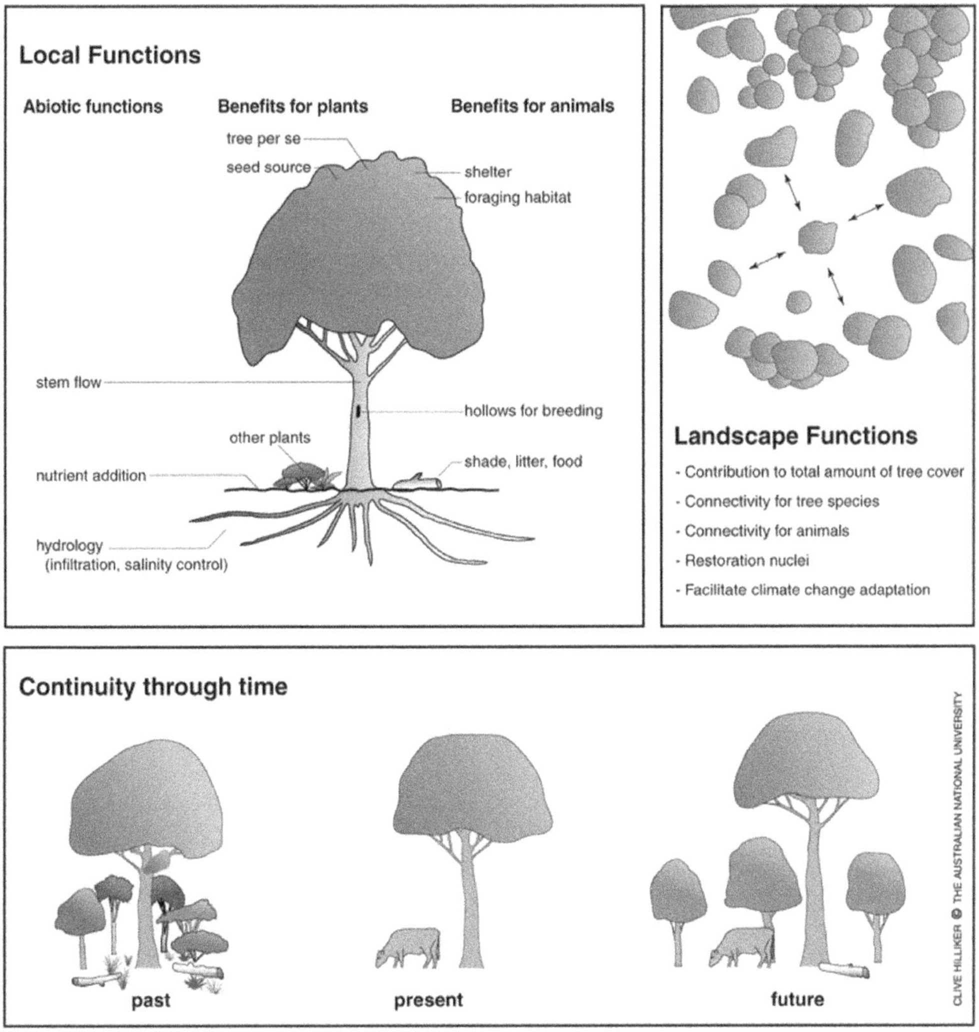

Figure 5.1 Schematic showing the key ecological functions of scattered trees at the local and landscape scales, and their importance for providing ecological continuity through time.

habitat in which they evolved. In response to climate change many species will shift their ranges. It is not clear that reserves, corridors or proposed ecological networks will be enough to facilitate such range shifts. Sparse scattered tree cover can complement ecological networks by creating gradual boundaries, thereby allowing the maximum number of species to move across and live in a given landscape (Manning *et al.* 2009).

Past and current policy and management practices have led to a decline in scattered trees. Changes to policy management will be crucial for the future survival and restoration of scattered tree landscapes in Australia.

7. We need to review existing policy settings critically to reflect current science

With few exceptions, current policy settings do not encourage the maintenance or regeneration of scattered paddock trees. Agricultural intensification, (e.g. through the use of central pivot irrigation systems or conversion of pastures to cropland), poses a major threat to scattered trees (Maron and Fitzsimmons 2007). Although clearing of scattered trees requires approval, existing regulations afford them relatively weak protection. For example, most successful applications for clearing in New South Wales in 2005/2006 were for the removal of scattered trees (Gibbons *et al.* 2009).

Current policy settings reflect an ethic of separating conservation and agricultural production, rather than integrating the two (underpinned by the patch-corridor-matrix landscape model; see first assumption listed in Table 5.2). An important policy instrument in several states is that of clearing offsets, where clearing in one place requires conservation activities to take place somewhere else, with the aim of achieving no net loss overall (Gibbons and Lindenmayer 2007). The 'no-net-loss principle' is sound in theory, but in practice, it could contribute to a pattern of landscape change whereby tree cover is increasingly confined to designated patches, while scattered paddock trees continue to disappear. Essentially, policy that does not protect and perpetuate scattered trees is a self-fulfilling prophecy where woodland patches become increasingly isolated 'islands' in a hostile 'sea' of agricultural production matrix. Similarly, many incentive schemes for conservation or restoration activities have focused on woodland patches or corridors, rather than on scattered trees. Despite tangible benefits of patches and corridors, a critical next step in conservation policy must be to develop additional mechanisms to also perpetuate scattered trees into the future (Fischer *et al.* 2010; Manning and Lindenmayer 2009).

8. We need to improve conditions for natural tree regeneration

Scattered trees need not be doomed to be the living dead (see Janzen 1986). Under certain conditions, their natural regeneration is possible, and examples of the natural regeneration of scattered paddock trees exist throughout Australia's temperate agricultural zone. Achieving natural regeneration at a given place and time can be very difficult. Natural regeneration is sporadic and uncertain, and a myriad of factors influence it – from seasonal conditions to seed predation by ants (Vesk and Dorrough 2006).

Although the success of management actions at a given place and time is uncertain, particular management strategies can greatly enhance the medium-term chances of natural regeneration at the regional scale (Fischer *et al.* 2009). In particular, low-input rotational livestock grazing promises greatly enhanced chances of natural regeneration compared to high-input, continuous grazing. In the Upper Lachlan catchment of New South Wales, high-intensity, short-duration rotational grazing increased the probability of natural regeneration in areas of scattered trees by up to four times (Fischer *et al.* 2009). Regeneration probabilities also were greatly enhanced in low-nutrient environments, implying that regular fertiliser use is incompatible with the process of natural tree regeneration (Fischer *et al.* 2009). The active encouragement of low-input rotational grazing throughout Australia's temperate agricultural zone could deliver major improvements in the regeneration of scattered paddock trees (as well as reduce farm input costs).

9. We need to plant and direct-seed scattered trees where natural regeneration is unlikely

Where natural regeneration is not possible (e.g. cropping situations), or where regeneration is needed with more certainty (e.g. to provide stock shelter or erosion control), two alternative methods of tree recruitment are possible: (1) planting tubestock; and (2) direct seeding. Planting tubestock is expensive, but useful where no seed trees exist, and could be used to specifically achieve a scattered arrangement of trees. Direct seeding, where trees are established in rows, is more cost-effective but trees may not exactly replicate naturally scattered woodland patterns. Direct seeding could be particularly valuable, however, in cropping areas where cultivation can occur between rows of trees. Seed stock used for these methods should originate from as close as possible to the regeneration site (Manning and Lindenmayer 2009).

Conclusions

Are scattered trees the 'living dead' or lifelines to the future? We argue that scattered trees can only be considered the living dead if this is the choice we make. Two critical factors that have contributed to the decline of paddock trees to date have been the passive 'benign' neglect of existing scattered trees, and the encouragement of management practices that prevent tree regeneration, and increase or encourage premature mortality. As we have outlined here, the assumptions that underpin the validity of these choices are largely false. Instead, we can help scattered trees provide a lifeline to the future. Scattered trees represent a unique (although rapidly diminishing) opportunity to create self-perpetuating woodland-agricultural systems that will provide valuable ecosystem services at the landscape scale, and genuinely integrate conservation and agricultural production. The choice is ours to make, but the time to make it is rapidly running out.

Bios

Adrian Manning is a Fellow at the Fenner School of Environment and Society at the Australian National University. His research interests are in conservation biology, landscape ecology and restoration ecology, and he has published widely in these areas. He is currently undertaking a large research project investigating ways of improving grassy eucalypt woodlands for biodiversity. Adrian has written extensively on scattered trees.

Joern Fischer is a Fellow at the Fenner School of Environment and Society at The Australian National University. His main research interest is biodiversity conservation in human-modified landscapes. He has published extensively in a wide range of journals, including empirical work on birds, reptiles and trees, as well as several conceptual papers. Scattered paddock trees have been of interest to Joern since 2000. In 2005, he co-published an influential review paper on scattered trees with Adrian Manning.

References

Burrows, G.E. (2000). Seed production in woodland and isolated trees of *Eucalyptus melliodora* (yellow box, Myrtaceae) in the South Western Slopes of New South Wales. *Australian Journal of Botany* **48**, 681–685.

Cunningham, R.B., Lindenmayer, D.B., Crane, M., Michael, D., MacGregor, C., Montague-Drake, R. and Fischer, J. (2008). The combined effects of remnant vegetation and tree planting on farmland birds. *Conservation Biology* **22**, 742–752.

Donald, P.F. and Evans, A.D. (2006). Habitat connectivity and matrix restoration: the wider implications of agri-environment schemes. *Journal of Applied Ecology* **43**, 209–218.

Fischer, J. and Lindenmayer, D.B. (2002). Small patches can be valuable for biodiversity conservation: two case studies on birds in southeastern Australia. *Biological Conservation* **106**, 129–136.

Fischer, J., Sherren, K., Stott, J., Zerger, A., Warren, G. and Stein, J. (2010). Towards landscape-wide conservation outcomes in Australia's temperate grazing region. *Frontiers in Ecology and the Environment* **8**, 69–74.

Fischer, J., Stott, J., Zerger, A., Warren, G., Sherren, K. and Forrester, R. (2009). Reversing a tree regeneration crisis in an endangered ecoregion. *Proceedings of the National Academy of Sciences USA* **105**, 10386–10391.

Franklin, J.F., Lindenmayer, D.B., Macmahon, J.A., Mckee, A., Magnusson, D.A., Perry, D.A., Waide, R. and Foster, D.R. (2000). Threads of continuity: ecosystem disturbances, biological legacies and ecosystem recovery. *Conservation Biology in Practice*, **1**, 8–6.

Gibbons, P. and Boak, M. (2002). The value of paddock trees for regional conservation in an agricultural landscape. *Ecological Management and Restoration* **3**, 205–210.

Gibbons, P., Briggs, S.V., Ayers, D., Seddon, J., Doyle, S., Cosier, P., McElhinny, C., Pelly, V. and Roberts, K. (2009). An operational method to assess impacts of land clearing on terrestrial biodiversity. *Ecological Indicators* **9**, 26–40.

Gibbons, P. and Lindenmayer, D.B. (2007). Offsets for land clearing: no net loss or the tail wagging the dog? *Ecological Management & Restoration* **8**, 26–31.

Janzen, D.H. (1986). The future of tropical ecology. *Annual Review of Ecology and Systematics* **17**, 305–324.

Killey, P., McElhinny, C., Rayner, I. and Wood, J. (in press). Modelling fallen branch volumes in a temperate eucalypt woodland: implications for large senescent trees and benchmark loads of coarse woody debris. *Austral Ecology*, doi:10.1111/j.1442-9993.2010.02107.x.

Manning, A.D., Gibbons, P. and Lindenmayer, D.B. (2009). Scattered trees: a complementary strategy for facilitating adaptive responses to climate change in modified landscapes? *Journal of Applied Ecology* **46**, 915–919.

Manning, A.D. and Lindenmayer, D.B. (2009). Paddock trees, parrots and agricultural production: an urgent need for large-scale, long-term restoration in south-eastern Australia. *Ecological Management & Restoration* **10**, 126–135.

Manning, A.D., Fischer, J. and Lindenmayer, D.B. (2006a). Scattered trees are keystone structures – implications for conservation. *Biological Conservation* **132**, 311–321.

Manning, A.D., Lindenmayer, D.B., Barry, S. and Nix, H.A. (2006b). Multi-scale site and landscape effects on the vulnerable superb parrot of south-eastern Australia during the breeding season. *Landscape Ecology* **21**, 1119–1133.

Manning, A.D., Lindenmayer, D.B. and Nix, H.A. (2004). Continua and *Umwelt*: novel perspectives on viewing landscapes. *Oikos* **104**, 621–628.

Maron, M. and Fitzsimons, J.A. (2007). Agricultural intensification and loss of matrix habitat over 23 years in the West Wimmera, south-eastern Australia. *Biological Conservation* **135**, 587–593.

McElhinny, C., Lowson, C., Schneemann, B. and Pachón, C. (2010). Variation in litter under individual tree crowns: implications for scattered tree ecosystems. *Austral Ecology* **35**, 87–95.

McIntyre, S. and Hobbs, R.J. (1999). A framework for conceptualizing human effects on landscapes and its relevance to management and research models. *Conservation Biology* **13**, 1282–1292.

Ottewell, K.M., Donnellan, S.C., Lowe, A.J. and Paton, D.C. (2009). Predicting reproductive success of insect- versus bird-pollinated scattered trees in agricultural landscapes. *Biological Conservation* **142**, 888–898.

Ozolins, A., Brack, C. and Freudenberger, D. (2001). Abundance and decline of isolated trees in the agricultural landscapes of central New South Wales, Australia. *Pacific Conservation Biology* **7**, 195–203.

Vesk, P.A. and Dorrough, J.W. (2006). Getting trees on farms the easy way? Lessons from a model of eucalypt regeneration on pastures. *Australian Journal of Botany* **54**, 509–519.

6

THE NORMAL, THE NOVEL AND THE NATURAL: THE CONTRIBUTION OF EMERGING PRODUCTION SYSTEMS TO THE MANAGEMENT OF WOODLAND REMNANTS IN AGRICULTURAL LANDSCAPES

Ted Lefroy

1. Whole of landscape change in land use and land management is required to manage many of the threats to woodland remnants in farming landscapes.
2. Emerging novel land-use systems have the potential to contribute to the conservation of woodland remnants on private land by moderating and reversing these threats.
3. Awareness, moral persuasion and public programs will never be sufficient to bring about change at the scale required.
4. Adoption of conservation farming practices requires that they deliver private as well as public good, are sufficiently modular to be trialled by landholders, and are sufficiently flexible to be adapted to different geographic and commercial situations.
5. The changing human face of Australia's rural landscapes means that meeting adoption criteria requires tailoring conservation technology, incentives and communication to a range of different audiences.
6. A marriage between the formal and informal cultures that acknowledges the role of science in developing components, and the role of land managers in developing and testing systems at larger scales, would accelerate this transition.

Introduction

This chapter reflects on 20 years of agricultural research and extension in the mixed farming zone of south-western Australia (see map on next page), designed to encourage the development of conservation farming. The aim was to manage rising groundwater, wind erosion, habitat loss and other side effects of agricultural development within productive farming landscapes. It is suggested that change at the scale necessary to influence these threats and protect

both the productive capacity of these landscapes and the remaining fragments of the original vegetation will never be achieved through awareness, moral persuasion and public programs. However, emerging land-use systems, many based on commercial uses for endemic plants, do have the potential to bring about a transition to states that are structurally and functionally more similar to the pre-cleared landscape. While agriculture has been a major source of threats to temperate woodland communities, the transition to farming systems better adapted to the Australian environment has the potential to improve their protection and persistence.

Lessons

1. Whole-of-landscape change in land use and land management is required to manage many of the threats to woodland remnants in farming landscapes

Managing threats to woodland communities requires action at the scales at which these threats arise. Much of Australia's southern woodlands occur on land that was considered to be well suited for agriculture and pastoralism from the earliest days of European settlement. As a consequence, they have been subject to grazing by sheep and cattle and clearing for crop and pasture production for much of the last 200 years. Woodland communities are now under-represented across the southern arc of Australia's 100 million hectares of farmland, particularly on valley floors and floodplains where their remnants commonly exist only as isolated fragments. These remaining fragments are vulnerable to their changed surroundings, particularly changes in surface and groundwater hydrology and the introduction of nutrients, disease, pesticides and exotic plants and animals through the agencies of wind, water and grazing animals. Disruption to the water cycle has been dramatic and well documented (Hatton and Nulsen 1999). Replacing woodlands with synthetic annual grasslands has led to reduced transpiration as a result of lower leaf area, increased runoff and soil evaporation, and reduced infiltration of water into surface soil as a result of reduced groundcover, increased groundwater recharge where water accumulates in the landscape, and consequently increased discharge of water and salt to streams. These changes have directly affected remnants of the original vegetation and reduced the capacity of these landscapes to buffer the effects of highly variable rainfall.

The larger scale threatening processes, particularly those influenced by changes to the water cycle, can never be managed by focusing on the woodland remnants themselves. Instead 'the boundaries of consideration must be isomorphic with the boundaries of causation' (Richard Levins, cited in Jackson 2002). The landscape-scale perspective that emerged from environmental planning as pioneered by McHarg (1971) and Integrated Catchment Management (Curtis and Lockwoood 2000), led to calls for major changes in Australian agricultural production systems to restore the water cycle to a state more closely matching those of the pre-cleared landscape, to both protect productive land and conserve remnants of the original vegetation.

2. Emerging novel land-use systems have the potential to contribute to the conservation of woodland remnants on private land by moderating and reversing these threats

Over the last 50 years, major changes have been advocated in production systems to adapt them better to Australia's ancient soils and highly variable climate. These fall into three broad types based on scale.

1. Whole-of-landscape change that was speculative, based on first principles, and proposed by scientists (e.g. Parker 1989; Lefroy *et al.* 1991; Stirzaker *et al.* 2000).
2. Novel farm-scale systems developed by land managers for specific landscape contexts such as Yeomans' (1964) 'Keyline' system of water harvesting and Andrews' (2006) 'Natural Sequence Farming' approach to rehydrating floodplains.
3. New paddock-scale production systems based on endemic and exotic woody and herbaceous perennial plants (Seis 2006; Goss 2007; Pannell 2008).

The first type was directed at improved conservation of soil, water and native vegetation and relied largely on advocacy (Figure 6.1). The second placed equal emphasis on productivity and resource conservation, has achieved significant levels of interest amongst land managers, but has been largely ignored by the formal culture (science and the bureaucracy). The third are products of both innovative land managers (Seis 2006) and public investment in research and development designed to correct market failure in management of natural resources, in many cases jointly funded by levies on rural industries (Goss 2007; Pannell 2008; Reid 2008; Lefroy 2009). These approaches are potentially complementary, but require synthesis in two ways. First, integration between the biophysical case for change at paddock, farm, catchment and landscape scales and the socio-economic drivers required to bring them about. Second, integration of scientific theory and empirical evidence from studies of novel systems developed *in situ*. It is suggested that the following four steps (Lessons #3 to #6) are required to bring that about.

3. Awareness, moral persuasion and public programs will never be sufficient to bring about change at the scale required

An example of the degree of landscape scale change that was considered desirable in the 1980s and 1990s is shown in Figure 6.1. The proposed strategies for improving management of soil, water and vegetation were to use revegetation to reduce runoff, wind erosion, soil evaporation and accession to water tables, and increase infiltration, soil water storage and transpiration. Design criteria were an increase in leaf area index to that approaching the pre-cleared vegetation over 10–20% of the landscape, aligning management boundaries with natural features

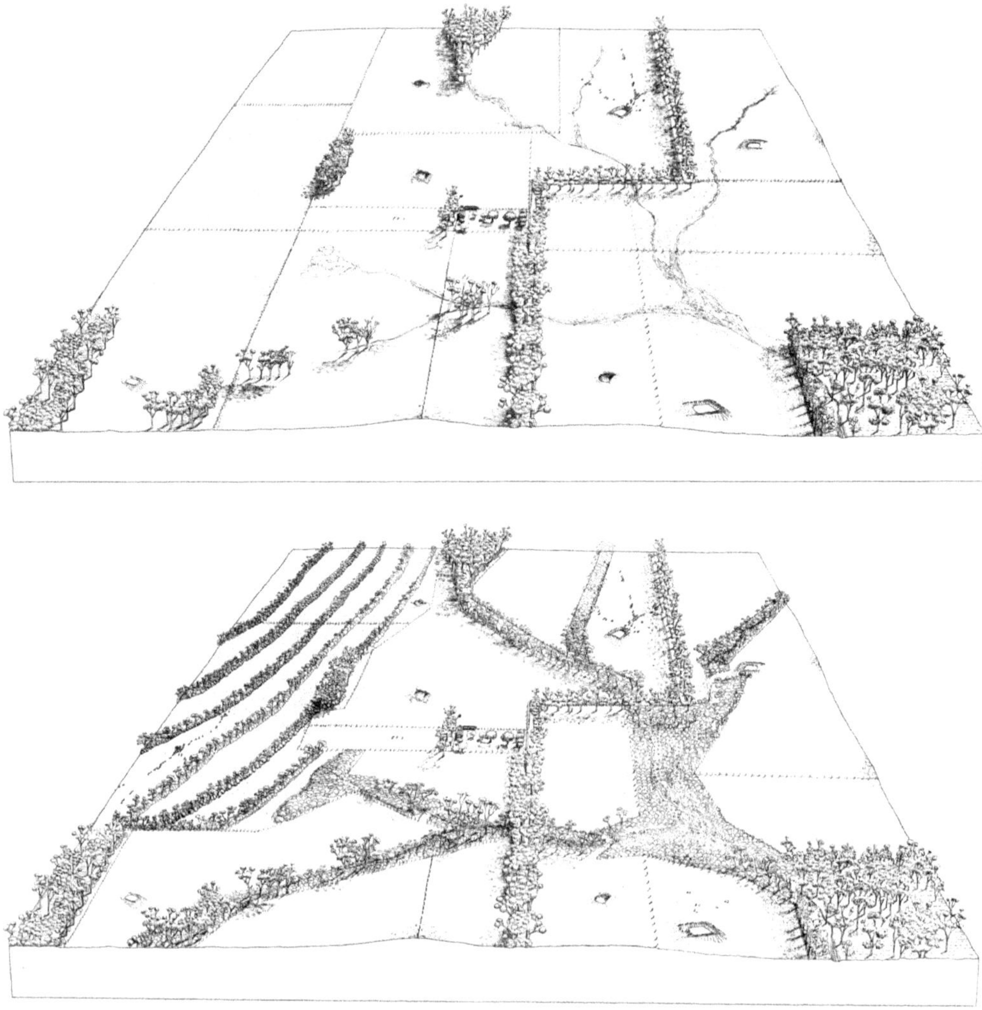

Figure 6.1 A restoration vision for the central wheatbelt of Western Australia featuring an increase in perennial plant cover, alignment of management boundaries along natural features and connecting remnants of native vegetation (Lefroy *et al.* 1991).

such as waterways and soil types, and arranging revegetation to connect and protect remnant vegetation. Such proposals were speculative, relying on advocacy and first principles rather than evidence to support assumed improvements in productivity and conservation. The goals were not explicit, neither was it clear whose goals they were – those of farmers, scientists or conservationists.

Since that time, there has been improved appreciation of the biophysical and social contexts of such change. Improved understanding of hydrology has suggested a greater proportion of the landscape would be required under perennial vegetation to manage the water cycle in most farming landscapes (Stirzaker *et al.* 2002). Studies of adoption of conservation practices by landholders have clarified the necessary conditions for change (Pannell *et al.* 2006). Emerging markets for timber, carbon and biodiversity have provided commercial incentives for some of this change to occur (Figure 6.2). Finally, there is wider acceptance that goal setting is a

Figure 6.2 The Coffey farm near Boyup Brook, Western Australia, in 2002, surrounded by Jarrah (*Eucalyptus marginata*) forest where commercial opportunities exist through high-value timber production to drive the transitional landscape envisaged for drier regions in Figure 6.1. (Photo: Department of Environment and Conservation, WA)

collaborative process that involves questions of value as much as science (Pannell 2004). On the questions of goals, change at landscape scale needs to meet the goals of different interests at different scales. The primary producer's emphasis is on productivity at paddock scale and economic viability at farm scale. Only at catchment and larger scales do the goals of conserving soil, water and biodiversity become meaningful and measurable.

4. Adoption of conservation farming practices requires that they deliver private as well as public good, are sufficiently modular to be trialled by landholders, and are sufficiently flexible to be adapted to different geographic and commercial situations

Improved understanding of the adoption of conservation practices over the last decade has suggested that if conservation farming is to emerge on any significant scale, it will be a product of the economic drivers, the social contexts within which they act, and personal motivations of those who own and manage land. A review of the adoption of conservation practices by landholders identified the requirements for successful practice change. These are awareness; a strong trigger, usually commercial, to stimulate interest in finding out more; the ability to trial the new technology or practice; and its adaptability to different physical, economic and social environments (Pannell *et al.* 2006).

5. The changing human face of Australia's rural landscapes means that meeting adoption criteria requires tailoring conservation technology, incentives and communication to a range of different audiences

In a demographic and social sense, Australia's rural communities have been in transition for some time. Studies by Mendham and Curtis (in press) suggest that 50% of rural properties in south-eastern Australia are likely to change hands in the next decade, with a declining number of those landholders describing themselves primarily as farmers. In areas described by Barr (2009) as 'transitional landscapes' (i.e. those within commuting distance of large regional centres), surveys by Mendham *et al.* (2007) indicate that non-farmers make up approximately 50% of landholders. Meanwhile in the traditional production landscapes of north-western Victoria, a longitudinal study indicates the proportion of farmers is 70% and declining (Curtis *et al.* 2008). The significance of this is underpinned by the fact that farmers report higher levels of investment in conservation activity such as fencing remnant vegetation than landholders who identify themselves as non-farmers. While non-farmers placed a higher value on native habitat, they carried out less activity to protect remnant vegetation on their properties. The level of conservation activity was instead positively correlated with the amount of income earned on-farm, technical knowledge, and membership of industry and Landcare networks, and inversely correlated with conservation values (Curtis *et al.* 2007).

6. A marriage between the formal and informal cultures that acknowledges the role of science in developing components, and the role of land managers in developing and testing systems at larger scales, would accelerate this transition

The failure of large-scale visions to take root contrasts with the adoption of a few notable novel systems that have been developed and championed by Australian farmers over the last 50 years. These have typically evolved as alternative approaches to land management in specific landscapes, such as Yeomans' (1964) approach to water harvesting on hill slopes in the Kiewa valley, Andrews' (2006) engineering of watercourses to rehydrate floodplains in the Hunter Valley, and Seis' (2006) polyculture system where winter-active cereals are oversown into dormant summer-active native pastures in central and northern NSW. They have all set out to simultaneously improve productivity and resource conservation. They have also overcome a major obstacle to adoption by being modular. That is, they can be tried at a manageable scale before being adopted over large proportions of farms and landscapes (Pannell *et al.* 2006; Lefroy *et al.* 2008).

That said, Australia has an unfortunate history of unproductive arguments between innovative practitioners and the formal culture on the question of how to best manage the Australian landscape. This is often because it's easy for a particular discipline, interest group or government agency to dismiss a practice as unacceptable on specific grounds, such as the use of exotic species in Andrews' (2006) Natural Sequence Farming. Few if any formal disciplines, however, are as integrated as managers must be, and are therefore poorly equipped to comment on whole of landscape approaches to management.

These farmer-led systems are strong on practice but light on theory and evidence, and usually fail to explain the basis for their advantage over conventional practice in terms either familiar or acceptable to scientists. Consequently, they are ignored or disparaged by the formal culture, not for their lack of apparent success, but their lack of theoretical underpinning. If science can see beyond unsatisfactory explanations for observed phenomena, and see past the components to the larger principles at work, these systems could offer great opportunities for cooperative learning.

Science has well established rules of evidence, and while applying these to test novel systems at large scales is challenging, it is not impossible. To initiate this process, a group of open-minded scientists from a range of disciplines (including geomorphology, hydrology, agronomy, landscape ecology and conservation biology) working with proponents of novel systems could:

1. Develop a shared conceptualisation of how these novel systems work by articulating a set of principles that could be applied generally.
2. Design a series of replicated case studies where these principles could be tested in landscapes with contrasting geomorphology, hydrology and land-use histories.
3. Have the designs for the case studies reviewed by a group of independent scientists and land managers to ensure they represent valid tests of the principles before they were implemented.
4. Assess the functioning of these systems in terms of water and nutrient cycling, energy flows, contribution to the conservation of selected native species, commercial value and adoptability.
5. Communicate the results in terms relevant to the demographic and social contexts of Australia's southern agricultural regions.

This exercise is essentially a marriage between agriculture and ecology, and would need to draw on the experience, language and performance measures of both fields. This concept is not new (Jackson and Piper 1989), but its application has the potential to advance both disciplines and the management of rural landscapes.

Conclusions

Managing woodland remnants in agricultural landscapes will not happen at sufficient scale through education, persuasion or public programs. Achieving conservation goals in these mixed-use landscapes requires integration at two levels. First, integration of human-centred and non-human-centred perspectives of landscapes, which means appreciating that economic drivers, the social context, and personal motivations will ultimately determine if and when conservation measures are adopted by land holders. Second, integration between the formal and the informal culture, between alternative approaches that have emerged from practice and passed the test of adoption, and the rules of evidence and theory required by science.

Agricultural research has traditionally focused at the paddock scale and has been very successful at increasing productivity, but has done so at the expense of the larger scale environmental consequences of simplified and homogenised landscapes. Novel systems that have emerged from practice rather than theory have the potential to bridge the gap between ecologically based visions of future landscapes, and new crops and farming systems in development. The result of such a marriage would be transitional landscapes that lie structurally and functionally between the present cleared and the pre-cleared states.

Acknowledgements

This chapter draws on research carried out in collaboration with Richard Stirzaker, John Pate, and Richard Hobbs funded by Land & Water Australia, the Grains R&D Corporation, the Rural Industries R&D Corporation, and the Cooperative Research Centre for Plant-Based Management of Dryland Salinity. Richard Stirzaker suggested the title 'The Normal, the Novel and the Natural', while two of the editors provided helpful comments on an earlier draft.

Bio

Ted Lefroy trained as an agricultural scientist and worked in rural extension in Queensland, Papua New Guinea and Western Australia before taking up a research career focused on the development of conservation farming techniques. In 2003 he led a team that was awarded a Eureka Prize for research into the sustainability of grain production, and in 2008 led a team of ecologists, farmers and extension workers that won a Banksia Award for a national study of the role of mixed farming in biodiversity conservation. Since 2005 he has been Director of the Centre for Environment at the University of Tasmania.

References

Andrews, P. (2006). *Back from the Brink. How Australia's Landscapes Can Be Saved.* ABC Books: Sydney.
Barr, N. (2009). *The House on the Hill. The Transformation of Australia's Farming Communities.* Halstead Press: Canberra.
Curtis, A. and Lockwood, M. (2000). Landcare and catchment management in Australia: lessons for state-sponsored community participation. *Society and Natural Resources* **13**, 61.
Curtis, A., McDonald, S., Sample, R. and Mendham, E. (2008). Understanding the social drivers for natural resource management in the Wimmera region. Institute for Land, Water and Society, Charles Sturt University: Albury, NSW.
Goss, K. (2007). Focus on future farm industries. Focus on Perennials October 2007, Future Farm Industries Co-operative Research Centre: Perth. http://www.futurefarmcrc.com.au/documents/FocusonFutureFarmIndustries.pdf. accessed September 2009.
Hatton, T.J. and Nulsen, R.J. (1999). Towards achieving functional ecosystem mimicry with respect to water cycling in southern Australian agriculture. In *Agriculture as a Mimic of Natural Ecosystems.* (Eds E.C. Lefroy, R.J. Hobbs, J.S. Pate and M.H. O'Connor) pp. 259–270. Kluwer: Dordrecht, The Netherlands.
Jackson, W. (2002). Poverty and agricultural policies. *Population and Environment* **24**, 55–67.
Jackson, W. and Piper, J. (1989). The necessary marriage between ecology and agriculture. *Ecology* **70**, 1591–1593.
Lefroy, E.C. (2009). Agroforestry and the functional mimcry of natural ecosystems. In *Agroforestry for Natural Resource Management.* (Eds I. Nuberg, B. George and R. Reid) pp. 23–35. CSIRO Publishing: Melbourne.
Lefroy, E.C., Bailey, K., Unwin, G. and Norton, T.W. (2008). *Biodiversity: Integrating Conservation and Production.* CSIRO Publishing: Melbourne.
Lefroy, E.C., Hobbs, R.J. and Atkins, L.J. (1991). 'Revegetation guide to the central wheatbelt'. Western Australian Department of Agriculture Bulletin 4231.
McHarg, I. L. (1971). *Design with Nature.* Doubleday: New York.
Mendham, E. and Curtis, A. (in press). Taking over the reins: trends and impacts of changes in rural property ownership. *Society and Natural Resources.*
Mendham, E., Millar, J. and Curtis, A. (2007). Landholder participation in native vegetation management in irrigation areas. *Ecological Management and Restoration* **8,** 42–48.
Pannell, D. J. (2004). Heathens in the chapel? Application of economics to biodiversity. *Pacific Conservation Biology* **10**, 88–105.
Pannell, D.J. (2008). Public benefits, private benefits, and policy intervention for land-use change for environmental benefits. *Land Economics* **84**, 225–240.

Pannell, D.J., Marshall, G.R., Barr, N., Curtis, A., Vanclay, F. and Wilkinson, R. (2006). Understanding and promoting adoption of conservation practices by rural landholders. *Australian Journal of Experimental Agriculture* **46**, 1407–1424.

Parker, C.A. (1989). Soil biota and plants in rehabilitation of degraded agricultural soils. In *Animals in Primary Succession: The Role of Fauna in Reclaimed Land.* (Ed. J.D. Majer) pp. 423–438. Cambridge University Press: Cambridge.

Reid, R. (2008). *Tree Change: The Australian Master Tree Grower Phenomenon.* Rural Industries Research and Development Corporation: Canberra.

Seis, C. (2006). Pasture cropping: techniques and equipment. *Australian Organic Journal*, Summer 2006, 44–45.

Stirzaker, R.J., Lefroy E.C., Keating B.A. and Williams J. (2000). 'A revolution in land use. Emerging land use systems for managing dryland salinity'. CSIRO Land & Water: Canberra.

Stirzaker, R.J., Ellis, T.J. and Lefroy, E.C. (2002). Mixing tree belts with agriculture. In *Trees, Water and Salt.* (Eds R.J. Stirzaker, R.J. Vertessey and A. Sarre) pp. 77–91. Rural Industries Research and Development Corporation: Canberra.

Yeomans, P.A. (1964). *Water for Every Farm.* Murray: Sydney.

PART B

Research: by state

7

ECOLOGY AND RESTORATION OF FRAGMENTED WOODLANDS: A WESTERN AUSTRALIAN PERSPECTIVE

Richard Hobbs

1. Understand the regeneration needs of woodland trees: the basics of a source of seeds, a suitable substrate, suitable growing conditions and lack of grazing are essential.
2. Understand the changes arising in an altered landscape: changing the landscape leads to changes in biological processes such as dispersal and pollination, and abiotic processes such as hydrology.
3. Figure out what needs to be done to repair damaged woodlands: aim for the minimum management input necessary for success, but don't settle for anything less.
4. Set priorities within a logical framework, but don't expect it to be an easy task.
5. Think about what climate change means for woodland conservation.
6. Use past work as a basis for understanding: don't reinvent the wheel.

Introduction

My appreciation and understanding of woodlands has developed over the 26 years I have lived and worked in Western Australia (see map on next page). I originally moved there to join a CSIRO research team investigating the dynamics and management of fragmented ecosystems in the Western Australian wheatbelt. Travelling through the wheatbelt, with its vast expanses of land cleared for wheat and sheep, the remaining patches of woodland (and other ecosystems such as the kwongan) provided a touchstone to what remained of the past nature of the area – not just for me, but for many of the locals and most people who worked there. I was surprised, therefore, when it came to considering how these woodlands worked and how they might be managed effectively, to encounter an almost complete lack of information on the basic ecology of even the dominant trees.

That realisation set me (and others) on a journey of finding out how the woodlands worked, how the dominant tree species regenerated, and ultimately, how the woodland systems might be managed and restored. This work was carried out predominantly with PhD students Colin Yates and Trish Fox, but many other people contributed to filling in our understanding of the

woodland systems and elements of how they worked, including Max Abensperg-Traun, Graeme Smith, Alison Hester, Denis Saunders and others with contributions in this book. Colin Yates continued on as a postdoctoral fellow and, as part of a Land & Water Australia-funded project, we ran the first workshop on woodland management and conservation in Australia, which resulted in a book in 2000 (Hobbs and Yates 2000).

Since then, the work has expanded to consider landscape-level issues such as salinity, and how climate change might interact with all the other environmental changes underway, asking the question of how we intervene meaningfully and effectively to ensure that woodlands and other ecosystems persist into the future.

Lessons

1. Understand the regeneration needs of woodland trees: the basics of a source of seeds, a suitable substrate, suitable growing conditions and lack of grazing are essential

Without trees there is no woodland. Hence it is important to understand how trees work – how they live, grow and reproduce. And yet because of their longevity and size, trees are some of the more difficult organisms to study effectively. A mature woodland tree may be several hundred years old and tall enough to make it very difficult to see what is going on at the top in terms of flowering, pollination, seed set, etc. There is a range of processes, however, which need to be understood to ensure that woodland trees grow to maturity and can successfully reproduce. Mature trees are the woodland of today – seeds and seedlings are the woodlands of tomorrow.

Mature trees in woodlands across Australia have been declining in both number and in condition. The phenomenon of rural tree dieback has been discussed for decades, and is a complex phenomenon with multiple causes (Jurskis 2005). Often trees persist in remnants which are degraded by stock grazing, weed invasion and/or altered fire regimes, and which are in a very different landscape matrix than previously. Add drought and increased prevalence of pests and diseases and you have a rich cocktail of elements which can cause problems for trees. In light of this, it is amazing that there still appear to be healthy trees persisting in some areas. Determining long-term trends in tree abundance and health is difficult without long-term observations, which unfortunately are still few and far between. Where researchers have persisted with observations or revisited study sites after some time, the results are both illuminating and daunting, with evidence of both declining tree health and numbers, with no regenerating seedlings to replace the mature trees which are dying (Saunders *et al.* 2003).

For woodland trees to successfully regenerate, they need to flower and set seed. That seed has to be present either in the canopy or in the soil when conditions are right for germination. Some tree species can regenerate at any time whereas others depend on disturbances such as fire, which open up spaces by killing adult trees or thinning the understory. Seeds thus have to be produced at the right time or be stored and avoid predation for long enough to encounter such an event. Assuming seeds are available, the soil substrate needs to be suitable in terms of water and nutrient supply to ensure that the seeds germinate successfully and that seedlings grow and survive, in some cases through a prolonged summer drought period. At the young stage, seedlings are particularly susceptible to grazing either by native animals or by stock. There are thus an array of hurdles to be overcome for successful regeneration to occur and the ability to assess whether conditions are likely to be conducive or not is an important part of woodland management (Hobbs and Yates 2000; Hobbs and Yates 2003).

2. Figure out what needs to be done to repair damaged woodlands: aim for the minimum management input necessary for success, but don't settle for anything less

The assessment of woodland condition and regeneration potential can elucidate whether there are problems which need to be attended to for successful regeneration to occur. Common problems in fragmented landscapes include a lack of viable seed through poor seed set or seed predation, adverse soil conditions which prevent seeds from germinating or seedlings from surviving, competition from weeds, and excess grazing pressure on seedlings. Long periods of stock grazing, in particular, can degrade woodland systems to the extent that regeneration can no longer occur. In some cases, stock can compact the soil, remove the litter layer and alter the soil microclimate and water infiltration (Hobbs and Yates 2000; Yates *et al.* 2000; Hobbs and Yates 2003). In these cases, simply removing grazing animals will not result in regeneration; instead, a woodland system is in a new alternative state which requires active management to overcome (Yates and Hobbs 1997; Standish *et al.* 2009).

Recognising the presence of barriers to regeneration is an essential element in deciding how to manage woodland remnants. This tells a manager what is needed to encourage regeneration. In some woodland systems, relatively little needs to be done. Regeneration will occur when the right climatic conditions or disturbances come along. In some cases, this process needs to be aided by the temporary removal of grazing pressure, particularly by stock. Where grazing has caused fundamental changes in the soil, however, fencing the woodland will have little effect: soil problems have to be dealt with as well. In addition, where weeds are a problem, these also have to be managed for successful regeneration to occur. Often there are nasty synergistic effects between weeds, grazing and disturbance which mean that conditions or events which would normally encourage regeneration instead promote weed growth which in turn thwarts seedling survival (Hobbs 2001; Standish *et al.* 2007). Hence, it is vital to consider all the possible factors affecting regeneration and dealing with all of them – doing a partial job will still take time and effort but not result in the desired outcome (Brown and Holt 2000; Hobbs and Yates 2000).

3. Understand the changes arising in an altered landscape: changing the landscape leads to changes in biological processes such as dispersal and pollination, and abiotic processes such as hydrology

Landscape modification through agricultural or urban development usually results in a reduction in the amount of native vegetation and its fragmentation into patches of varying sizes and degrees of isolation. Considerable effort has gone into trying to understand the effects of these changes, and over time a better appreciation of the effects of both degree of vegetation loss and changes in condition of the remaining patches has developed (McIntyre and Hobbs 1999; Lindenmayer and Fischer 2006). Deciding what to do where, when managing fragmented landscapes depends on an initial assessment of landscape condition and the opportunities and constraints this provides (Salt *et al.* 2004). In addition, a better understanding has developed of the importance of considering not only the remaining vegetation patches but also the surrounding altered matrix and its properties. In some cases, especially when considering fauna habitat and connectivity or permeability, it may be better to consider fragmented landscapes using a gradient model or a mosaic model that recognises an array of different kinds of patches (or land uses) of varying quality rather than a simple binary habitat/non-habitat model (Fischer *et al.* 2004; Fischer and Lindenmayer 2006; Lindenmayer *et al.* 2008).

As landscapes become fragmented, processes such as seed dispersal and pollination are at risk of disruption (Hobbs and Yates 2003). Some recent studies indicate, however, that pollination processes may remain relatively intact even in highly fragmented systems and highlight the importance of even quite small remnants in maintaining connectivity across fragmented populations (Byrne *et al.* 2008). The impacts of fragmentation on processes such as pollination and seed dispersal will depend on how fragmentation affects the seed and pollen vectors, particularly bird, mammal and insect communities. Such communities change in complex ways in response to the altered landscape matrix and also as a result of introduced species: these issues are considered in other chapters in this book.

Abiotic processes also change as a result of landscape fragmentation and the broadscale removal of native vegetation. Of particular concern in south-western Australia and the eastern agricultural regions is the change in hydrology resulting from altered evapotranspiration as perennial native vegetation is replaced by annual crops and pastures. While considerable effort has gone into understanding the process of secondary salinisation resulting from these hydrological changes from an agricultural perspective, until relatively recently very little attention was paid to the effects of salinisation on native vegetation. Woodlands occupying lower and mid-slope landscape positions are particularly at risk from rising water tables and salinity, and woodland management has to consider not only the processes within the woodland itself but also the broader landscape processes underway (Cramer and Hobbs 2002; Cramer and Hobbs 2005). Management of these processes of necessity has to extend beyond the remnant boundaries and tackle the broader hydrological processes using revegetation, engineering works and novel agricultural systems as appropriate.

4. Set priorities within a logical framework, but don't expect it to be an easy task

There is always going to be more to do in conserving and restoring woodlands, and, barring a radical change in societal and governmental priorities, there are always going to be limited resources available to do the work. Hence, some form of priority setting is inevitably necessary.

What needs to be done where and in which order? Numerous schemes for setting priorities have been developed, including some that are outlined in other chapters in this book, and it is clear that there is not a universally agreed approach to this. We have attempted to develop a set of approaches to setting priorities for woodland management based on the assessment of within-patch condition described above and a consideration of the landscape context. Correct diagnosis of what, if anything, is currently wrong is an essential first step. Further considerations include the degree of threat, the feasibility of the management required, the likelihood of it succeeding, and the relative value of the different patches or remnants being considered (Cramer and Hobbs 2002; Hobbs *et al.* 2003; Hobbs and Kristjanson 2003).

Considering all these factors results in a matrix within which management categories can be recognised based on the degree of urgency and level of intervention necessary. While this scheme might seem logical, there are, in fact, endless complications involved in actually applying it. Considering more than one or a few threats at a time (e.g., salinity, invasive species) rapidly makes the assessment very complex. Assigning degrees of urgency and estimating likelihood of success are tricky enterprises, as is the assignation of value. Finally, making some areas or actions a priority over others immediately leads to there being winners and losers – consigning some areas to a low priority status has inevitable economic, social and political consequences which may be unpalatable to the local community and untenable politically. Hence, it is essential to consider how to set priorities, but socially and politically tricky to do so effectively. There appears to be no easy answer to this conundrum.

5. Think about what climate change means for woodland conservation

Most thoughtful individuals who have examined the facts rather than the hype on climate change recognise that the climate is in a state of rapid flux, probably at least partially as a result of human activities. The impacts of climate change on biodiversity are uncertain but potentially huge (Steffen *et al.* 2009). In south-western Australia, there is clear evidence for declining rainfall over the past 40 years, with ensuing drought impacts on both natural and agricultural systems (Indian Ocean Climate Initiative 2002). Observed declines in tree health across many woodland types are notoriously hard to pin on particular causes, but prolonged climatic stress is increasingly seen as being an important factor, either directly or as a precursor to attack by pests and pathogens, both in Australia (Rice *et al.* 2004; Hooper and Sivasithamparam 2005; Jurskis 2005), and worldwide (Allen *et al.* 2010).

Future climate scenarios are relatively unhelpful in considering regional and local biotic and management responses. In the case of south-western Australia, however, ongoing reductions in rainfall amounts and changes in seasonal rainfall distributions are predicted. How projected climatic changes will affect woodlands and other ecosystems is still largely the stuff of conjecture, and the task is made more difficult by the likelihood that many regions may experience climatic conditions completely unlike anything currently in place. If there is no current analog for the climate projected for a particular area, standard methods of predicting species distributions on the basis of their current climate envelope become unusable (Fitzpatrick and Hargrove 2009).

Climate change also has important implications for restoration efforts. A directionally changing climate leads to important questions concerning what the target for any particular restoration project should be, given that the current local climate is likely to change (Harris *et al.* 2006). In addition, restoration practitioners will have to make increasingly difficult decisions about what species or genotypes to use in restoration projects: the standard protocol of using local species and provenances may no longer be considered optimal and some consideration of using material from a wider range of sources may need to be considered (Broadhurst *et al.* 2008). There is, however, as yet no new consensus on how to deal with this issue effectively, and all restoration activities need to be tackled in an experimental way, perhaps using a variety of options where possible.

In addition, the likely impacts of climate change cannot be considered in isolation, but instead must be viewed in the context of other ongoing changes, including fragmentation, altered fire regimes and invasive species (Hobbs and Cramer 2008). Tomorrow's world is likely to be even more complex and unpredictable than today's. This is not a reason to give up – rather it is a reason to look for more effective and durable solutions. Simply putting into practice what is already known about how woodlands should be managed would go a long way to preparing woodland systems, and us, for the future.

6. Use past work as a basis for understanding: don't reinvent the wheel

People know stuff. Particularly people who have been working in an area for a long time, be they scientists, farmers or conservation managers. It is important when embarking on a new project, either a research project or a management or restoration effort, to check to see what is known already, who has worked in the area, and what might be learned from what others have done in similar situations elsewhere. Resources and manpower are always limited in both research and management, and hence it is essential to ensure that we spend our time doing things that will make a difference. We can't afford to spend time reinventing the wheel.

As an example, in the final chapter of the 2000 woodlands book, Colin Yates and I pulled together a checklist of how to assess the current condition of a woodland, diagnose what was wrong, and initiate suitable restorative actions (Hobbs and Yates 2000). This checklist was then translated into extension material and made available to practitioners (Brown and Holt 2000), as well as being covered in a number of national practitioner-based periodicals. Since then, however, both the book and the extension materials have become relatively inaccessible, and hence the information is in danger of being lost. Indeed, I have recently been involved in discussions regarding the need to develop such materials as part of newer research initiatives. The checklist developed in 2000 remains relevant today, and a good starting point from which to incorporate newer information. The lessons from this are two-fold. First, it is apparent that senior researchers like me need to keep plugging the same message repeatedly (hopefully with greater finesse as new insights become available) and not assume that because it has been published and promoted in the past, it will remain current. Second, it is clear that young researchers and people new to management roles need to seek advice from the senior people in their area to find out what is already known and what guidance might be available. Institutional memory is fragile and once it is lost, it can be hard to re-establish.

Conclusions

We have learned a lot since the publication of the 2000 woodlands book, but we still have a lot more to learn. In Western Australia, we are still lacking information on the basic ecology of many of the dominant tree species and most of the other components. We are now grappling with the added complication of climate change as we decide how best to manage the remaining woodlands and restore woodlands to cleared and damaged areas. This is not an excuse for doing nothing or waiting for more research. We know enough to do 'best-bet' approaches which can be monitored and modified as necessary. We need to get on with the job lest woodlands become a haunting image from the past rather than an iconic symbol for today and the future.

Acknowledgements

I thank all the people who have contributed to the work on woodlands referred to here and all the students and colleagues who have participated over the years. I also thank the Australian Research Council, RIRDC and particularly Land & Water Australia for supporting this research.

Bio

Richard Hobbs is currently Professor of Restoration Ecology in the School of Plant Biology at the University of Western Australia, where he holds an ARC Australian Laureate Fellowship, and leads the Ecosystem Restoration Laboratory, a dynamic research group engaged in a variety of projects. Originally from Scotland, he spent three years in California and has been in Western Australia since 1984, working with CSIRO and at Murdoch University before joining UWA in 2009. His particular interests are in vegetation dynamics and management, invasive species, ecosystem restoration, conservation biology and landscape ecology. He is the author of over 300 scientific publications, many magazine articles and other publications, and author/editor of 18 books. He is currently Editor in Chief of the journal Restoration Ecology

References

Allen, C.D., Macalady, A.K. *et al.* (2010). A global overview of drought and heat-induced tree mortality reveals emerging climate change risks for forests. *Forest Ecology and Management* **259**, 660–684.

Broadhurst, L.M., Lowe, A., Coates, D.J., Cunningham, S.A., McDonald, M., Vesk, P.A. and Yates, C. (2008). Seed supply for broadscale restoration: maximizing evolutionary potential. *Evolutionary Applications* **1**, 587–597.

Brown, L. and Holt, C. (2000). 'Eucalypt woodlands: a guide to management. Miscellaneous Publication 17/00.' Agriculture WA: Perth.

Byrne, M., Elliott, C.P., Yates, C.J. and Coates, D.J. (2008). Maintenance of high pollen dispersal in *Eucalyptus wandoo*, a dominant tree of the fragmented agricultural region in Western Australia. *Conservation Genetics* **9**, 97–105.

Cramer, V.A. and Hobbs, R.J. (2002). Ecological consequences of altered hydrological regimes in fragmented ecosystems in southern Australia: impacts and possible management responses. *Austral Ecology* **27**, 546–564.

Cramer, V.A. and Hobbs, R.J. (2005). Assessing the ecological risk from secondary salinity: a framework addressing questions of scale and threshold responses. *Austral Ecology* **30**, 537–545.

Fischer, J. and Lindenmayer, D.B. (2006). Beyond fragmentation: the continuum model for fauna research and conservation in human-modified landscapes. *Oikos* **112**, 473–480.

Fischer, J., Lindenmayer, D.B. and Fazey, J. (2004). Appreciating ecological complexity: habitat contours as a conceptual landscape model. *Conservation Biology* **18**, 1245–1253.

Fitzpatrick, M.C. and Hargrove, W.W. (2009). The projection of species distribution models and the problem of non-analog climate. *Biodiversity Conservation* **18**, 2255–2261.

Harris, J.A., Hobbs, R.J., Higgs, E. and Aronson, J. (2006). Ecological restoration and global climate change. *Restoration Ecology* **14**, 170–176.

Hobbs, R.J. (2001). Synergisms among habitat fragmentation, livestock grazing and biotic invasions in south-western Australia. *Conservation Biology* **15**, 1522–1528.

Hobbs, R.J. and Cramer, V.A. (2008). Restoration ecology: interventionist approaches for restoring and maintaining ecosystem function in the face of rapid environmental change. *Annual Review of Environment and Resources* **33**, 39–61.
Hobbs, R.J., Cramer, V.A. and Kristjanson, L.J. (2003). What happens if we can't fix it? Triage, palliative care and setting priorities in salinising landscapes. *Australian Journal of Botany* **51**, 647–653.
Hobbs, R.J. and Kristjanson, L.J. (2003). Triage: how do we prioritize healthcare for landscapes? *Ecological Management and Restoration* **4 Suppl.**, S39–S45.
Hobbs, R.J. and Yates, C.J. (2000). Priorities for action and management guidelines. In *Temperate Eucalypt Woodlands in Australia: Biology, Conservation, Management and Restoration.* (Eds RJ Hobbs and CJ Yates) pp. 400–414. Surrey Beatty & Sons: Chipping Norton, NSW.
Hobbs, R.J. and Yates, C.J. (2003). Impacts of ecosystem fragmentation on plant populations: generalizing the idiosyncratic. *Australian Journal of Botany* **51**, 471–488.
Hooper, R.J. and Sivasithamparam, K. (2005) Characterization of damage and biotic factors associated with the decline of *Eucalyptus wandoo* in southwest Western Australia. *Canadian Journal of Forest Research* **35**, 2589–2602.
Indian Ocean Climate Initiative (2002). 'Climate variability and change in south west Western Australia.' Department of Environment, Western Australia: Perth.
Jurskis, V. (2005). Eucalypt decline in Australia, and a general concept of tree decline and dieback. *Forest Ecology and Management* **215**, 1–20.
Lindenmayer, D.B. and Fischer, J. (2006). *Landscape Change and Habitat Fragmentation.* Island Press: Washington D.C.
Lindenmayer, D., Hobbs, R.J. *et al.* (2008). A checklist for ecological management of landscapes for conservation. *Ecology Letters* **11**, 78–91.
McIntyre, S. and Hobbs, R.J. (1999). A framework for conceptualizing human impacts on landscapes and its relevance to management and research. *Conservation Biology* **13**, 1282–1292.
Rice, K.J., Matzner, S.L., Byer, W. and Brown, J.R. (2004). Patterns of tree dieback in Queensland, Australia: the importance of drought stress and the role of resistance to cavitation. *Oecologia* **139**, 190–198.
Salt, D., Lindenmayer, D.B. and Hobbs, R.J. (2004). 'Trees and biodiversity: a guide for improving biodiversity outcomes in tree plantings on farms.' RIRDC: Canberra.
Saunders, D.A., Smith, G.T., Ingram, J.A. and Forrester, R.I. (2003). Changes in a remnant of salmon gum *Eucalyptus salmonophloia* and York gum *E. loxophleba* woodland, 1978 to 1997. Implications for woodland conservation in the wheat-sheep regions of Australia. *Biological Conservation* **110**, 245–256.
Standish, R.J., Cramer, V.A., Wild, S. and Hobbs, R.J. (2007). Seed dispersal and recruitment limitation are barriers to native recolonisation of old-fields in Western Australia. *Journal of Applied Ecology* **44**, 435–445.
Standish, R.J., Cramer, V.A. and Yates, C.J. (2009). A revised state-and-transition model for the restoration of woodlands in Western Australia. In *New Models for Ecosystem Dynamics and Restoration.* (Eds R.J. Hobbs and K.N. Suding) pp. 169–188. Island Press: Washington D.C.
Steffen, W.L., Burbidge, A.A., Hughes, L., Kitching, R.L., Lindenmayer, D.B., Musgrave, W., Stafford Smith, D.M. and Werner, P.A. (2009). *Australia's Biodiversity and Climate Change.* CSIRO Publishing: Melbourne.

Yates, C.J. and Hobbs, R.J. (1997). Woodland restoration in the Western Australian wheatbelt: a conceptual framework using a state and transition model. *Restoration Ecology* **5**, 28–35.

Yates, C.J., Hobbs, R.J. and Atkins, L. (2000). Establishment of perennial shrub and tree species in degraded *Eucalyptus salmonophloia* remnant woodlands: effects of restoration treatments. *Restoration Ecology* **8**, 135–143.

Yates, C.J., Norton, D.A. and Hobbs, R.J. (2000). Grazing effects on plant cover, soil and microclimate in fragmented woodlands in south-western Australia: implications for restoration. *Austral Ecology* **25**, 36–47.

8

THE EUCALYPT WOODLANDS OF WESTERN AUSTRALIA: LESSONS FROM THE BIRDS

Harry F. Recher, Jonathan D. Majer and William E. Davis, Jr.

1. Vegetation structure and floristics are the primary determinants of the composition of woodland bird communities, with species affected seasonally and from year to year by the abundance and kind of food resources available.
2. The most productive woodlands support the richest bird communities.
3. Within woodlands there are micro-habitats that act as refuges during drought and other times of food scarcity.
4. Birds use and require many different resources throughout the year for food and nesting.
5. Insectivorous birds affect the abundance, kinds and size of canopy arthropods, but do not prevent outbreaks.
6. The majority of woodland bird species are nomadic or migratory, but the most abundant species are sedentary.
7. Increased temperatures and changed patterns of rainfall associated with climate change have significant effects on woodland birds.
8. The conservation and management of woodland birds need to be integrated on a landscape scale.

Introduction

Since 1986, we have studied birds in the eucalypt woodlands of south-western Western Australia at Dryandra near Narrogin, at Mt Gibson Station north of the wheatbelt, and in the Great Western Woodlands at Yellowdine, Widgiemootha, and Norseman (Majer *et al.* 1989; Recher *et al.* 1991, 1996, 2002; Recher and Davis 1998, 2002; Recher and Majer 2006) (see map on next page and Figure 8.1). At each place, we record the abundance of birds, their foraging habits, the foods they take and where they nest. We work during the spring and early summer (July–October) when breeding and nesting activities are greatest. In association with Jonathan Majer (see Majer *et al.* this book), we measured the abundance and richness of canopy and bark arthropod faunas (e.g. Majer *et al.* 1989; Recher *et al.* 1996), which are primary food sources for birds, and we have investigated the impact of birds feeding on the arthropods associated with

the tree canopy (Evelegh *et al.* 2001; Recher and Majer 2006). Our aim has been to describe the bird assemblages of the woodlands and the resources (food and nest sites) they require for their long-term survival.

Despite the clearing of the wheatbelt, Western Australia retains the most extensive and least disturbed temperate woodlands in southern Australia. The Great Western Woodlands falls beyond the clearing line and is a 16 million hectare region of eucalypt woodland associated with mallee, heaths, and salt lakes (Watson *et al.* 2008). Although the Great Western Woodlands has a long history of disturbance by mining, roads, logging, and wildfires, a core of more than 6 million hectares has never been grazed by domestic stock, with large areas unaffected by mining or logging. The lack of grazing is particularly significant, as domestic grazers compact the soil, remove the litter layer and change the structure and species composition of the vegetation. Grazing may not be as severe an impact as clearing for farming (but see Majer *et al.* this book), a practice which continues in Western Australia (Fulton and Majer 2006), but ungrazed woodlands are a unique and precious heritage in the 21st century. Their management and conservation should be a national priority. Thus the goal of our studies has been to expand our knowledge of woodland birds and their food resources, and use that knowledge to develop guidelines that will improve Australia's conservation management. In this, we are strongly influenced by the concept of WildCountry and the need for habitat connectivity (Recher 2003; Soule *et al.* 2004; Mackey *et al.* 2007).

Lessons

1. Vegetation structure and floristics are the primary determinants of the composition of woodland bird communities, with species affected seasonally and from year to year by the abundance and kind of food resources

Vegetation structure and floristics are the primary determinants of avian species composition within continuous eucalypt woodlands, with composition and abundances affected seasonally

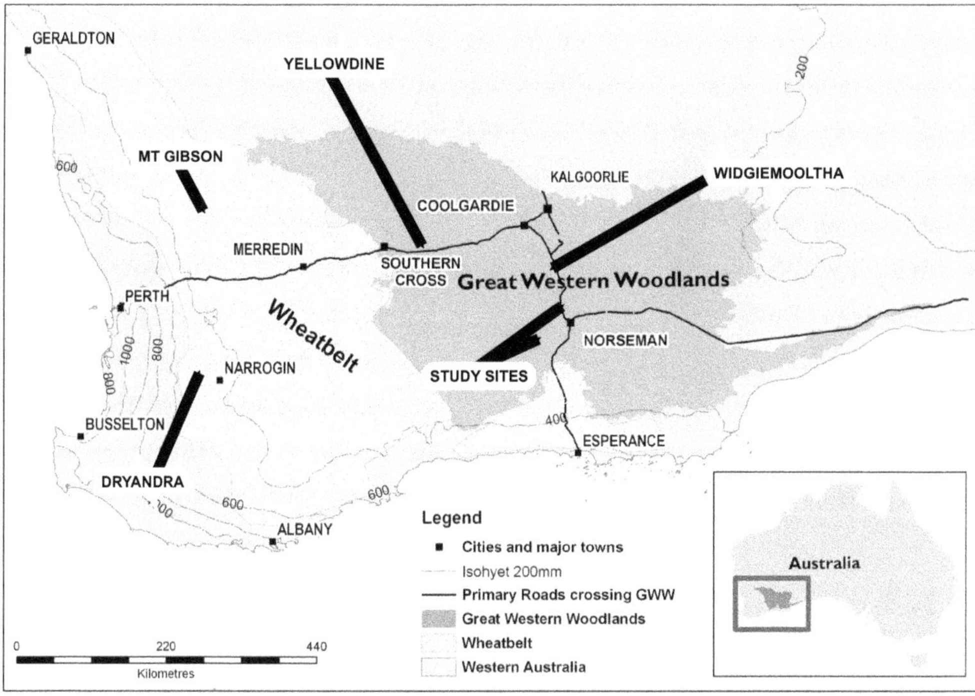

Figure 8.1 Map of south-western Western Australia showing principal towns and roads where birds were studied at Dryandra (near Narrogin); on Mt Gibson Station along the mulga-eucalypt line; and, in the Great Western Woodlands at Yellowdine east of Southern Cross; at Widgiemooltha between Coolgardiae and Norseman; and, immediately north and west of Norseman.

and from year to year by the abundance and kinds of food resources available. Omnivorous, fruit-eating, nectar-feeding, seed-eating and insect-eating birds are affected, with the most pronounced short-term changes in abundance occurring in migratory and nomadic species. Differences in food availability between years are primarily related to variation in rainfall, with spatial patterns determined by site productivity and floristics, while rainfall and temperature affect seasonal abundances (Recher *et al.* 1996). As best we can determine, ungrazed woodlands, such as those in the Great Western Woodlands centred on Norseman, retain an intact avifauna, with no evidence of either species or trophic level extinctions, although migratory and nomadic species have declined in abundance. Where woodlands are grazed by domestic animals or have been fragmented and isolated by agricultural clearing, such as within the wheatbelt, only a dysfunctional avifauna characterised by the decline and loss of many species, and the increased abundance of others, remains.

2. The most productive woodlands support the richest bird communities

As with canopy and bark arthropods (Majer *et al.* 1989, 1992; Recher *et al.* 1996), birds are most abundant and there is the greatest number of species in woodlands on fertile, well-watered, and well-drained soils. Often these are at the bottom of slopes or are 'run-on' areas that receive water and nutrients from adjacent habitats through surface flow and wind (Recher *et al.* 1991). These are the habitats that have been preferentially cleared for farming and are most affected by logging and tourism, as well as by human settlement. The most productive woodlands are often floristically and structurally diverse, thereby offering a wider array of foraging and nesting opportunities than less productive and/or more simply structured

Figure 8.2 Salmon Gum (*Eucalyptus salmonophloia*) woodland with a bluebush (*Maireana* sp.) understorey in the Great Western Woodlands near Norseman, Western Australia (H.F.Recher, photographer).

habitats, and have less seasonal and/or annual variation in resource availability. Birds are abundant and species are numerous because such habitats provide more food with more variety in the form of insects and other animals, as well as food from plants (e.g. nectar, fruit). By contrast, extensive areas of what are probably some of the most productive woodlands are dominated by the Yellow-plumed Honeyeater (*Lichenostomus ornatus*), a semi-colonial species, which excludes other, mainly smaller birds aggressively, creating what we have come to call 'honeyeater deserts'.

3. Within woodlands there are micro-habitats that act as refuges during drought and other times of food scarcity

Within woodlands there are micro-habitats that act as refuges during drought and other times of food scarcity. Even slight difference in soil and position in the landscape, as well as the lingering effects of storm, fire and other disturbances, create a mosaic of habitats that differ in vegetation structure and plant species. Birds respond to these differences, with the result that each patch of habitat within a woodland sustains a different mix of bird species. Habitat patches, even ones seemingly structurally and floristically identical, differ in productivity, and can be so small that they support only one or a few breeding units of a bird species. Our (unpublished) observations suggest that the most productive patches enable successful reproduction even through severe and prolonged drought, while birds leave or fail to breed elsewhere. Presently, other than by monitoring breeding outcomes, there is no way to identify these micro-refuges, yet they are likely to be important for the survival of woodland birds as climate change progresses. In itself, this is a lesson; often observations of natural phenomena require years or decades of study to be confirmed, but this does not mean that they should not be considered when managing natural areas. If there are 'micro-refuges' that shelter woodland birds during drought, management actions, such as hazard reduction burning that may reduce habitat complexity, should be avoided or used in ways that ensure a mosaic of micro-habitats is conserved.

4. Birds use and require many different resources throughout the year for food and nesting

Knowing what foods birds use, where they feed and how their foraging habits differ between habitats, as well as seasonally and from year to year, is important for the effective conservation of birds (and other fauna). Woodland birds use a wide range of foods, changing where and how they feed in response to daily, seasonal and year-to-year shifts in the abundances and kinds of food available (Recher 1989; Wilson and Recher 2001). Unless it is large, with a diversity of habitats, a conservation reserve may not provide the full range of foods needed by birds through time. Where this occurs, connections with other reserves are necessary for the survival of species, especially nomadic birds that chase food resources such as nectar over large geographic regions.

Many woodland birds depend on carbohydrates for energy. Some are seed-eaters, although finches, doves and the seed-eating parrots are not common in most of the woodlands where we work. In the Great Western Woodlands, the most common parrots are omnivores, feeding on leaf and flower buds, young plant growth, and green seeds, as well as taking nectar and insects. The Purple-crowned Lorikeet (*Glossopsitta porphyrocephala*) forages almost exclusively on the nectar and pollen of eucalypts and is joined by a dozen or so species of honeyeaters (Meliphagidae). Honeyeaters can be described as 'short' or 'long' billed, with the short-billed species less dependent on nectar, relying instead on alternative carbohydrates for energy, such as lerp (sugary excretions of sap-sucking psyllid insects) taken from eucalypt foliage, or honeydew (more sugary excretions of insects) obtained by probing under bark. The abundance of nectar, lerp and honeydew changes through time, with many honeyeaters and lerp-eaters moving long distances between patches of nectar-rich flowers or outbreaks of psyllid insects. Managing these spatially and temporally variable resources requires large areas and good connectivity between different habitats and reserves.

Most species of woodland birds are insect-eaters and require different management strategies from nectar-feeders. Insectivorous species differ in body size, the height above the ground at which they feed, and the way they catch their prey (Recher *et al.* 1985; Recher and Davis 1998, 2002). Some pounce on their prey from perches, while others catch insects in the air and their foraging requirements can be very specific. The ground-pouncing robins (e.g. *Petroica, Eopsaltria*), for example, do best in habitats with an abundance of perches and a mosaic of bare soil, litter, ground vegetation and woody debris, all of which provides good habitat for ground-dwelling insects and spiders, while not being so thick as to interfere with the birds seeing and catching prey (Recher *et al.* 2002; Cousin 2004). The lesson here is that changed fire regimes, grazing by domestic animals and weeds alter the mosaic of soil, litter and ground vegetation, which in turn affects the birds that feed there. Insectivores feeding above the ground are equally affected by any management action that changes the density and composition of the vegetation, or that affects the abundance and kinds of arthropods available in the shrub and canopy layers.

Food is not the only resource needed for survival by birds. Being able to nest is also important. Unfortunately, too much conservation emphasis is placed on tree hollows and the birds that nest in tree hollows. All birds have specific nesting requirements and each species selects a unique site for its nest (Recher 2004). Only a few woodland birds in Western Australia rely on tree hollows for nesting and tree hollows are not the only nest sites associated with mature and large trees that birds require. Cuckoo-shrikes, for example, place their nests on large, horizontal branches, usually where a fork in the branch provides extra support. Dead branches, whose colour blends with the spiderweb and dead plant material used to build the nest, are preferred and the branch must be in the open so that access is easy. Similarly, the Black-capped Sittella (*Daphoenositta pileata*) builds a nest of spiderweb, lichen and bark on a vertical dead branch against which the nest is virtually invisible. Such branches are common only on mature trees, but may not be provided by the same species of trees that develop hollows suitable for nesting. Nest-building materials are also important resources, with many woodland birds using spiderweb to bind and hold their nests (Recher 2004). The risk is that by emphasising hollows and hole-nesting birds, these other nest requirements may be neglected and that the birds that rely on them will be affected adversely.

5. Insectivorous birds affect the abundance, kinds and size of canopy arthropods, but do not prevent outbreaks

By excluding birds from the canopy foliage of eucalypts, we have shown that insectivorous birds affect the abundance, size, and kinds of arthropods in the canopy, although birds are unlikely to prevent outbreaks of insects (Evelegh *et al.* 2001; Recher and Majer 2006). In eastern Australia, however, the Bell Miner (*Manorina melanophrys*), a colonial honeyeater, encourages outbreaks of psyllid insects by excluding other birds from the colony, which in turn leads to the decline and death of the trees affected by the psyllids (Wardell-Johnson *et al.* 2005). Predation by birds on canopy arthropods results in more small insects and spiders and fewer large ones. The impact on spiders is particularly striking, with more and significantly larger spiders found when birds are excluded from the canopy. Although spiders benefit from the increased numbers of insects in canopies from which birds are excluded, spiders are a preferred bird food. Our trials were limited, but no arthropod increased to outbreak proportions when bird predation was removed (Evelegh *et al.* 2001; Recher and Majer 2006). The lesson learned from our exclusion studies and the interactions between the Bell Miner and psyllid insects is to show how small changes in the environment or abundance of a particular species can have cascading effects on other groups of animals and even affect the health of entire ecosystems. Resource managers need to be aware of this whenever they consider changing fire regimes or even eradicating exotic species.

6. The majority of woodland bird species are nomadic or migratory, but the most abundant species are sedentary

Nectar-eaters are not the only woodland birds to disperse widely in search of food. Birds that depend on insects and spiders for food also move long distances as their food supply changes seasonally and across the landscape. Migrants, which rely on an abundance of arthropods in spring and summer to raise young, are the most familiar, but there are also many insect-eaters that move nomadically in response to changing food availability. Among these are pardalotes (*Pardalotus*), which rely on an abundance of psyllid insects for lerp, trillers (*Lalage*), whistlers (*Pachycephala*), cuckoo-shrikes (*Coracina*), and woodswallows (*Artamus*), which track outbreaks of insects associated with good rainfall or the flush of growth following fires. Altogether, more than 60% of the bird species in Western Australia's woodlands are migratory or nomadic, although some individuals within each species may be sedentary. Migrants and nomads often give an illusion of abundance as large flocks aggregate at rich patches of food (e.g. nectar-rich flowers, insect outbreaks), but such birds are not spread evenly across the landscape and their total numbers are probably less than appears, and less than many sedentary species. There are really two lessons in these observations. The first is that abundance can be illusory and a species may decline significantly while still aggregating in flocks and appearing common. In such cases, conservation action may be delayed and recovery made more difficult. The second lesson is that a system of conservation reserves cannot adequately protect the majority of birds unless reserves are large, diverse and linked by habitat through which migrants and nomads can move and find food.

7. Increased temperatures and changed patterns of rainfall associated with climate change has significant impacts on woodland birds

Human-induced climate change has dominated conservation and ecological debate during the first decade of the 21st century. There are good reasons for this. Average annual rainfall in south-western Australia has declined by ~20% since the 1970s. Since 1986, when we first began to study birds in the woodlands of Western Australia, it has become progressively drier. From 2006 through 2008, the Great Western Woodlands has been particularly dry. Progressively as the drought lengthened, we observed the impact that the lack of rain had on the woodlands and on birds in particular. The first, most visible effects were on the drying of the ground and litter layers followed by the failure of ephemeral winter and spring ground vegetation (grasses, herbs, ferns) to grow. This was accompanied by declines in the numbers of ground-foraging birds and evidence that many birds were failing to breed. What nesting occurred was successful, but restricted to a few pairs of each species in our presumed micro-refuges (see Lesson #3 above) where food seemingly remained abundant.

By 2007, local beekeepers in the vicinity of Norseman failed to harvest any honey, a sign of reduced flowering and little nectar. In the same year, there was a noticeable decrease in the abundance of nectar-feeding birds. Some, like the Purple-crowned Lorikeet, were dispersed and found only in small flocks. With these went the hawks, eagles, kites and falcons that preyed on the flocks. Yellow-plumed Honeyeaters deserted areas of woodland that they had colonised only a few years earlier when it was wetter. By 2008, even fewer birds nested and migratory and nomadic insect-eaters failed to arrive or were found only in small numbers. As the amount and frequency of cloud cover declined in association with less rain, it was evident that birds ceased foraging earlier in the morning despite cool temperatures. Birds do not forage when it is hot as they need to conserve moisture and keep body temperatures within their normal range, which is high (~40°C). Not foraging when it is still cool but the sunlight is intense, suggests that moisture conservation may be a significant constraint on foraging during drought, especially if food is limited. Woodland birds obtain almost all their water from the food they eat.

Climate change is predicted to increase average temperatures in south-west Western Australia, with more frequent and longer-lasting droughts and more hours of sunshine. The lesson we need to learn is that, while woodland bird species may survive the worsening conditions, numbers will decline, with populations progressively restricted to drought refuges. In these circumstances, the survival of species will be increasingly at risk from wildfire and the effects of habitat fragmentation from mining, tourism and the associated development of roads and other infrastructure.

8. The conservation and management of woodland birds need to be integrated on a landscape scale

Although the avifauna of the woodlands outside the Western Australian wheatbelt appears to be intact, there have been significant and mostly undocumented changes. As highlighted by the effects of the current drought, it is likely that dispersive species have declined in abundance on a regional scale as a result of land clearing and development outside the intact woodlands. Most such development has taken place in higher rainfall areas, which previously offered dispersive species a refuge from drought. Increased human population, activity, and infrastructure throughout the South West, and the Great Western Woodlands in particular, threaten the sustainability of the avifauna through internal habitat fragmentation from roads and infrastructure and through increased fire frequency. Establishing a comprehensive system of conservation reserves in the South West will assist in conserving part of the avifauna, but no system of relatively small and isolated reserves can provide the complex array of spatially and temporally varying resources required by dispersive species. Nor is a reserve system likely to adequately provide the micro-refuges that seem so important for the survival of sedentary species during drought and other times of environmental stress. The essential ingredient for the conservation of regional and continental biodiversity, birds and all other biota, is continuity of habitat. This requires all land, irrespective of tenure and use, to be managed as part of a regional conservation framework (Recher 2003; Soule *et al.* 2004; Mackey *et al.* 2007).

Conclusions

The study of birds in their natural environments provides detailed information on the resources they require for survival and how they go about obtaining those resources. What we have learned is that each species requires and uses a complex array of resources differing spatially from the scale of the landscape or region to micro-refuges of hundreds or tens of square metres. This spatial complexity changes hour by hour, day by day, season by season, and from year to year. Integrating these needs across the entire biota requires more than a few reserves scattered across the landscape; the reserve system needs to be planned to meet the needs of all species, those that are sedentary and resident, as well as those that are nomadic and migratory. This can only be done by treating and managing the entire landscape, regardless of tenure, as a single ecological entity, with an overarching objective of conserving continental biodiversity and ecosystem function, both of which are needed for human life-support systems and sustainable economies.

Acknowledgements

R.T. Holmes assisted with fieldwork in the Great Western Woodlands during 2005. The research in the Great Western Woodlands from 2005 to 2007 was supported by Australian Research Council grant LP0455163 for which The Wilderness Society was an Industry Partner. Our other work on birds in Western Australian woodlands was supported at various times by Curtin University, Edith Cowan University and the University of New England.

Bios

'Ted' Davis was Professor of Science and Mathematics at Boston University, Boston, USA. Although trained as a geologist, Professor Davis's main research interests have been with birds. He has worked extensively on avian foraging ecology in Australia, but is also an historical ornithologist and co-editor of several books on the history of ornithology in North America. With Walter Boles and Harry Recher, he is editing a series of books on the history of Australian ornithology. Professor Davis has also published a book on the natural history of Tasmania and written others about nuclear power and how to play chess. Professor Davis lives near Cape Cod in Massachusetts.

Jonathan Majer was until recently Head of the Department of Environmental and Aquatic Sciences at Curtin University of Technology, Bentley, Western Australia, and is now Professor of Invertebrate Conservation. Jonathan was born in Sussex, England, and obtained his undergraduate degree from Bristol University and his PhD from the University of Ghana, where he studied tree ants in Ghana. After completing his doctorate, Professor Majer took a one-year position at WAIT (Western Australian Institute of Technology, now Curtin University) that quickly developed into a permanent post. His special research interests remain ants, with a particular emphasis on the role of ants and other fauna in mine site rehabilitation. Jonathan has been President of the Entomological Society of Australia, travels regularly to Brazil where he continues his interests in tropical ecology first developed in Ghana. Professor Majer lives in Perth, Western Australia.

Harry Recher was Foundation Professor of Environmental Management at Edith Cowan University, Perth, Western Australia. Prior to that he held appointments at the University of Sydney, the Australian Museum, and the University of New England. An ecologist, Professor Recher has studied the ecology of birds, mammals, and terrestrial arthropods throughout Australia. He has special interests in forest management and the ecology of fire in Australian ecosystems. Professor Recher works from his home on Dangar Island in the Hawkesbury River just north of Sydney, New South Wales.

References

Cousin, J.A. (2004). Pounce site characteristics of the Western Yellow Robin *Eopsaltria griseolgularis*: the importance of assessing foraging microhabitat. *Pacific Conservation Biology* **10**, 21–27.

Evelegh, N.C., Majer, J.D. and Recher, H.F. (2001). The effects of reducing bird predation on canopy arthropods of Marri *Eucalyptus calophylla* saplings on the Swan Coastal Plain, Western Australia. *Journal of the Royal Society of Western Australia* **84**, 13–21.

Fulton, G.R. and Majer, J.D. (2006). The effect of recent chaining on birds in the eastern wheatbelt of Western Australia. *Pacific Conservation Biology* **12**, 168–174.

Mackey, B.G., Soule, M.E., Nix, H.A., Recher, H.F., Lesslie, R.G., Williams, J.E., Woinarski, J.C.Z., Hobbs, R.J. and Possingham, H.P. (2007). Applying landscape-ecological principles to regional conservation: the WildCountry Project in Australia. In *Key Topics in Landscape Ecology.* (Eds J. Wu and R. J. Hobbs). pp. 192–213. Cambridge University Press: Cambridge.

Majer, J.D., Recher, H.F. and Ganeshanandam, S. (1992). Variation in the foliar nutrients and its relation to arthropod communities on Eucalyptus trees in eastern and Western Australia. *Australian Journal of Ecology* **17**, 383–394.

Majer, J.D., Recher, H.F., Perriman, W.S. and Achuthan, N. (1989). Spatial variation of invertebrate abundance within the canopies of Australian eucalypt forests. *Studies in Avian Biology* **13**, 65–72.

Recher, H.F. (1989). Foraging segregation of Australian warblers (Acanthizidae) in open-forest near Sydney, New South Wales. *Emu* **89**, 204–215.

Recher, H.F. (2003). WildCountry. *Pacific Conservation Biology* **8**, 221–222.

Recher, H.F. (2004). Eucalypt forest birds: the role of nesting and foraging resources in conservation and management. In *Conservation of Australia's Forest Fauna*. 2nd ed. (Ed D. Lunney.) pp. 23–35. Royal Zoological Society (NSW): Mosman.

Recher, H.F. and Davis, W.E. (1998). Foraging profile of a wandoo woodland avifauna during spring. *Australian Journal of Ecology* **23**, 514–528.

Recher, H.F. and Davis, W.E. (2002). Foraging profile of a Salmon Gum woodland avifauna in Western Australia. *Journal Royal Society of Western Australia* **85**, 103–111.

Recher, H.F., Davis, W.E. and Calver, M. (2002). Resource partitioning and the comparative foraging ecology of five species of ground-pouncing birds in Western Australian eucalypt woodlands. *Ornithological Science* **1**, 29–40.

Recher, H.F., Holmes, R.T., Schulz, M., Shields, J. and Kavanagh, R. (1985). Foraging patterns of breeding birds in eucalypt forest and woodland of south-eastern Australia. *Australian Journal of Ecology* **10**, 399–420.

Recher, H.F., Kavanagh, R.P., Shields, J.M. and Lind, P. (1991). Ecological association of habitats and bird species during the breeding season in southeastern New South Wales. *Australian Journal of Ecology* **16**, 337–352.

Recher, H.F. and Majer, J.D. (2006). Effects of bird predation on canopy arthropods in wandoo *Eucalyptus wandoo* woodland. *Austral Ecology* **31,** 349–360.

Recher, H.F., Majer, J. and Ford, H.A. (1991). Temporal and spatial variation in the abundance of eucalypt canopy invertebrates: the response of forest birds. *Proceedings International Ornithological Congress* **20**, 1568–1575.

Recher, H.F., Majer, J.D. and Ganeshanandam, S. (1996). Birds, insects and eucalypts, on the relation between foliar nutrients and species richness. *Forest Ecology and Management* **85**, 177–196.

Soule, M.E., Mackey, B.G., Recher, H.F., Williams, J.E., Woinarski, J., Driscoll, D., Dennison, W. and Jones, M. (2004). The role of connectivity in Australian conservation. *Pacific Conservation Biology* **10**, 266–279.

Wardell-Johnson, G., Stone, C., Recher, H. and Lynch, A.J.J. (2005). A review of eucalypt dieback associated with bell miner habitat in south-eastern Australia. *Australian Forestry* **68**, 231–236.

Watson, A., Judd, S., Watson, J., Lam, A. and Mackenzie, D. (2008). *The Extraordinary Nature of the Great Western Woodlands*. The Wilderness Society: Perth.

Wilson, K. and Recher, H.F. (2001). Foraging ecology and habitat selection of the Yellow-plumed Honeyeater *Lichenostomus ornatus* in a Western Australian woodland. *Emu* **101**, 89–94.

9

THE WHEATBELT WOODLANDS OF WESTERN AUSTRALIA: LESSONS FROM THE INVERTEBRATES

Jonathan D. Majer, Harry F. Recher and Anita Lyons

1. In terms of extent and magnitude of decline, biodiversity integrity is lower in the Western Australian wheatbelt than in any other botanical district of the State.
2. The trunks and canopies of trees are important sources of invertebrate biodiversity, with arboreal invertebrate richness exhibiting high species turnover over short distances and being higher in the wheatbelt woodland than in the higher rainfall forests to the west.
3. Isolated trees in paddocks and trees in corridors are important refuges of biodiversity.
4. Restoration through tree planting or seeding has maximum value for biodiversity if local species are used.
5. Oil Mallees planted in paddocks for economic reasons and salinity control have significant potential for enhancing the persistence of many native species in agricultural landscapes.

Introduction

The majority of our studies of woodland invertebrates have been conducted in the Wandoo (*Eucalyptus wandoo* and *E. capitata*) woodland region of Western Australia (see map on next page). This region, which is immediately to the east of the Jarrah/Marri (*E. marginata*/*Corymbia calophylla*) forest, gives way to lower mallee formations, before blending into the Great Western Woodlands, which lie beyond the agricultural clearing line to the east. Due to the adequate rainfall and suitability of the soil for agriculture, this Wandoo/Mallee area is known as the Western Australian wheatbelt, and over 90% of it has been cleared for agricultural production (Environmental Protection Authority 2007).

Lessons

1. In terms of areal extent and magnitude of decline, biodiversity integrity is lower in the Western Australian wheatbelt than in any other botanical district of the State

Massive clearing has had profound effects on the biodiversity of this region. This region contains the highest number of plant species listed as threatened or extinct for any region of Western Australia (Environmental Protection Authority 2007). Another way to look at this is to consider the degree of reduction of biodiversity by various land uses and the percentage of the area that is occupied by each land use. Majer and Beeston (1996) applied this concept to the entire State of Western Australia, using Beard's (1990) botanical districts as the individual land units. On the basis of previous studies, and using ant assemblages as the response group, they quantified the change in biodiversity caused by agricultural clearing, rangeland grazing, transport corridors, urbanisation and mining. The percentage extent of each land use, plus the amount of unaffected land in each botanical district, was measured and this value was multiplied by the corresponding reduction or change in biodiversity. The values were then summed for each botanical district to produce a Biodiversity Integrity Index. The results, shown in Figure 9.1, indicate that the wheatbelt of Western Australia has the lowest Biodiversity Integrity Index of any botanical district in the State. The procedure was subsequently applied to the Avon River Basin (Majer 2000), which traverses the central region of the wheatbelt, using shire boundaries as the area for calculations. It was found that lowest Biodiversity Integrity Index values centred around the shires of Tammin and Cunderdin (Figure 9.1 inset), two of the oldest settled areas of the wheatbelt. The Biodiversity Integrity Index was considerably higher in the Great Western Woodlands to the east, suggesting a more optimistic future for this region.

Figure 9.1 Biodiversity Integrity Index values, ranging from low (white) to high (black) for Western Australia, divided into Beard's (1990) botanical districts. The inset shows equivalent information for the Avon River Basin, divided into shire boundaries.

2. The trunks and canopies of trees are important sources of invertebrate biodiversity, with arboreal invertebrate richness exhibiting high species turnover over short distances and being higher in the wheatbelt woodland than in the higher rainfall forests to the west

With such a threat to biodiversity in the wheatbelt, it is essential that we characterise what is there, how it is threatened, and the measures that might be taken to arrest any decline. Majer and Recher have investigated these questions since 1986, initially focusing on tree canopy invertebrates, and continuing with investigations on bark fauna. The impetus for these investigations was not only for biodiversity conservation, but also because we need to know more about these pivotal 'drivers' of ecosystem functions and processes (Wilson 1987), and because they are important food resources for birds and other insectivorous vertebrates (see Chapter 8).

The initial work was performed in Dryandra Forest near Narrogin, where Wandoo (*E. wandoo*) and Powderbark Wandoo (*E. accedens*) are common, but enclaves of Jarrah and Marri occur on higher lateritic ground; and also at Karragullen in the western part of the Jarrah forest. Majer and Recher (1988) sampled the canopies of these trees by chemical knockdown and were able to quantify invertebrate abundances and biomass. At Dryandra, the abundance and biomass of invertebrates on Wandoo were at least double the values on Jarrah, and at Karragullen values were much higher on Marri than on Jarrah. These differences appeared to be positively associated with levels of foliar nitrogen, potassium and phosphorus, which varied greatly between tree species (Recher *et al.* 1996). The work at Karragullen was subsequently expanded to examine invertebrate richness, and canopies of 40 Jarrah and Marri trees were found to harbour at least 422 and 417 species, respectively, totalling 561 species when overlap was taken into account (Majer *et al.* 2002). The canopy diversity work was never extended into the Wandoo or Powderbark Wandoo woodlands but, on the basis of arthropod abundances and biomass, as well as tree trunk faunas, canopy species richness on these species is likely to be as high, or higher, than on the Jarrah forest tree species.

Trunk-associated invertebrates were sampled by placing intercept traps to catch specimens alighting on trunks, and bark traps to catch invertebrates living on or passing up the trunk. Jarrah and Marri were sampled at Karragullen and Jarrah, Marri, Wandoo and Powderbark Wandoo were sampled at Dryandra. Additionally, Marri trees were sampled at 10 km intervals along a transect between the two sites (Majer *et al.* 2003a, 2006). A total of 1234 species of invertebrates was found on the trunks and the majority of these were additional to those previously sampled in the canopy. Turnover of species along the geographic transect of Marri trees was considerable (Majer *et al.* 2003b). Species richness on Wandoo and, to a lesser extent Powderbark Wandoo, exceeded that on Jarrah and Marri (497, 422, 320, 398 invertebrates on 10 trees, respectively, at Dryandra). No conclusive explanation could be advanced for the higher species richness on the two smooth-barked Wandoos than on the rough-barked Jarrah and Marri, since nutrient levels at the various sites did not differ. One possibility is that the Wandoos occupy soils which are lower in the landscape and which may have better water-holding capacities. Another explanation is that species richness may be higher in the Wandoo woodland area than in the Jarrah/Marri forest, despite the former being in a lower rainfall area. There is strong evidence for this, since species richness on Marri trunks sampled along the geographic transect rose sharply at the point where the Marri trees were surrounded by Wandoo woodland rather than Jarrah/Marri forest (Majer *et al.* 2006).

Regardless of the reasons for differences in canopy and bark faunas between tree species or between Karragullen and Dryandra, the trends reported here highlight the importance of canopy and bark of each species of eucalypt as a habitat and resting place for a large array of invertebrates. Tree canopies and trunks should therefore be seen as important sources of woodland biodiversity. This has particular implications for the way woodlands are managed, since management practices such as prescribed burning (affecting canopy and bark structure), selective logging (affecting tree-age profile and tree-species composition), and clearfelling (affecting tree-age profiles), have the potential to influence canopy and bark-associated invertebrates and to restrict the movement of invertebrates through the woodland ecosystem.

3. Isolated trees in paddocks and trees in corridors are important refuges of biodiversity

With such a large percentage of wheatbelt vegetation having been cleared, and the knowledge that each tree species (and probably each shrub and herb species) is an important harbour of invertebrate richness, it is important to understand the contribution to conservation of the remaining vegetation.

The value of roadside corridors to the conservation of ant communities, themselves good indicators of the variety and composition of other invertebrate groups (Andersen and Majer 2004), was investigated in the Wubin to Perenjori area of the northern wheatbelt by Keals and Majer (1991). Ant species richness was around eight species per 100 m transect in farm paddocks compared to values in the mid-twenties in remnant vegetation. Five metre-wide road verges with little or no native vegetation supported around 12 species, although this value climbed to 18 species if native vegetation remained on the verge. In the areas where 20 m wide road verges remained, ant species richness was 24, which is similar to that of the remnant vegetation. It appears that, at least in the short term, vegetated road verges have a reasonable capacity to conserve invertebrate species richness, especially if they are wide.

In addition to remnant vegetation along transport corridors, there are also trees in paddocks, most of which can be isolated from remnants or from each other by distances exceeding 100 m. Using a modification of the chemical knockdown procedure (branchlet shaking), necessitated by strong winds, Majer *et al.* (1999) compared the invertebrate fauna in Wandoo trees in nature reserves, roadside corridors and in farm paddocks in the Dowerin to Wyalkatchem area of the central wheatbelt. Invertebrate abundances were often as high, or even higher, in the canopies of trees in corridors or paddocks than in the reserves. Analysis of nutrients in the soil in these growing situations revealed elevated levels of nitrogen and phosphorus under the corridor and paddock trees. It seems that the trends in invertebrate numbers may, in part, be explained by the interplay between the progressive isolation of trees from colonisation sources and the elevation of soil nutrients beneath the corridor and paddock trees, associated with fertiliser application and the excreta of farm animals.

Subsequent identification of the sucking bugs (Homoptera), beetles (Coleoptera) and ants (Formicidae) to species level indicated that species richness on the isolated trees could be high, but generally not as high as in the reserves. Also, species composition of corridor/paddock trees was rather different to that on reserve trees (Bae *et al.* 1998), probably because a number of the species on the isolated trees were cosmopolitan species, introduced species, or disturbance specialists. It thus seems that isolated trees have value for conservation of biodiversity, but their utility is inferior to that of trees remaining in large tracts of vegetation.

4. Restoration through tree planting or seeding has maximum value for biodiversity if local species are used

In response to the recognised losses in biodiversity, and also to rising salinity levels, there are numerous revegetation programs designed to ameliorate degraded farmland and salt-affected land or to revegetate road corridors. These revegetation efforts do not always involve native vegetation.

Bhullar and Majer (2000) examined the canopy fauna on street trees planted in the Perth metropolitan area, an area once vegetated with *Banksia* woodland. Tree species examined included local Marri, Bald Island Marlock (*E. conferruminata*) from the south coast of Western Australia, Bolivian Glory Tree (*Tipuana tipu*) from South America and Camphor Laurel (*Cinnamomum camphora*) from Asia. In that order, invertebrate levels per tree fell from 453 to 383 to 117 to 29 individuals, suggesting that the more alien the tree species is to the area, the lower the variety of invertebrates that occur on it.

This choice of non-native tree species does not generally apply to revegetation efforts within the wheatbelt, but, in a study of canopy invertebrates on revegetated trees in the Northam area, Majer *et al.* (2001) found that a mixture of local, non-local Western Australian, northern Australian and eastern Australian eucalypts were being used. They sampled invertebrates on examples of each of these categories and found, with exceptions, that even within the genus *Eucalyptus*, more abundant and more diverse invertebrate assemblages were found on the local

Figure 9.2 York Gum (mallee form, *Eucalyptus loxophleba* subsp. *lissophloia*) alley in the Western Australian wheatbelt.

species, with abundance and diversity declining with distance from the source area of the tree species. It is clear from this that maximum benefits to biodiversity can be obtained if local species are planted or seeded.

5. Oil Mallees planted in paddocks for economic reasons and salinity control have significant potential for enhancing the persistence of many native species in agricultural landscapes

Recently, Oil Mallees have been developed for use in the reduction and prevention of salinity, and in the production of oil, activated carbon and electricity (Figure 9.2). While some work has been done on aspects of mallee ecology to maximise productivity, little is known about their usefulness for biodiversity in natural and agricultural systems.

Lyons (2009) examined the potential contribution of these mallees to 'lost' biodiversity in the paddocks. Sampling was conducted in alleys of Oil Mallee vegetation and in remnant vegetation in the Narrogin-Wickepin region. Three mallee species, Blue-leaved Mallee (*E. polybractea*), Oil Mallee (*E. kochii* subsp. *borealis*) and the mallee form of York Gum (*E. loxophleba* subsp. *lissophloia*), plus two native remnant species, Wandoo and Mallet (*E. astringens*), were used in the study. Trees were sampled by chemical knockdown and samples were sorted to the ordinal level in the laboratory; Coleoptera (beetle) specimens were identified to the species level. Invertebrate ordinal richness and abundance were not significantly different between the tree species, whether they were planted mallees or native trees in remnants. It appears that arthropod abundance, in particular, is related to soil and leaf nutrient levels, sometimes positively, at other times negatively.

The roles of leaf essential oils and other secondary plant compounds, including cineole and pinene, in determining arthropod abundance and ordinal richness were also examined. A range of relationships, including deterrent and attractant effects, were observed between arthropod assemblages and the various secondary plant compounds. It seems that complex interactions between these chemicals cause changes in nutritional quality, palatability and digestibility of foliage, influencing the feeding behaviour, development, distribution and abundance of herbivores, in turn affecting predator densities and feeding behaviour.

Of all the tree species, Blue-leaved Mallee had both the highest beetle species richness, and high levels of beetle abundance. Strong similarities were apparent between beetle assemblages resident on the same tree species. No influence of soil or leaf nutrients on either beetle species richness or abundance was observed; however, beetle abundance was found to have a negative relationship with leaf pinene. Cineole levels were lower in the remnant species (which tended to have similar assemblages), while pinene was high in York Gum and the remnant species. The other major leaf chemical examined, sideroxylonal, had no significant impact on beetle species richness or abundance. Statistical analysis selected cineole as the single factor best explaining the pattern of beetle assemblages observed, although this result should be treated with caution due to possible confounding of the results as a consequence of interaction between the factors.

Lyons' (2009) work indicates that Oil Mallees support high levels of arthropod biodiversity, a finding mirrored in a parallel study of ground-active beetles at one of the same sites (Leng 2006). Overall, the mallees had a level of species richness not dissimilar to that of high quality remnant vegetation. When planted in alleys across agricultural fields, they represent a significant change in the vegetative and architectural diversity of the landscape, and can have a positive influence on the environment by supporting beneficial arthropods and other native animals, reducing dryland salinity and improving the aesthetics of the wheatbelt.

Conclusions

The invertebrate studies described here highlight the diverse nature of the wheatbelt fauna, and also the extreme threat it is facing from excessive clearing. The high turnover of species on an east-west gradient across the wheatbelt has important implications for climate change and reserve design in this highly fragmented landscape, and indicates the importance of conserving all native vegetation that remains. Vegetation along corridors and even isolated trees in paddocks play some role in biodiversity conservation. This will not be sufficient, however, to conserve invertebrate biodiversity, and of course that of other organisms, so greater attention should be paid to revegetation, using native plant species in strategic locations in order to restore connectivity between fragments.

Acknowledgements

The authors thank the numerous people who have assisted with these projects and the Australian Research Council, the Gordon Reid Foundation, the WA Department of Environment and Conservation, and CSIRO Sustainable Ecosystems for funding various components of this work.

Bios

Jonathan Majer was until recently Head of the Department of Environmental and Aquatic Sciences at Curtin University of Technology, Bentley, Western Australia, and is now Professor of Invertebrate Conservation. Jonathan was born in Sussex, England, and obtained his undergraduate degree from Bristol University and his PhD from the University of Ghana, where he studied tree ants in Ghana. After completing his doctorate, Professor Majer took a one-year position at WAIT (Western Australian Institute of Technology, now Curtin University) that quickly developed into a permanent post. His special research interests remain ants, with a particular emphasis on the role of ants and other fauna in mine site rehabilitation. Jonathan has been President of the Entomological Society of Australia, travels regularly to Brazil where he continues his interests in tropical ecology first developed in Ghana. Professor Majer lives in Perth, Western Australia.

Harry Recher was Foundation Professor of Environmental Management at Edith Cowan University, Perth, Western Australia. Prior to that he held appointments at the University of Sydney, the Australian Museum, and the University of New England. An ecologist, Professor Recher has studied the ecology of birds, mammals, and terrestrial arthropods throughout Australia. He has special interests in forest management and the ecology of fire in Australian ecosystems. He works from his home on Dangar Island in the Hawkesbury River just north of Sydney, New South Wales.

Anita Lyons is a Quarantine Entomologist responsible for supporting biosecurity through taxonomic triage and remote diagnostics. She also acts as a liaison between AQIS entomologists and CSIRO taxonomists. Anita is based at the Australian National Insect Collection in Canberra, ACT. She obtained her PhD at Curtin University where she studied arthropod biodiversity of Oil Mallee plantings. In her liaison role, Anita gives AQIS the ability to utilise the national collection and gain advice from taxonomic specialists, which are invaluable tools in making quarantine decisions and ultimately protecting Australia's biosecurity. She lives in Canberra.

References

Andersen, A.N. and Majer, J.D. (2004). Ants show the way Down-Under: invertebrates as bioindicators in land management. *Frontiers in Ecology and the Environment* **2**, 291–298.

Bae, T.W., Majer, J.D. and Recher, H.F. (1998). Canopy arthropod faunas in fragmented agricultural landscapes: patterns in the distribution of beetles. *Research Bulletin of the Institute of Agricultural Resources, Dong-A University* **35**, 431–444.

Beard, J.S. (1990). *Plant Life of Western Australia*. Kangaroo Press: Kenthurst.

Bhullar, S. and Majer, J.D. (2000). Street trees for wildlife and for people. *Pacific Conservation Biology* **6**, 171–173.

Environmental Protection Authority (2007). *State of the Environment Report*. Department of Environment and Conservation: Perth.

Keals, N. and Majer, J.D. (1991). The conservation status of ant communities along the Wubin-Perenjori Corridor. In *Nature Conservation 2: The Role of Corridors*. (Eds D. A. Saunders and R. J. Hobbs.) pp. 387–393. Surrey Beatty & Sons: Chipping Norton.

Leng, M.C. (2006). Beetles and belts: Biodiversity of ground dwelling invertebrates (Coleoptera) associated with oil mallee plantings in the Western Australian wheatbelt. Honours thesis, Curtin University of Technology: Perth.

Lyons, A. (2009). Oil mallee plantings and arthropod biodiversity in the Western Australian wheatbelt: effects of host species, nutrition and leaf chemistry. PhD thesis, Curtin University of Technology: Perth.

Majer, J.D. (2000). The biodiversity integrity index: an example using ants in the Avon River Basin. *Ecological Management and Restoration* **1**, 232–234.

Majer, J.D. and Beeston, G. (1996). Biodiversity integrity index: an illustration using ants in Western Australia. *Conservation Biology* **10**, 65–73.

Majer, J.D. and Recher, H.F. (1988). Invertebrate communities on Western Australian eucalypts: a comparison of branch clipping and chemical knockdown procedures. *Australian Journal of Ecology* **13**, 269–278.

Majer, J.D., Recher, H.F., Graham, R. and Gupta, R. (2003a). Trunk invertebrate faunas of Western Australian forests and woodlands: influence of tree species and season. *Austral Ecology* **28**, 629–641.

Majer, J.D., Recher, H.F., Graham, R. and Watson, A. (2001). 'The potential of revegetation programs to encourage invertebrates and insectivorous birds'. Curtin University School of Environmental Biology Bulletin No. 21.

Majer, J.D., Recher, H.F. and Heterick, B.E. (2003b). Trunk invertebrate faunas of Western Australian forests: Implications for global warming. *Ecological Management & Restoration* **4**, 143–145.

Majer, J.D., Recher, H.F. and Keals, N. (1999). Canopy arthropod faunas in fragmented agricultural landscapes. In *Conservation of Australia's Temperate Woodlands.* (Eds R. J. Hobbs and C. Yates) pp. 235–247. Surrey Beatty & Sons: Chipping Norton.

Majer, J.D., Recher, H.F., Graham, R. and Gupta, R. (2006). Trunk invertebrate faunas of Western Australian forests and woodlands: seeking causes of patterns along a west-east gradient. *Austral Ecology* **31**, 503–511.

Majer, J.D., Recher, H.F., Heterick, B.E. and Postle, A.C. (2002). The canopy, bark, soil and litter fauna of the Darling Plateau and adjacent woodland near Perth, Western Australia, with reference to the diversity of forest and woodland invertebrates. *Pacific Conservation Biology* **7**, 229–239.

Recher, H.F., Majer, J.D. and Ganesh, S. (1996). Eucalypts, insects and birds: on the relationship between foliar nutrients and species richness. *Forest Ecology and Management* **85**, 177–195.

Wilson, E.O. (1987). The little things that run the world (the importance and conservation of invertebrates). *Conservation Biology* **1**, 344–346.

10

MANAGING THE WOODLANDS OF THE MT LOFTY REGION, SOUTH AUSTRALIA

David C. Paton, Nigel Willoughby, Daniel J. Rogers, Matthew J. Ward, Joel R. Allan and Andrew West

1. The woodlands of the Mt Lofty Ranges continue to deteriorate and remain management-dependent.
2. On-ground capacity exists, but spatial coverage is inadequate.
3. A mismatch exists between the scale of on-ground works and the scale of ecological processes.
4. Adequate, integrated and iterative planning for the recovery of the woodlands of the region is lacking.
5. A significant gap for managers relates to how systems will respond to recovery activity.
6. While research has influenced strategic policy, translating this into effective on-ground action has been difficult.

Introduction

We have been working within university and/or government sectors on research and planning aimed at maintaining South Australian species, communities and ecosystems. More recently we have been trying to bring these two components of conservation (research and planning) closer. This chapter documents some of the key points that have arisen from these efforts.

This chapter concerns some of the research and application lessons learnt from working on the conservation of woodland systems in the Mt Lofty region of South Australia (see map on next page). The Mt Lofty region provides a diverse range of environmental settings on which woodlands develop (see Specht 1972; Armstrong *et al.* 2003), resulting in an equally diverse set of ecosystems. Furthermore, these systems are isolated from other large areas of woodland by semi-arid shrubland and mallee to the east and north.

Lessons

1. The woodlands of the Mt Lofty Ranges continue to deteriorate and remain management-dependent

Within the Mt Lofty region, as with many other areas around the world, woodlands on the soils better for agriculture have been particularly heavily cleared (Paton *et al.* 1999). By the late 1970s, there were major concerns about the areas of habitat remaining and broadacre clearance of native vegetation without a permit ceased in the 1980s. In spite of the removal of this fundamental threat, the condition of the remaining woodlands within the Mt Lofty region continues to deteriorate (Paton *et al.* 1999; Paton *et al.* 2004; Ward 2005), along with the local (patch-scale) extinction, and regional decline of fauna species (Paton 2001). Ongoing threats to the woodlands of the region include grazing by domestic and native animals, competition by introduced plants and animals, and others typical for a highly developed area (e.g. Saunders *et al.* 1991). In spite of attempts to address these threats, all remain in the region (AMLR NRM Board 2007). Both the ongoing decline of woodland systems, and the continued presence of exacerbating threatening processes, suggest that the systems remain management-dependent in order to avoid loss of species and ecological function.

2. On-ground capacity exists, but spatial coverage is inadequate

Within the Mt Lofty region, a suite of government agencies and non-government organisations are actively involved in addressing some of the threats affecting components of the region's woodlands. These include the Adelaide and Mt Lofty Ranges Natural Resource Management Board, the SA Department for Environment and Heritage, the SA Department of Water, Land and Biodiversity Conservation, Trees-for-Life, the Nature Conservation Society of South Australia, Friends of Parks groups, Conservation Council of SA, Greening Australia (SA) and other specialist societies (orchids, native grasses, etc.). Both paid and volunteer workers are involved in a range of activities that undoubtedly improve the condition of some patches (e.g. Jury and Croft 2008). The actual area treated is still very small relative to the area that needs to be managed, however, and the management actions often need to be continually applied or re-applied. This suggests that the condition of the woodland systems of this region remains management dependent. Without a significant increase in resources (funding and people), the condition of all but a small component of the remnant woodlands of this region will continue to deteriorate. Further, monitoring the outcomes from these on-ground activities is far from adequate. If conducted, monitoring is usually focused on auditing the on-ground works that have been implemented and not whether they have delivered desired conservation goals.

Collectively this on-ground activity suggests that there is the will and motivation, and possibly considerable resources, to undertake the work required. Ideally there would be better funding, a better knowledge base and better planning, but the Mt Lofty region is probably better equipped than other regions of Australia, to deliver regional recovery programs for woodlands.

3. A mismatch exists between the scale of on-ground works and the scale of ecological processes

There is a large mismatch between the scales of on-ground activities and the scales at which the Mt Lofty region functions ecologically. Within the Mt Lofty region, the on-ground works focus at the patch scale, and often at a sub-patch scale. While this may be appropriate at times, such as for some threatened species or communities with limited distributions and low abundance, other components operate at greater scales, and some patches are limited in their response due to processes occurring at other scales (e.g. Cale 2007). Many ecological processes operate at a landscape scale (e.g. changes in groundwater) or a regional scale. Since management actions at the patch scale are usually implemented in the absence of landscape or regional contexts, they largely ignore the scale at which many species and threats operate. Undoubtedly, conservation action needs to occur with the intent of influencing outcomes at a range of scales, depending on the conservation goal. For example, recent work on home ranges of declining woodland bird species in the Mt Lofty region has shown that individuals of the Restless Flycatcher (*Myiagra inquieta*), Varied Sittella (*Daphoenositta chrysoptera*) and Black-chinned Honeyeater (*Melithreptus gularis*) move over areas of at least 100 ha during periods of 10 or so days of radio-tracking (Willoughby 2005, Allan unpublished). There are currently no programs working to restore habitat at a scale appropriate to meet these spatial requirements.

One existing example provides a compelling argument for restoring habitat at larger scales. In the 1970s, the South Australian Government undertook a broadscale revegetation program in the vicinity of Monarto on the eastern edge of the Mt Lofty region. The purpose of this work was to provide aesthetic and erosion control functions for a proposed satellite city (eventually abandoned). Over 1800 ha of overstorey and limited understorey species from all over Australia were planted by tubestock, approximately doubling the vegetation cover in that landscape. Today this vegetation is an open woodland that provides habitat for a range of declining woodland birds. Some, such as the Restless Flycatcher, Hooded Robin (*Melanodryas cucullata)* and Varied Sittella have home ranges entirely within revegetated areas, while others, such as Diamond Firetail (*Stagonopleura guttata*) and Brown Treecreeper (*Climacteris picumnus*) use revegetated areas in conjunction with remnant areas (Paton *et al.* 2004; Ankor 2005; Gillespie 2005; Barker 2007; Northeast 2007; Allan, Rogers and Paton unpublished data). All of these species have disappeared from the woodlands of the Mt Lofty region adjacent to Monarto.

Components of the Mt Lofty region also operate at a regional scale. For example, movements of honeyeaters occur into and out of different woodland systems within the region with individual birds moving up to 100 km (Paton *et al.* 1999; Paton 2000), over periods as little as 58 days (D.C. Paton unpublished data). Within this system, many honeyeaters move away from woodlands and forests with a heathy understorey during spring as flowering wanes in these areas. Traditionally other woodland systems provided resources for the birds outside winter and spring, including some of the grassy eucalypt woodland systems. Knowledge of the plant communities that once existed on the better quality land indicate that Silver Banksia (*Banksia marginata*), several eucalypts including South Australian Blue Gum (*Eucalyptus leucoxylon*), Grey Box (*E. microcarpa*) and Peppermint Box (*E. odorata*), and Box Mistletoe *Amyema miquelii* provided nectar resources for honeyeaters during this period (Ford and Paton 1977; Ford 1979; Paton 1986). These latter habitats have been disproportionately cleared leading to a greater regional restriction of floral resources over summer and autumn. Honeyeater populations have declined as a result and are subsequently unable to fully service the pollination requirements of winter-flowering plants during subsequent winters (Paton 2000). Based on this scenario, re-establishing the complementary summer–autumn habitats becomes important for maintaining the birds, as well as improving pollination outcomes for winter-flowering plants in the region.

A significant note here is that the distances over which honeyeaters have been detected moving in the Mt Lofty region ranged from 1 km to over 100 km (see Paton *et al.* 2004), with more than 10% of the detected movements exceeding 20 km (D.C. Paton unpublished data). Clearly, most honeyeater species are capable of moving through heavily cleared landscapes (and their movements may be greater now than in the past). Given their apparent ease of movement, the re-establishment of summer and autumn habitats may not need to be juxtaposed with the areas used in winter, and a regional approach to recovering this system can be adopted – if honeyeaters and the services they provide are a conservation goal.

4. Adequate, integrated and iterative planning for the recovery of the woodlands of the region is lacking

There are several reasons why conservation and restoration activities have failed to account for the scales at which this ecological system operates. Among these are the logistical difficulties, particularly for single agencies, to implement effective on-ground works at landscape and regional scales. Another major limitation is that there is no process in place that allows ecological knowledge to be integrated into the planning of on-ground works, no integration between the various groups doing on-ground work, and no effective monitoring of outcomes (e.g. Australian Government 2008). Monitoring the outcomes of on-ground works is also fraught with difficulty, since the expected context-specific conservation outcomes are rarely stated explicitly beforehand so that they can then be monitored. Ultimately, both goal-setting and monitoring need to be incorporated into a single planning process that directly links explicit conservation outcomes with both on-ground activities and monitoring programs (that monitor both activity *and* outcomes). This lack of integration can be extended beyond monitoring to research, which has largely been driven by the individual researchers' interests rather than addressing the gaps in information needed to manage these systems.

The Monarto area again provides a valuable example for the need for explicit conservation goals, established through better planning processes. This area is now compositionally different to remaining vegetation in the region, including the few remaining pieces of native vegetation on similar landforms (J.R. Allan and D.C. Paton unpublished data). Thus, despite providing essential habitat for a range of bird species, the revegetated area now conserves few other compositional values 'natural' to the area. While the original goals (aesthetic and erosion control) may have been met, the lesson for current conservation and management is that the recovery trajectory set by restoration work is primarily a function of the initial actions taken. Goal-setting must take this into account, specifying in as much detail as possible the desired conservation outcomes from specific restoration actions. Retrofitting conservation goals to the Monarto example – if the initial goal of the program had been to provide habitat important to declining woodland birds of the Mt Lofty region, then success has been achieved. If the goal was to conserve woodland composition, then the goal has not been met.

5. A significant gap for managers relates to how systems will respond to recovery activity

Patches that have lost compositional, structural or functional diversity may or may not recover these attributes with the removal of stressors (depending on a range of factors, e.g. Beisner 2003; Suding *et al.* 2004). Determining when a threshold has been crossed (preventing the desired recovery) is difficult and specific to a particular context. In the Mt Lofty region, some woodland areas degraded through grazing still retain appropriate propagules for recovery (e.g. components of a heathy understorey re-established after removal of grazing pressure for Sandy Creek Conservation Park; Paton unpublished). For other woodlands, the removal of key threats like excessive grazing may not result in any recovery because of depleted or lost propagule banks, or due to changes in the physical environment that prevent recovery. At present it is not clear when a degraded woodland system within the Mt Lofty region still has the capacity to recover functionality. Given this, the restoration of these woodlands will require an adaptive management approach that assesses the relative responses of degraded woodlands to restoration activity.

Setting appropriate restoration trajectories is still a massive gap in our current knowledge about how to make restoration actions achieve the outcomes sought. This applies across all spatial scales. Questions include the relative benefits or tradeoffs of repairing degraded systems versus building additional woodlands, and whether the additional woodlands should be designed to have added resilience to climate change or replicate the original woodland systems. Other questions relate to the ability to recover from stresses such as elevated grazing and browsing pressure from kangaroos, rabbits and koalas, and how to implement programs that manage these threats at landscape and regional scales as opposed to just patch scales. As suggested above, however, there needs to be an iterative process in place that ensures that information being generated is used to improve delivery of conservation outcomes (while also informing the process that sets these goals), and that the information is context-specific such that it is directly useful for management.

6. While research has influenced strategic policy, translating this into effective on-ground action has been difficult

The research and knowledge generated in the Mt Lofty region and elsewhere in South Australia has undoubtedly influenced policy and there are now a range of government plans and policies that theoretically empower action at appropriate scales. The South Australian Strategic Plan (Government of South Australia 2007) has two policies relevant to biodiversity conservation and management – 'NatureLinks' and 'Lose No Species' (see also DEH 2003, 2007) and these are translated into regional plans (e.g. AMLR NRM Board 2008). These policy statements have rarely been translated into activity that has demonstrably achieved conservation outcomes, however, as they are generally too vague, and rely too heavily on general enhancement (*sensu* Lambeck 1999) to guide management activity. The challenge for both planning and research is to turn these aspirational statements into tangible, quantifiable and context-specific targets.

Conclusions: managing the woodlands of the Mt Lofty region in the future

Time is running out to arrest the observed declines in composition and function of the woodland systems in the Mt Lofty region. Future generations will not level blame on our generation for losses in the integrity of some of these woodland systems provided we implement significant regional programs that have a chance of recovering these ecosystems. We are unlikely to ever acquire the knowledge needed to design and implement a perfect regional-scale recovery program. This, however, should not prevent taking action. Instead what is now needed is a truly functional and iterative planning and implementation process. This will require significant integration and collaboration from policy makers, planners, developers, ecologists, managers and implementers. Importantly, there is a need for a long-term and ongoing commitment to woodland conservation since many of the processes within these woodland systems operate over long time periods (decades or more) and there will be significant lags between implementing on-ground action and measuring the outcome (e.g. Vesk and Mac Nally 2006). Failure to secure the woodlands of the Mt Lofty region to date reflects a failure to establish an adequate and iterative planning process to deliver appropriate conservation activity at ecologically relevant scales.

Acknowledgements

This chapter grew from discussion during an Environmental Futures Network workshop on Habitat Reconstruction and Restoration, funded by the Australian Research Council. Funding for this work was partially provided by Caring for Our Country, the South Australian Murray–Darling Basin NRM Board, the Adelaide and Mt Lofty Ranges NRM Board, the South Australian Department for Environment and Heritage and the University of Adelaide.

Bios

David Paton is an ecologist at the University of Adelaide with a broad interest in the ecology of birds, and how this can assist in their management and conservation. This has included studies of pollination biology, interactions between honeybees and native flora and fauna, patterns of floral resource production and how these influence movements and population dynamics of honeyeaters. Other activities relate to responses of biota to fires and droughts in semi-arid shrublands, ecology of aquatic birds in the Coorong, and behavioural and demographic studies on a range of bird species. He is involved in several whole-farm habitat restoration programs.

Nigel Willoughby is an ecologist with the Department for Environment and Heritage, South Australia. His current work centres on determining ecological priorities for restoration and/or conservation at landscape scale. The practical application of this work occurs in the South Australian Murray Mallee with a focus on bridging the 'research–implementation gap' by getting work implemented against the priorities and monitoring the ecological outcomes of implementation.

Daniel Rogers holds the position of Ecologist (Restoration Ecology) at the South Australian Department for Environment and Heritage. His primary role is to develop information to determine landscape-scale conservation priorities across South Australia's diverse landscapes that will be used to support conservation investment strategies and implementation across the state. He has previously worked in a diverse range of settings, including wetlands, semi-arid shrublands, temperate woodlands and semi-natural grasslands, with a focus on understanding the responses of avifauna to key environmental drivers in these systems.

Matthew Ward's current role is Regional Ecologist for the Aboriginal Lands with the Department for Environment and Heritage, South Australia. His role is to deliver conservation, research and natural resource management outcomes on the vast Aboriginal Lands of South Australia, by facilitating indigenous land management and knowledge in a contemporary scientific framework. Previously he has conducted research on mistletoe proliferations in woodlands of the Mount Lofty Ranges, and fire management and avifauna conservation in heath and mallee systems of South Australia.

Joel Allan is a PhD candidate in the School of Earth and Environmental Sciences at the University of Adelaide. His research interests revolve around determining the habitat requirements of woodland birds at multiple spatial scales, with particular focus on practical translation to restoration activities. Joel's PhD research involves using the reconstructed woodland system at Monarto to provide habitat reconstruction guidelines for the Mount Lofty Ranges.

Andrew West is currently the Manager of Biodiversity Conservation with the Adelaide and Mount Lofty Ranges Natural Resources Management Board. He previously worked in the South Australian Department for Environment and Heritage in a number of different roles, focussing on landscape restoration, threatened species and ecological community recovery and rural biodiversity extension programs.

References

AMLR NRM Board (2007). 'Creating a Sustainable Future. A natural resources management plan for the Adelaide and Mount Lofty Ranges region. Volume A – State of the Region Report'. Adelaide and Mount Lofty Ranges Natural Resources Management Board: Eastwood, South Australia.

AMLR NRM Board (2008). 'Creating a Sustainable Future. A natural resources management plan for the Adelaide and Mount Lofty Ranges region. Volume B – Ten Year Plan for the Region'. Adelaide and Mount Lofty Ranges Natural Resources Management Board: Eastwood, South Australia.

Ankor, E. (2005). Is seed availability a limiting factor for diamond firetails in the Monarto region, South Australia? B.Sc. Honours thesis, Department of Environmental Biology, The University of Adelaide: South Australia.

Armstrong, D.M., Croft, S.J. and Foulkes, J.N. (2003). A Biological Survey of the Southern Mount Lofty Ranges, South Australia 2000–2001. Department for Environment and Heritage, South Australia.

Australian Government (2008). Natural Resource Management, Monitoring, Evaluation, Reporting and Improvement Framework. Australian Government: Canberra.

Barker, A. (2007). Brown treecreeper at Monarto: habitat suitability and management. B.Sc. Honours thesis, Department of Environmental Biology, The University of Adelaide: South Australia.

Beisner, B.E. (2003). Alternative states in ecology. *Frontiers in Ecology and the Environment* **1**, 376–382.

Cale, P. (2007). A contribution to the development of a conceptual framework for landscape management: a landscape state and transition model. In *Managing and Designing*

Landscapes for Conservation: Moving from Perspectives to Principles. (Eds D. B. Lindenmayer and R. J. Hobbs) pp. 527–545. Blackwell Publishing: Oxford.
DEH (2003). NatureLinks. Implementing the WildCountry philosophy in South Australia. Department for Environment and Heritage: South Australia.
DEH (2007). No Species Loss. A Nature Conservation Strategy for South Australia 2007–2017. Department for Environment and Heritage: South Australia.
Ford, H.A. (1979). Interspecific competition in Australian honeyeaters – depletion of common resources. *Australian Journal of Ecology* **4**, 145–164.
Ford, H.A. and Paton, D.C. (1977). The comparative ecology of ten species of honeyeaters in South Australia. *Australian Journal of Ecology* **2**, 399–407.
Gillespie, C. (2005). Foraging behaviour and habitat use of hooded robins (*Melanodryas cucullata*) and red-capped robins (*Petroica goodenovii*) at Monarto. B.Sc. Honours thesis, Department of Environmental Biology, The University of Adelaide: South Australia.
Government of South Australia (2007). South Australia Strategic Plan 2007.
Jury, T. and Croft, S. (2008). Comparison of managed with unmanaged grassy woodland at Belair National Park. *Xanthopus* **26**, 16–19.
Lambeck, R.J. (1999). 'Landscape planning for biodiversity conservation in agricultural regions. A case study from the wheatbelt of Western Australia'. CSIRO Division of Wildlife and Ecology, Biodiversity Technical Paper, No. 2.
Northeast, P. (2007). Habitat use by foraging red-capped robins *Petroica goodenovii* (Petroicidae) and hooded robins *Melanodryas cucullata* (Petroicidae) within the Monarto Woodlands, South Australia. B.Sc. Honours thesis, School of Earth and Environmental Sciences, The University of Adelaide: South Australia.
Paton, D.C. (2000). Disruption of bird-plant pollination systems in southern Australia. *Conservation Biology* **14**, 1232–1234.
Paton, D.C. (2001). Death of a treecreeper and we only celebrate. *Xanthopus* **19**, 8–12.
Paton, D.C. (1986). Honeyeaters and their plants in south-eastern Australia. In *The Dynamic Partnership. Birds and Plants in Southern Australia*. (Eds H. A. Ford and D. C. Paton) pp. 9–19. Government Printer: South Australia.
Paton, D.C., Prescott, A.M., Davies, R.J.-P. and Heard, L.M. (1999). The distribution, status and threats to temperate woodlands in South Australia. In *Temperate Eucalypt Woodlands in Australia: Biology, Conservation, Management and Conservation*. (Eds R. J. Hobbs and C. J. Yates) pp. 57–85. Surrey Beatty & Sons: Chipping Norton, NSW.
Paton, D.C., Rogers, D.J. and Harris, W. (2004). Birdscaping the environment: restoring the woodland systems of the Mt Lofty region, South Australia. In *Conservation of Australia's Forest Fauna*. (Ed. D. Lunney) pp. 331–358. Royal Zoological Society of New South Wales: Mosman, NSW.
Saunders, D.A., Hobbs, R.J. and Margules, C.R. (1991). Biological consequences of ecosystem fragmentation: a review. *Conservation Biology* **5**, 18–31.
Specht, R.L. (1972). *The Vegetation of South Australia*. Government Printer: South Australia.
Suding, K.N., Gross, K.L. and Houseman, G.R. (2004). Alternative states and positive feedbacks in restoration ecology. *Trends in Ecology and Evolution* **19**, 46–53.
Vesk, P.A. and Mac Nally, R.C. (2006). The clock is ticking – revegetation and habitat for birds and arboreal mammals in rural landscapes of southern Australia. *Agriculture, Ecosystems and Environment* **112**, 356–366.
Ward, M.J. (2005). Patterns of box mistletoe *Amyema miquelii* infection and pink gum *Eucalyptus fasciculosa* condition in the Mount Lofty Ranges, South Australia. *Forest Ecology and Management* **213**, 1–14.

Willoughby, N. (2005). The comparative ecology, and conservation, of the *Melithreptus* genus in the Southern Mount Lofty Ranges, South Australia. PhD thesis, Discipline of Environmental Biology, School of Earth and Environmental Sciences, The University of Adelaide: South Australia.

11

CONSERVING GRASSY WOODLAND IN TASMANIA

Jamie B. Kirkpatrick

1. To help prevent trees from dying out, exclude or kill Common Brushtail Possums (*Trichosurus vulpecula*) (Tasmania only).
2. To help prevent trees from dying out, destock or stock lightly until a new generation establishes.
3. Burning can help prevent trees from dying out in the short term.
4. Both burning and grazing are necessary to prevent native tree and shrub thickening.
5. Vertebrate grazers can be used to control some species of woody weeds.
6. Mowing or slashing can be used to prevent tree and shrub thickening.
7. To maintain significant species, keep on doing what has been done until you prove that it is better to do otherwise.
8. Old, small, isolated remnants with high perimeter: area ratios can be very important for nature conservation. Corridors may not be particularly important.
9. Latitude and high relative relief make most of the Tasmanian grassy woodland biota resilient to climate change.
10. Regulation combined with stewardship payments seems to be the only way to stop more clearance and fertilisation of grassy woodlands.
11. People with local ecological expertise and experience can usually provide good management advice, whereas tool kits and models cannot cope with contingencies.

Introduction

Tasmanian woodlands range from the coast to the climatic treeline, from some of the poorest soils in the world to some of the richest, from waterlogged ground to sharply drained dry ridges. They are unusual in temperate Australia in that they support a full complement of native marsupial herbivores, which can occur in high densities. This chapter covers the subset of these woodlands with understoreys potentially dominated by native grasses and herbs, henceforth called grassy woodlands. These grassy woodlands have been dramatically reduced

and degraded in Tasmania (Kirkpatrick *et al.* 1988), but still cover large areas of private land, largely utilised for wool production (Kirkpatrick *et al.* 2007b) (see map above). They also occur on some smaller areas of public land, largely utilised for recreation and nature conservation. The remaining areas of grassy woodland on both types of tenure are under severe threat as the State Government embarks upon a misguided program to turn the dry Midlands of Tasmania into an irrigated food bowl.

One key to the conservation of those grassy woodlands that are fortunate enough to survive lies in a middle path, avoiding both the thinning and thickening of trees. The other key lies in preventing a native grass and herb-dominated understorey from losing significant species or transforming into heath, scrub or vegetation dominated by exotic plants. The prevention of the last of these transitions relates to land-use decisions, and therefore to the economic, social and political context. The other transitions involve ecological interactions between climate, soils, fire regimes and grazing regimes. While climate change cannot be altered in the short term, fire and grazing regimes, two of the most effective vegetation management tools, are fortunately more malleable.

The task given for this chapter was to report the conservation implications of ecological and social studies of Tasmanian grassy woodland in which I have been involved (see map). Particular reference is made to the role of vegetation remnants in conservation and the impacts of recent climate change.

Lessons

1. Exclude or kill Common Brushtail Possums (*Trichosurus vulpecula*)

The drying out of eastern Tasmania that took place after 1978 has been associated with widespread morbidity and mortality of woodland trees (Kirkpatrick *et al.* 2000b; Calder and

Kirkpatrick 2008). At two out of three experimental sites in the Tasmanian Midlands, dying woodland trees recovered when the Common Brushtail Possum was excluded (Kirkpatrick *et al.* 2000). The only populations of Cider Gum (*Eucalyptus gunnii* ssp. *divaricata*) that remained healthy after decades of drought had the good fortune to be the locations of shooters' camps or had nearby residents who shot possums (Calder and Kirkpatrick 2008).

Noisy Miner (*Manorina melanocephala*) breeding colonies are concentrated on the margins of grassy woodland native vegetation remnants, where they are statistically associated with tree dieback (McDonald and Kirkpatrick 2003). The breeding colonies may be as much of an effect as a cause of dieback, however, as remnant edges also have elevated populations of the Common Brushtail Possum because of the availability of adjacent crops and improved pastures.

The killing of the Common Brushtail Possum as a woodland management tool is not recommended for mainland Australia, where, outside cities and their suburbs, the species may soon lose its first adjective. In Tasmania, such action should only be taken with appropriate permits and in areas where populations of the species are known to be substantial.

2. Destock or stock lightly until a new generation establishes

Tree regeneration occurs in lightly stocked (<0.5 dry sheep equivalents), rocky hill country, usually in tangles of fallen branches, or within prickly shrubs, but is absent on the margins of improved pastures and on more heavily stocked country. High levels of past and present stocking were associated with an absence of regeneration and a substantial age gap in the Cider Gum woodlands of the Central Plateau (Calder and Kirkpatrick 2008), and eucalypt seedlings were located only within fenced plots in the dieback recovery experiment of Kirkpatrick *et al.* (2000). If seed remains on woodland trees, some scarification or burning around them, combined with exclusion of stock, should ensure a new generation of trees in most circumstances, although it may fail where European Rabbit (*Oryctolagus cuniculus*) or marsupial herbivore populations are particularly high, as at one of the experimental sites of Kirkpatrick *et al.* (2000). Exclusion of these animal populations through rabbit-proof and possum-proof fencing will then be required. If no seed remains on the woodland trees, restoration through planting or direct seeding is the only solution. This is difficult and expensive, with watering of seedlings often being necessary to ensure establishment success (Kirkpatrick *et al.* 2007e).

3. Burning can help prevent trees from dying out in the short term

Tree and shrub dieback in dry eucalypt forests and woodlands during drought in Tasmania has been observed not to occur where sites have been recently burned (Kirkpatrick and Marks 1985). Eucalypt dieback in the grassy woodlands in the Domain, a park in the centre of Hobart, was greatest where the total basal area was highest (Kirkpatrick *et al.* 2007e), suggesting that burning allows trees to survive because it reduces total transpiration. To be effective in preventing dieback of the larger trees in grassy woodland, burns should have a scorch height sufficient to defoliate middle storey and understorey shrubs, but insufficient to scorch the crowns in the dominant layer. Care needs to be taken to make sure that fire does not ingress to basal hollows in the older trees, as this can cause death through mechanical collapse. Fire can be a short-term solution to one problem, while generating another; in this case, tree thickening (see below).

4. Both burning and grazing are necessary to prevent native tree and shrub thickening

Tree thickening has probably been more extensive than tree thinning in the grassy woodlands of Tasmania during the last half-century. This has been documented for the hills to the east of the Midlands graben and for the Domain in Hobart (Kirkpatrick 2004; Kirkpatrick *et al.* 2007e). The trees that thicken are typically not eucalypts. Drooping Sheoak (*Allocasuarina verticillata*), Silver Wattle (*Acacia dealbata*), Black Wattle (*A. mearnsii*) and Prickly Box (*Bursaria spinosa*) are the most common proliferating species. A high fire frequency on the Domain, where the only vertebrate herbivore in the ground-layer is the Common Brushtail Possum, has not resulted in the maintenance of a grassy woodland structure. Rather, the more frequent the fires, the greater the thickening of the small tree layer. In the hills to the east of the Midlands graben, stock and native animal grazing, in the absence of burning, has not prevented tree thickening (Kirkpatrick *et al.* 2007e). In both cases it might be tempting to blame climate change, but the grassy woodland country that is still both burned and grazed, by either a mixture of wild herbivores and stock, or by wild herbivores alone, has not suffered tree thickening. Thus burning followed by grazing is one way to prevent tree and shrub thickening in these ecosystems.

Fire may not need to be highly frequent. A reconstruction of fire history in the grassy woodland and open forests of the Eastern Tiers using cross-sections of eucalypt stumps has indicated that the mean fire frequency under indigenous management was once every 10–20 years (von Platen 2008). The normal level of grazing in lowland grassy woodland country creates a lawn which accumulates enough fuel to burn in the occasional years when low animal grazing pressure (induced by mortality or destocking in response to drought) is combined with rapid grass growth, induced by good spring rainfall. The highland grassy woodland with a Tussock Grass (*Poa*) grassland understorey can burn only when the dead grass content of the tussocks is highest, usually in late winter to early spring (Leonard 2009). Given that there is likely to be considerable variation in the rate of growth of thickening trees and shrubs between and within lowlands and highlands, the appropriate fire frequency needs to be judged on the basis of its potential to set woody plants back to below head height of grazing animals.

5. Vertebrate grazers can be used to control some species of woody weeds

While burning might facilitate the elimination of exotic shrub seedlings and resprouts by grazing animals, at least one highly invasive shrub species, Boneseed (*Chrysanthemoides monilifera* ssp. *monilifera*), has been shown to be susceptible to exclusion by stock and Tasmanian Pademelon (*Thylogale billardierii*) grazing in the absence of fire (Scurr *et al.* 2008). Several common woody weed species of grassy woodland planted in an experiment on the outskirts of Hobart were eaten to the ground at the sites where macropods survived (T. Zagorski, personal communication 2007). The experiment did not include Gorse (*Ulex europaeus*), the most widespread woody weed in grassy woodland, but did include Cotoneaster (*Cotoneaster* spp.), Spanish Heath (*Erica lusitanica*) and Broom (*Genista monspessulana*). Gorse is topiarised, but not eliminated, by stock and macropods (Kirkpatrick *et al.* 2007b). The deliberate introduction of the Fox (*Vulpes vulpes*) to Tasmania, combined with the sustained undermining of the control program by some politicians and conservationist-haters, promises to reduce the effectiveness of native animals in controlling these woody weeds.

6. Mow or slash

Mowing has been effective in maintaining grassy woodland on the Domain (Kirkpatrick 2004). Annual mowing is sufficient to prevent tree seedlings from becoming unmowable. Mowing is not a viable option on rocky ground, where slashing needs to be used.

7. If you have them, keep on doing what has been done until you prove that it is better to do otherwise

Most of the rare or threatened vascular plant species of Tasmania's grassy woodlands have this status either because most of their potential habitat has been converted into crops and improved pasture (Kirkpatrick *et al.* 2007c), or because the disturbance regimes that they require have become rare since the European occupation (Kirkpatrick 2007a, b). Different species require different disturbance regimes. Heavy grazing by stock and repeated mechanical disturbance may be critical for the future of particular species. Protection from stock and a lack of repeated mechanical disturbance may be critical for others. A change in disturbance regime in an area may render a rare or threatened species locally extinct, while the rare or threatened species that are suited to the new disturbance regime cannot access the area because it is outside their dispersal range. The vegetation remnants in the best condition, as indicated by low exotic plant cover, high native species richness and intact structure, are rarely those with concentrations of rare or threatened species (Kirkpatrick and Gilfedder 1995). Attempts to improve the condition of such remnants need to be incremental and cautious, as the rare or threatened species may have survived because of the very management regime that has degraded the 'integrity' of the remnant. In both remnants and continuous vegetation, experiments with management regimes may be worthwhile, as survival of significant species under a particular management regime may be no guarantee of future survival.

8. Old, small, isolated remnants with high perimeter: area ratios do matter – corridors probably do not

The vegetation, vascular plant flora, environment, disturbance history and management of one hundred vegetation remnants in subhumid Tasmania were recorded in 1993 (Kirkpatrick and Gilfedder 1995; Gilfedder and Kirkpatrick 1998). The surviving remnants were resurveyed in 1999 (Kirkpatrick *et al.* 2007c) when mosses and liverworts were also recorded (Pharo *et al.* 2005). A subset of the remnants plus some control areas were surveyed for their avifauna in the late 1990s (McDonald and Kirkpatrick 2003). The remnants were 2–200 ha in size, mostly grassy woodland and varied in age, shape and degree of isolation from other remnants. The geometric and age qualities that have been assumed to make remnants poor candidates for conservation efforts had only minor effects on the biota, which were most influenced by environmental conditions and management regimes (Kirkpatrick *et al.* 2007c), as in less fragmented country (Leonard and Kirkpatrick 2004). Some of the smallest, least compact and oldest remnants were the most important for the conservation of vascular plants. There was little encouragement for the establishment of corridors or the abandonment of small, contorted remnants to their fate in these data. It would be wise to assume the same for elsewhere in Australia, until data are available to demonstrate otherwise. Sophisticated and elegant theoretical and simulation models do not necessarily reflect reality.

9. Latitude and relative relief make most of the Tasmanian grassy woodland biota resilient to climate change

The contrast in radiation receipt between north and south facing slopes is maximal in the mid-latitudes (Kirkpatrick and Nunez 1980), where Tasmania is situated. There are few parts of Tasmania that are flat, and these have been comprehensively cleared, or inundated. Tree dieback caused by climate change has already occurred in the grassy woodlands of Tasmania, but the island is rich in very local refugia. Some populations of significant species in small remnants, however, may require management intervention if they are to survive further drying and warming (see above).

10. Regulation combined with stewardship payments seems to be the only way to stop more clearance and fertilisation of grassy woodlands

There is a conflict between maintaining grassy woodlands and maximising cash flows from the agricultural enterprise (Kirkpatrick *et al.* 2007d). Legislation to protect grassy woodlands, as in the case of those dominated by Black Gum (*Eucalyptus ovata*) in Tasmania, has proven ineffective in constraining both the State Government and private landowners from engaging in vegetation clearance, inundation or fertilisation. The ongoing development of offsetting arrangements has exacerbated the problem. The many schemes that have been designed to get private landowners to engage in conservation of native vegetation have managed to conserve some grassy woodland, but have alienated many of the most conservation-minded landowners who feel that the processes and restrictions are bureaucratic and often unfit for purpose (Kirkpatrick *et al.* 2007a). A stewardship payment system based on outcomes, within a framework of regulation, appears to be the only long-term option for the conservation of significant vegetation and species on broadacre grazing land (Kirkpatrick *et al.* 2007d).

11. People with local ecological expertise and experience can usually provide good management advice, whereas tool kits and models cannot cope with contingencies

Any comparison of results between different experiments and observations for the same vegetation type (e.g. Kirkpatrick *et al.* 2005) is likely to reveal a high degree of contingency in the responses of taxa to variation in management regimes at different times in different places. Guides, guidelines and rules for vegetation management, otherwise known as 'tools', are about as useful for most particular situations as a medical website is for diagnosing cancer. Some long-term primary producers, land managers and ecologists have the accumulated experience and knowledge of their places to be able to diagnose most ills and suggest a cure that might work. People employed on short-term contracts in an unfamiliar place find that their training in ecological theory and models is of little help in either diagnosis or cure. Although successful conservation in the rural context is as much about empathising with people as it is with in-depth knowledge of ecosystems, you should not hire a brain surgeon because of their smile, nor an ecologist for their social skills. The grumpy local pastoralist may be the only one who can help.

Acknowledgements

I am immensely grateful to my students and other colleagues who have worked so enthusiastically with me on the dire problems of grassy ecosystem conservation. The work summarised herein has been partly supported by grants from many sources, the more significant of which were from Land & Water Australia and its earlier manifestations. The writing of the present paper was supported by an Australian Research Council Discovery Grant (DP0665083).

Bio

Jamie Kirkpatrick is Distinguished Professor of Geography and Environmental Studies at the University of Tasmania. He enjoys doing research that might improve the conservation of nature, or that amuses him. Too much of his life has been spent in Government and NGO committees. This has resulted in a sock drawer partly filled with medals and statuettes, as well as some minor improvements in the conservation of nature in Australia. Several of his more than 50 graduated research postgraduate students are in the process of filling their own sock drawers.

References

Calder, J.A. and Kirkpatrick, J.B. (2008). Climate change and other factors influencing the decline of the Tasmanian cider gum (*Eucalyptus gunnii*). *Australian Journal of Botany* **56**, 684–692.

Gilfedder, L. and Kirkpatrick, J.B. (1998). Factors influencing the integrity of remnants in subhumid Tasmania. *Biological Conservation* **84**, 89–96.
Kirkpatrick, J.B. (2004). Vegetation change in an urban grassy woodland 1974–2000. *Australian Journal of Botany* **52**, 597–608.
Kirkpatrick, J.B. (2007a). Sheep and nature on the run country. In *People, Sheep and Nature Conservation: The Tasmanian Experience.* (Eds J.B. Kirkpatrick and K.L. Bridle) pp. 139–160. CSIRO Publishing: Melbourne.
Kirkpatrick, J.B. (2007b). Collateral benefit: unconscious conservation of threatened plant species. *Australian Journal of Botany* **55**, 221–224.
Kirkpatrick, J.B., Bridle, K.L., Edwards, J. and Vercoe, J. (2007a). Conserving on the run country. In *People, Sheep and Nature Conservation: The Tasmanian Experience.* (Eds J.B. Kirkpatrick and K.L. Bridle) pp. 99–124. CSIRO Publishing: Melbourne.
Kirkpatrick, J.B., Bridle, K.L. and Leith, P. (2007b). Managing the run country for production. In *People, Sheep and Nature Conservation: The Tasmanian Experience.* (Eds J.B. Kirkpatrick and K.L. Bridle) pp. 45–98. CSIRO Publishing: Melbourne.
Kirkpatrick, J.B., Gilfedder, L., Mendel, L. and Jenkin, E. (2007c). Run country on the run. In *People, Sheep and Nature Conservation: The Tasmanian Experience.* (Eds J.B. Kirkpatrick and K.L. Bridle) pp. 161–182. CSIRO Publishing: Melbourne.
Kirkpatrick, J.B., Jensen, A. and Bridle, K.L. (2007d). The future of the run country. In *People, Sheep and Nature Conservation: The Tasmanian Experience.* (Eds J.B. Kirkpatrick and K.L. Bridle) pp. 183–208. CSIRO Publishing: Melbourne.
Kirkpatrick, J.B., Wilson, D., Meiss, A., Mollon, A. and Bridle, K.L. (2007e). Trees on the run. In *People, Sheep and Nature Conservation: The Tasmanian Experience.* (Eds J.B. Kirkpatrick and K.L. Bridle) pp. 125–138. CSIRO Publishing: Melbourne.
Kirkpatrick, J.B. and Gilfedder, L. (1995). Maintaining integrity compared with maintaining rare and threatened taxa in remnant bushland in subhumid Tasmania. *Biological Conservation* **74**, 1–8.
Kirkpatrick, J.B., Gilfedder, L., Bridle, K.L. and Zacharek, A. (2005). The positive and negative conservation impacts of sheep grazing and other disturbances on the vascular plant species and vegetation of lowland subhumid Tasmania. *Ecological Management and Restoration* **6**, 51–60.
Kirkpatrick, J.B., Gilfedder, L. and Fensham, R. (1988). *City Parks and Cemeteries: Tasmania's Grassy Heritage.* Tasmanian Conservation Trust: Hobart.
Kirkpatrick, J.B. and Marks, F. (1985). Observation of drought damage to some native plant species in eucalypt forests and woodlands near Hobart, Tasmania. *Papers and Proceedings of the Royal Society of Tasmania* **119**, 15–21.
Kirkpatrick, J.B. and Nunez, M. (1980). Vegetation-radiation relationships in mountainous terrain: eucalypt-dominated vegetation in the Risdon Hills, Tasmania. *Journal of Biogeography* **7**, 197–208.
Kirkpatrick, J.B., Zacharek, A. and Chappell, K. (2000). Testing methods for mitigation of tree dieback in Tasmanian dry eucalypt forests and woodlands. *Pacific Conservation Biology* **6**, 94–101.
Leonard, S.W.J. (2009). Effects of vertebrate grazing and environmental factors on fuel and fire potential in Tasmanian native grasslands. PhD thesis, School of Geography and Environmental Studies, University of Tasmania.
Leonard, S.W.J. and Kirkpatrick, J.B. (2004). Effects of grazing management and environmental factors on native grassland and grassy woodland, Northern Midlands, Tasmania. *Australian Journal of Botany* **52**, 529–542.

MacDonald, M. and Kirkpatrick, J.B. (2003). Explaining bird species composition and richness in eucalypt-dominated remnants in subhumid Tasmania. *Journal of Biogeography* **30**, 1415–1426.

Pharo, E.J., Kirkpatrick, J.B., Gilfedder, L., Mendel, L. and Turner, P.A.M. (2005). Predicting bryophyte diversity in grassland and eucalypt-dominated remnants in subhumid Tasmania. *Journal of Biogeography* **32**, 2015–2024.

Scurr, G., Kirkpatrick, J.B., Daniels, G.D. and McQuillan, P.B. (2008). Biotic resistance to boneseed (*Chrysanthemoides monilifera* ssp. *monilifera*) in Tasmania. *Austral Ecology* **33**, 941–950.

von Platen, J. (2008). A history and interpretation of fire frequency in dry eucalypt forests of eastern Tasmania. PhD thesis, School of Geography and Environmental Studies, University of Tasmania.

12

LESSONS IN WOODLAND MANAGEMENT FROM NORTHERN AND CENTRAL VICTORIA

Ralph Mac Nally

1. Increase native vegetation cover.
2. Use planning and optimisation tools for guiding replanting
3. Development of habitat resources is very slow.
4. Fallen wood is a critical ecological resource.

Introduction

I have worked with colleagues and students in central and northern Victoria since the middle 1980s, mostly in box and ironbark forests and River Red Gum (*Eucalyptus camaldulensis*) floodplain forests (see map on next page). These studies have been built on several major collaborations over that period. The first focused on box and ironbark forests (Figure 12.1), mainly in the 1990s, with a team of researchers from Victoria's Department of Sustainability and Environment (Arthur Rylah Research Institute led by G.W. Brown and L.F. Lumsden), Deakin (A.F. Bennett) and Monash Universities, and the Museum of Victoria (A. Yen). The work explored impacts of habitat fragmentation and degradation on many biotic elements (birds, arboreal and volant mammals, reptiles, terrestrial and airborne invertebrates), and provided baseline data (for that time) on distributions of fauna in different ecological vegetation classes (Mac Nally *et al.* 2002a). Another group of Victorian box-ironbark researchers was assembled in the mid-2000s looking more explicitly at landscape-scale biodiversity questions. That team consisted of personnel from Monash (myself, J.R. Thomson), Deakin University (A.F. Bennett and J.Q. Radford), the University of Melbourne (P.A. Vesk), and the Department of Sustainability and Environment (D. Duncan, J. Dorrough). The other main collaboration I have had is with other Monash workers looking at stand condition (Cunningham *et al.* 2009) and factors affecting biodiversity in River Red Gum forests on the Murray floodplain and its Victorian tributaries (Ballinger *et al.* 2005; Lada and Mac Nally 2008; Mac Nally and Horrocks 2007). I list my lessons from issues relating to the largest spatial scales and strategic character through to site-based, tactical issues.

Lessons

1. Increase native vegetation cover

There is too little native woodland in landscapes, and most of it is on the driest and most infertile land. The deleterious impacts on the woodlands since broad-scale agriculture began (and extensive mining in the 1850s) have made them very vulnerable to external forces

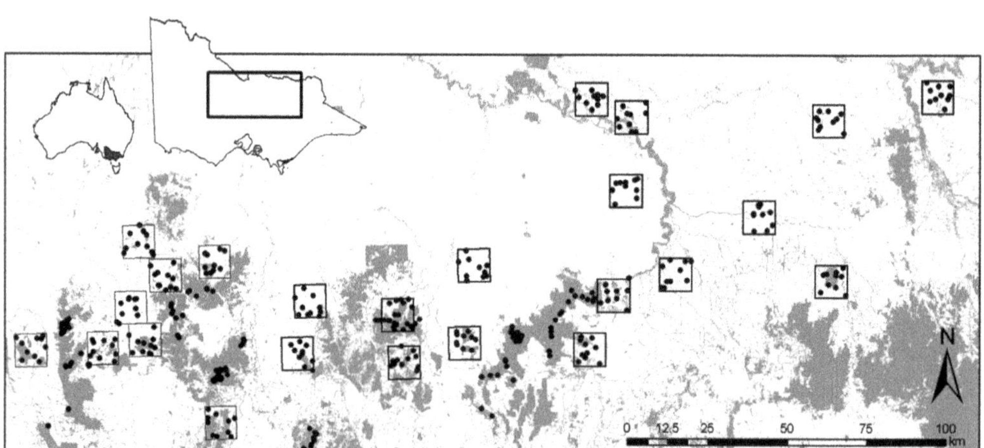

Figure 12.1 Location of study locations across the box and ironbark region of north-central Victoria, Australia, with remnant vegetation in grey shade. Study sites in boxed regions are the landscapes of the landscape-scale study of Radford *et al.* 2005.

(Camilleri *et al.* 2010). Climate change in this region is perhaps as pronounced as anywhere in Australia (Ummenhofer *et al.* 2009), leaving the existing woodlands exposed to effects of aridification. In both the box-ironbark and River Red Gum forests, drying has seriously reduced the condition of tree stands (Cunningham *et al.* 2009). Across the box and ironbark estate, drying appears to have had profoundly negative consequences for birds (Mac Nally *et al.* 2009a) and for amphibians (Mac Nally *et al.* 2009b). The enforced restriction of the box-ironbark remnants to the dry, infertile soils means that the vegetation has little chance to deal with the harsher conditions it is experiencing; flowering seems much diminished (Mac Nally *et al.* 2009a). Many of the animals and plants evolved to use a mosaic of habitat quality across continuous landscapes of millions of hectares (Mac Nally and McGoldrick 1997), but the woodland clearances have reduced the extent, connections and variation of habitat quality. The balance of areas of different ecological vegetation classes has been much altered since the 18th century.

Management changes needed – (1) There needs to be at least 40% appropriate native vegetation in landscapes at all scales up to bioregion (Radford and Bennett 2007). (2) The 'narrow corridor' and minor tinkering approaches, which place very little new habitat in landscapes and mostly as narrow linear strips (Lambeck 1997), appear to have failed. We need a strategic 'biolink' paradigm as its replacement (Brereton *et al.* 1995), where a biolink is a much broader swathe of vegetation (tens of kilometres) that joins up remnant vegetation both of the same general type (e.g. open woodlands) but also with other forms of vegetation (e.g. plains open woodlands with floodplain forests). Biolinks also offer the promise of allowing species' migration in response to climate change.

2. Use planning and optimisation tools for guiding replanting

While landscape management is a compromise activity that needs to address ecological as well as social and production objectives, use of validated spatial models (Thomson *et al.* 2007) and smart algorithms (Thomson *et al.* 2009) can provide biodiversity managers with compelling, defensible tools to argue their cases in how and where replantings should occur and be scheduled.

Research needs – Existing tools largely are based on schedules of habitat suitability that would change through time in replantings, taking into account the extent and distribution of extant vegetation and replantings (Thomson *et al.* 2009). A gap lies in the translation of these maps into projections for concurrent population *viabilities* of multiple species, although some steps are proceeding in this direction (Drielsma and Ferrier 2009).

Management changes needed – These approaches provide blueprints for amounts and schedules for landscape reconstruction but probably will be compromised by tactical decision-making – auction methods tend to produce such disjointed outcomes (Stoneham *et al.* 2003). Scheduling must be strategic and guided by the optimised plans. The main management challenge is the swaying of decision makers to invest heavily enough in the pressing challenges facing woodlands, especially in south-eastern Australia.

3. Development of habitat resources is very slow

A consequence of the dryness and infertility of the remnant woodland vegetation is that growth rates of trees are very slow, and likely to slow even further under climate change. From a calibrated chronosequence of replanted sites, we know definitely that the provision of key ecological resources, such as tree hollows, is a multi-decadal process (Vesk *et al.* 2008).

Management changes needed – The ongoing exploitation of the woodlands through timber harvesting, which affects tree sizes and availabilities of the key ecological resources, needs to be curtailed immediately. It also suggests that there needs to be a 'buy back,' analogous to the acquisition of water in the Murray–Darling Basin, of lands of higher fertility upon which accelerated growth of woodlands can be accomplished (Vesk and Mac Nally 2006).

4. Fallen wood is a critical ecological resource

While the work was undertaken in River Red Gum forests, there now is little doubt that fallen-wood loads of at least 40 t ha^{-1} evoke positive biodiversity responses from birds (Mac Nally and Horrocks 2007) and mammals, at least for the widespread Yellow-footed Antechinus (*Antechinus flavipes)* (Lada *et al.* 2007). Diversity and numbers of birds increase dramatically at this level or above. Numbers of antechinuses, numbers of young produced and female survivorship all are higher above 40 t ha^{-1} (Mac Nally and Horrocks 2008). The importance of fallen wood has been established for other woodland types (Antos *et al.* 2008), but comparable experimental evidence to the Red Gum results is difficult to obtain due to cost and logistics.

Research needs – Fallen wood is one of the few precisely manipulable elements in habitat structure in woodlands, and a determination of the transferability of the River Red Gum outcomes to, say, box and ironbark forests, would provide greater confidence that key thresholds are similar in a range of woodland types. Experimentation over broad scales is needed on degrees of heterogeneity of loads that best contributes to biodiversity outcomes.

Management changes needed – Fallen wood needs to be managed as a key ecological resource rather than as an energy supply (firewood) or as a fire risk. If the River Red Gum results are transferable to other woodland types, which I believe likely, then there are huge shortfalls of fallen wood over most of the extant woodland areas of northern and central Victoria (Mac Nally *et al.* 2002b).

Conclusions

Management of woodland assets – especially biodiversity – has had a history of minimalism, essentially addressing the question, what is the least biodiversity management that can be done, rather than what is necessary to maintain viability of woodland biodiversity? This stance has led to potentially catastrophic declines in woodland flora and fauna, which almost certainly has been exacerbated by a lack of ecological elasticity across the woodland estate due to habitat loss, habitat fragmentation and loss of quality when exposed to major drivers such as aridification and higher temperatures. The primary management issue is investment – its level (woefully inadequate at present), its suffocation by a complex network of institutional and policy strictures, its scheduling over the near to medium future, and its mode of delivery (rarely strategic and mostly piecemeal).

Acknowledgements

Work described here was supported by the Australian Research Council, Land & Water Australia, Environment Australia, the Murray–Darling Basin Authority and the Hermon Slade Foundation. I acknowledge much collaborative involvement with the Goulburn-Broken (especially Tim Barlow), North-central (Geoff Park) and North-east Catchment Management Authorities in Victoria, and the Victorian Departments of Sustainability and Environment (David Duncan, Josh Dorrough, Geoff Brown, Lindy Lumsden, Kim Lowe, David Parkes, Matt White) and Primary Industry (Richard MacEwan). This is contribution 197 from the *Australian Centre for Biodiversity* at Monash University.

Bio

Ralph Mac Nally holds a PhD and a DSc from the University of Melbourne, and had postdoctoral experience at McGill University in Montréal, Canada. He has been at Monash since 1993, and currently is Director of the Australian Centre for Biodiversity at Monash and was the inaugural Theme Leader in Biodiversity & Ecosystem Processes in the Monash Sustainability Institute. His main areas of research interest are landscape ecology, conservation biology, invasion biology and spatial ecology. He also has worked extensively in modelling ecological change in California and Nevada. He has a special reverence for the Reverend Bayes.

References

Antos, M.J., Bennett, A.F. and White, J.G. (2008). Where exactly do ground-foraging woodland birds forage? Foraging sites and microhabitat selection in temperate woodlands of southern Australia. *Emu* **108**, 201–211.

Ballinger, A., Mac Nally, R. and Lake, P.S. (2005). Immediate and longer-term effects of managed flooding on floodplain invertebrate assemblages in south-eastern Australia: generation and maintenance of a mosaic landscape. *Freshwater Biology* **50**, 1190–1205.

Brereton, R., Bennett, S. and Mansergh, I. (1995). Enhanced greenhouse climate change and its potential effect on selected fauna of south-eastern Australia – a trend analysis. *Biological Conservation* **72**, 339–354.

Camilleri, S., Thomson, J. and Mac Nally, R. (2010). The interaction between land use and catchment physiognomy: understanding avifaunal patterns of the Murray-Darling Basin, Australia. *Journal of Biogeography.*

Cunningham, S., Mac Nally, R., Read, J., Baker, P., White, M., Thomson, J.R. and Griffioen, P. (2009). A robust technique for mapping vegetation across a major river system. *Ecosystems* **12**, 207–219.

Drielsma, M. and Ferrier, S. (2009). Rapid evaluation of metapopulation persistence in highly variegated landscapes. *Biological Conservation* **142**, 529–540.

Lada, H. and Mac Nally, R. (2008). Decline and potential recovery of Yellow-footed Antechinus in parts of south-eastern Australia: a perspective with implications for management. *Ecological Management and Restoration* **9**, 120–125.

Lada, H., Thomson, J.R., Mac Nally, R., Horrocks, G. and Taylor, A.C. (2007). Evaluating simultaneous effects of three anthropogenic effects on a floodplain-dwelling marsupial Antechinus flavipes. *Biological Conservation* **134**, 527–536.

Lambeck, R.J. (1997). Focal species: a multi-species umbrella for nature conservation. *Conservation Biology* **11**, 849–856.

Mac Nally, R., Bennett, A.F., Brown, G.W., Lumsden, L.F., Yen, A., Hinkley, S., Lillywhite, P. and Ward, D. (2002a). How well do ecosystem-based planning units represent different components of biodiversity? *Ecological Applications* **12**, 900–912.

Mac Nally, R., Bennett, A.F., Thomson, J.R., Radford, J.Q., Unmack, G., Horrocks, G. and Vesk, P. (2009a). Collapse of an avifauna: climate change appears to exacerbate habitat loss and degradation. *Diversity & Distributions* **15**, 720–730.

Mac Nally, R. and Horrocks, G. (2007). Inducing whole-assemblage change by experimental manipulation of habitat structure. *Journal of Animal Ecology* **76**, 643–650.

Mac Nally, R. and Horrocks, G. (2008). Longer-term responses of a floodplain-dwelling marsupial to experimental manipulation of fallen timber loads. *Basic & Applied Ecology* **9**, 458–465.

Mac Nally, R., Horrocks, G., Lada, H., Lake, P.S., Thomson, J.R. and Taylor, A.C. (2009b). Distribution of anuran amphibians in massively altered landscapes in south-eastern Australia: effects in climate change in an aridifying region. *Gobal Ecology and Biodiversity* **18**, 575–585.
Mac Nally, R. and McGoldrick, J.M. (1997). Landscape dynamics of bird communities in relation to mass flowering in some eucalypt forests of central Victoria, Australia. *Journal of Avian Biology* **28**, 171–183.
Mac Nally, R., Parkinson, A., Horrocks, G. and Young, M. (2002b). Current loads of coarse woody debris on south-eastern Australian floodplains: evaluation of change and implications for restoration. *Restoration Ecology* **10**, 627–635.
Radford, J.Q., Bennett, A.F. and Cheers, G.J. (2005). Landscape-level thresholds of habitat cover for woodland-dependent birds. *Biological Conservation* **124**, 317–337.
Radford, J.Q. and Bennett, A.F. (2007). The relative importance of landscape properties for woodland birds in agricultural environments. *Journal of Applied Ecology* **44**, 737–747.
Stoneham, G., Chaudhri, V., Ha, A. and Strappazzon, L. (2003). Auctions for conservation contracts: an empirical examination of Victoria's BushTender trial. *Australian Journal of Agricultural and Resource Economics* **47**, 477–500.
Thomson, J., Mac Nally, R., Fleishman, E. and Horrocks, G. (2007). Predicting bird species distributions in reconstructed landscapes. *Conservation Biology* **21**, 752–766.
Thomson, J.R., Moilanen, A., Vesk, P., Bennett, A.F. and Mac Nally, R. (2009). Where and when to revegetate: a quantitative method for scheduling landscape reconstruction. *Ecological Applications* **19**, 817–828.
Ummenhofer, C.C., England, M.H., McIntosh, P.C., Meyers, G.A., Pook, M.J., Risbey, J.S., Gupta, A.S. and Taschetto, A.S. (2009). What causes southeast Australia's worst droughts? *Geophysical Research Letters* **36**, L04706.
Vesk, P. and Mac Nally, R. (2006). The clock is ticking – revegetation and habitat for birds and arboreal mammals in rural landscapes of southern Australia. *Agriculture, Ecosystems and Environment* **112**, 356–366.
Vesk, P., Nolan, R., Thomson, J., Dorrough, J. and Mac Nally, R. (2008). Time lags in provision of habitat resources through revegetation. *Biological Conservation* **141**, 174–186.

13

LANDSCAPE PERSPECTIVES ON WILDLIFE CONSERVATION IN WOODLANDS

Andrew F. Bennett and James Q. Radford

1. Scale is important! Complementary conservation actions are needed at different scales of management.
2. Ecological responses may not be directly proportional to the amount of environmental change: look for nonlinear and threshold responses
3. Incremental changes and cumulative impacts have profound consequences in woodland ecosystems
4. Networks of linear wooded vegetation (roadsides, streams) are critical components in modified landscapes
5. Woodland species show diverse functional responses to landscape change
6. Land-use history influences current landscape values and conservation issues

Introduction

Temperate woodland ecosystems in Victoria have been profoundly modified since the first European explorers and settlers traversed the State in the late 1830s. Indeed, Major Thomas Mitchell foresaw such change when, in 1836, he looked out from Mt Hope across the plains of northern Victoria:

> *As I stood, the first intruder on the sublime solitude of these verdant plains as yet untouched by flocks and herds, I felt conspicuous of being the harbinger of mighty change there; for my steps would soon be followed by the men and animals for which it seemed to have been made.*

The ensuing changes to vegetation, soils, hydrology and native fauna (ECC 1997; Lunt and Bennett 2000) have been extraordinarily rapid on an ecological time scale – less than the lifespan of a living organism, as evidenced by the continued presence of trees of pre-European settlement origin.

Faunal response to landscape change in woodland ecosystems has been a central theme of some of our recent research. We have worked mainly in agricultural landscapes in north-central Victoria, with woodland birds as a particular focus of investigation (see map on next page).

We also draw on our experience from previous work in woodlands and dry box and ironbark forests in northern and north-western Victoria. Radford investigated the conservation biology of the White-browed Treecreeper (*Climacteris affinis*) in dry woodlands in semi-arid north-western Victoria (Radford 2002). Bennett has undertaken collaborative research on a range of taxa in woodlands, including arboreal mammals (e.g. van der Ree and Bennett 2003), bats (Lumsden *et al.* 2002), birds (Mac Nally *et al.* 2000) and reptiles (Brown *et al.* 2008).

Lessons

1. Scale is important! Complementary conservation actions are needed at different scales of management

Our research has focused on the effect on fauna of landscape change at large scales, using 'whole landscapes' (10 × 10 km in size) as the unit of study (Radford *et al.* 2005). This allows inference about the properties of landscapes that most influence the status of native fauna. The composition of the landscape, particularly the overall amount of remnant wooded vegetation compared with other land uses, was found to be of paramount importance. At this scale, protecting existing vegetation and increasing the total extent of wooded vegetation by restoration is a key priority for conservation. At the scale of individual patches, however, studies of woodland birds typically identify aspects of habitat configuration (e.g. patch size, context, isolation) as having significant influences on which, and how many, species occur. At an even finer scale, local habitat condition as indicated by attributes such as ground-layer complexity, shrub cover, tree density, fallen timber and tree hollows, also influence where species occur (Mac Nally *et al.* 2002; Antos *et al.* 2008).

Insights from all scales are important. Effective conservation requires balancing different kinds of actions across different scales in a complementary manner. Re-establishing understorey plants, for example, may enhance local habitat quality for a species, but this needs to be complemented by strategies at the landscape scale that determine its overall population size and viability. Revegetation projects to increase connectivity between patches should be nested within broader strategies designed to increase the total extent of native vegetation in the landscape.

2. Ecological responses may not be directly proportional to the amount of environmental change: look for nonlinear and threshold responses

A striking result from our work was the identification of a threshold response in the species richness of woodland birds to the amount of wooded cover in the landscape (Radford *et al.* 2005). Above 10% cover there was only gradual decline in richness with decreasing cover; but below 10% cover there was a rapid decline in species richness with decreasing cover. Numerous species, such as the Eastern Yellow Robin (*Eopsaltria australis*), Grey Shrike-thrush (*Colluricincla harmonica*) and Crested Shrike-tit (*Falcunculus frontatus*), also showed disproportionate declines in occurrence (but not necessarily abrupt thresholds) at different levels of wooded cover in the landscape (Radford and Bennett 2007). Many other nonlinear responses also occur in woodland ecosystems. For example, the use of clumps of trees in paddocks by the Squirrel Glider (*Petaurus norfolcensis*) declined inversely with the gap distance to be crossed, and were rarely used if >75 m from a source (van der Ree *et al.* 2003). The White-browed Treecreeper did not occupy suitable woodland patches in agricultural landscapes if isolated from other populations by more than a 3 km threshold distance (Radford and Bennett 2004).

We can not assume that plant and animal communities (and species) will respond to environmental change in a directly proportional way. It is vital to become better at recognising response 'shapes', especially the critical points or stages at which change occurs rapidly, so that intervention can be undertaken before populations crash, species disappear, or irreversible changes occur to ecosystem processes. It may also allow more strategic decisions about where to invest in landscape restoration by identifying landscapes where the greatest conservation gains can be made.

3. Incremental changes and cumulative impacts have profound consequences in woodland ecosystems

Incremental change and the 'death of a thousand cuts' from the cumulative effects of many seemingly small actions, are major issues for woodland conservation. Large, relatively undisturbed woodlands are scarce in our study areas: woodlands typically occur in remnant blocks, both large and small, with differing levels of isolation and subject to a range of human land uses. Many small changes occur and accrue across the landscape: loss of paddock trees, clearing on roadsides, subdivision of forest blocks, firewood collection, and the introduction of intensive land uses (e.g. vineyards, pivot irrigation). The benefits of positive actions also are cumulative: revegetation along gullies and streams, fencing to control stock access, and protecting roadside vegetation.

Changes to individual elements in a landscape and the cumulative impact of many changes do matter; not only because of their local effects but because they affect the structure of the landscape as a whole. We found that 75% of the variation in numbers of woodland bird species could be attributed to the properties of the whole landscape and its geographic location (Radford *et al.* 2005). Thus, in land-use planning and impact assessment, much greater scrutiny is needed of the wider consequences of individual actions across the landscape in which they are located. For example, in what ways will this proposal or land use affect other parts of the landscape? What will be the cumulative impact of this land use if it is repeated multiple times across the landscape?

4. Networks of linear wooded vegetation (roadsides, streams) are critical components in modified landscapes

Strips of remnant wooded vegetation along roadsides, drainage lines and streams are an invaluable component of rural landscapes, particularly where they form interconnected systems of habitat. They may represent a large proportion of the remaining wooded vegetation in some districts, they contribute to finer-grained and more spatially heterogeneous landscapes, and they enhance structural connectivity for woodland species. In northern Victoria, roadside networks provide critical habitat for threatened species such as the Squirrel Glider, Brush-tailed Phascogale (*Phascogale tapoatafa*) and Grey-crowned Babbler (*Pomatostomus temporalis*) (e.g. van der Ree and Bennett 2002), especially due to the presence of large old (pre-European) trees as a source of tree hollows and foraging substrates. Riparian vegetation extending across our study landscapes in dendritic patterns also has great value for terrestrial fauna. Riparian vegetation supports a large and distinctive assemblage of terrestrial birds, and riparian sites in these landscapes typically have a higher species richness and density of birds than other wooded sites (Radford and Bennett unpublished).

Linear habitats are sensitive to cumulative effects from disturbance; such as 'tidying up' roadsides, firewood removal, weed invasion, and incremental loss of trees from road works. They also offer many opportunities for restoration and improved management in agricultural environments, particularly of riparian vegetation, as part of a broader landscape approach. Restoration of riparian zones is increasingly being undertaken (fencing, revegetation) and has multiple benefits for soil, water and biodiversity conservation.

5. Woodland species show diverse functional responses to landscape change

Many workers, including ourselves, have divided species into broad categories such as 'woodland dependent', 'open country' and 'edge' species, as a basis for examining their patterns of occurrence and responses to land use. Such an approach is a compromise: it recognises that there are differences among species within assemblages but that treating each species individually in conservation planning often is impractical. It still is essential, however, to be aware of differences among species, even within these categories. We found that among species classed as 'woodland dependent', there were diverse responses to landscape change (Radford and Bennett 2007). The incidence of the Rufous Whistler (*Pachycephala rufiventris*), for example, was related primarily to the total amount of wooded cover in a landscape, whereas the Grey Shrike-thrush and Spotted Pardalote (*Pardalotus punctatus*) responded to both the amount of wooded cover and its configuration (favouring aggregated blocks). The Brown Treecreeeper (*Climacteris picumnis*), in contrast, responded most strongly to the diversity of vegetation types and the condition of the vegetation in each landscape.

Such differences are not surprising, given the variation in ecological attributes (e.g. body size, mobility, food and foraging locations, migratory status) encompassed within a group such as woodland birds. Even greater variation occurs when comparing responses to landscape change across higher taxonomic groups (mammals, reptiles, frogs, spiders etc). While composite indices provide useful summaries for aggregations of species, the implications of land-use change will often be taxa-specific. Where priority species (e.g. threatened, endemic, keystone) or species known to be susceptible to the proposed change persist, evaluation of species-specific responses is appropriate.

6. Land-use history influences current landscape values and conservation issues

Temperate woodlands differ, both within and among regions, in the composition and structure of their vegetation and biota. Such natural variation in relation to soils, climate and topography, is compounded by land-use history. In northern Victoria there are marked differences in land-use history, particularly between woodlands of the alluvial plains (Northern Plains) and those of the inland hills of sedimentary origin ('Box-Ironbark' forests and woodlands) (ECC 1997). These differences give rise to different kinds of conservation issues. On the plains, higher soil fertility has resulted in extensive clearing for agriculture. Key issues relate to the extent of overall loss of woodlands, the small sizes of remaining fragments, effects of grazing by stock, values of roadside and stream networks, and the future of scattered old trees in farmland. The inland hills, in contrast, have poorer soils and larger tracts of wooded vegetation remain. Gold mining and timber extraction have been dominant historical land-uses; and their legacy includes a current scarcity of large trees in public land blocks, dense stands of regrowth trees, and a degraded ground-layer from soil disturbance and firewood removal (ECC 1997).

These differences highlight that a 'one size fits all' solution is not appropriate for restoring woodland landscapes. General principles are useful as a baseline, but they need to be adapted in the light of differing circumstances. An historical perspective, as clearly articulated by Lunt and Spooner (2005), is invaluable both for understanding why modified landscapes are as they are today, and for identifying specific issues to address in restoration and conservation planning.

Conclusions

In an era of increasing reliance on statistical and spatial models, remote sensing and GIS, we emphasise the important role of field-based observations and research. Getting a 'feel' for an ecosystem and its function, and understanding the natural history of its plants and animals should not be underestimated. There are many gaps in empirical knowledge that have profound consequences for conservation planning, and many important discoveries are yet to be made. For example, little is known about the scale, frequency, timing and pathway of dispersal movements of most species. Yet such knowledge is essential for understanding (and modelling) the status of populations and metapopulations, and for more effective landscape design for restoration.

Woodland ecosystems in northern Victoria and elsewhere face ongoing changes in land use, an increasing human population and cumulative impacts from diverse sources of disturbance. A challenge for conservation planning and management is to move from a reactive, case by case, mode of operation to a longer-term, proactive perspective. Conservation outcomes ultimately depend on the choices that society makes, often through the political process; but these choices need to be well informed. There is a widespread, but inaccurate, perception in society that continued growth and intensity of human use of the environment is entirely compatible with maintaining natural ecosystems without loss of species. One option is for conservation biologists to participate in a process of developing alternative future scenarios, based on realistic projections in land use. By documenting the likely consequences for biodiversity of each scenario, society may more clearly see the ecological trade-offs that will result from the different choices we face.

Acknowledgements

For funding and support we gratefully acknowledge the Department of Sustainability and Environment, Department of Primary Industries, North-Central and North-Eastern CMAs, the Australian Research Council, Land and Water Australia, and Deakin University. We have benefited from shared experiences, discussion and insights into woodland ecology with many colleagues: from the Department of Sustainability and Environment (Geoff Brown, Josh Dorrough David Duncan, Kim Lowe, Lindy Lumsden, Peter Morison, Rob Price), Department of Primary Industries (Alan Yen, Richard McEwen), North-Central CMA (Geoff Park, Ian Higgins), Monash University (Ralph Mac Nally, Jim Thomson, Greg Horrocks, Peter Vesk), La Trobe University (Mike Clarke, Merilyn Grey) and Deakin University (Mark Antos, Glenn Ehmke, Angie Haslem, Greg Holland, Lindy MacRaild, Chris Tzaros, Rodney van der Ree, Jenny Wilson). Thanks to Gary Cheers for assistance with many bird surveys. We warmly thank many landholders for access to their land, and salute those leading the way in rural conservation by protecting natural environments on their property.

Bio

Andrew Bennett has more than 25 years research experience in wildlife ecology and conservation biology. After 12 years with the Arthur Rylah Institute, Department of Sustainability and Environment, he joined Deakin University in 1996 where he teaches ecology and landscape ecology, and leads an active research group in landscape ecology. He has conducted wildlife research in many regions of Victoria, and led teams investigating factors that influence the conservation of species and assemblages (mammals, birds, reptiles) in woodland and mallee ecosystems. A particular focus, both in Australia and internationally, is wildlife conservation in agricultural landscapes and the effect of human land-use and landscape change on nature conservation. He holds a Personal Chair in Landscape Ecology at Deakin University.

Jim Radford completed his PhD on the ecology of the White-browed Treecreeper at Deakin University in 2002. He then took up a Post-doctoral Fellowship at Deakin University to research the effects of land-use change on avifaunal communities at landscape scales. Jim's research interests include understanding the interplay between behavioural responses to landscape change and the ecological processes that determine the distribution and abundance of fauna. He is passionate about transferring this understanding into constructive landscape management principles and guidelines for ecologically sustainable practices. Jim Radford currently works for Bush Heritage Australia, where he manages the ecological monitoring program for their conservation reserves.

References

Antos, M.J., Bennett, A.F. and White, J.G. (2008). Where exactly do ground-foraging woodland birds forage? Foraging sites and microhabitat selection in temperate woodlands of southern Australia. *Emu* **108**, 201–211.

Brown, G.W., Bennett, A.F. and Potts, J. (2008). Regional faunal decline – reptile occurrence in fragmented rural landscapes of south-eastern Australia. *Wildlife Research* **35**, 8–18.

ECC (1997). 'Box-Ironbark Forests and Woodlands Investigation. Resources and Issues Report'. Environment Conservation Council: Victoria.

Lumsden, L.F., Bennett, A.F. and Silins, J.E. (2002). Locations of roosts of the lesser long-eared bat *Nyctophilus geoffroyi* and Gould's wattled bat *Chalinolobus gouldii* in a fragmented landscape in south-eastern Australia. *Biological Conservation* **106**, 237–249.

Lunt, I. and Bennett, A.F. (2000). Temperate woodlands in Victoria: distribution, composition and conservation. In *Temperate Eucalypt Woodlands in Australia: Biology, Conservation, Management and Restoration.* (Eds. RJ Hobbs and CJ Yates) pp. 17–31. Surrey Beatty and Sons: Chipping Norton, NSW.

Lunt, I.D. and Spooner, P.G. (2005). Using historical ecology to understand patterns of biodiversity in fragmented agricultural landscapes. *Journal of Biogeography* **32**, 1859–1873.

Mac Nally, R., Bennett, A.F. and Horrocks, G. (2000). Forecasting the impact of habitat fragmentation. Evaluation of species-specific predictions of the impact of habitat fragmentation on birds in the box-ironbark forests of central Victoria, Australia. *Biological Conservation* **95**, 7–29.

Mac Nally, R., Soderquist, T.R. and Tzaros, C. (2000). The conservation value of mesic gullies in dry forest landscapes: avian assemblages in the box-ironbark ecosystem of southern Australia. *Biological Conservation* **93**, 293–302.

Radford, J.Q. (2002). Conservation ecology and breeding biology of the White-browed Treecreeper *Climacteris affinis*. PhD Thesis, Deakin University, Melbourne.

Radford, J.Q. and Bennett, A.F. (2004). Thresholds in landscape parameters: occurrence of the white-browed treecreeper *Climacteris affinis* in Victoria, Australia. *Biological Conservation* **117**, 375–391.

Radford, J.Q. and Bennett, A.F. (2007). The relative importance of landscape properties for woodland birds in agricultural environments. *Journal of Applied Ecology* **44**, 737–747.

Radford, J.Q., Bennett, A.F. and Cheers, G.J. (2005). Landscape-level thresholds of habitat cover for woodland-dependent birds. *Biological Conservation* **124**, 317–337.

van der Ree, R. and Bennett, A.F. (2003). Home range of the squirrel glider (*Petaurus norfolcensis*) in a network of remnant linear habitats. *Journal of Zoology* **259**, 327–336.

van der Ree, R., Bennett, A.F. and Gilmore, D.C. (2003). Gap-crossing by gliding marsupials: thresholds for use of isolated woodland patches in an agricultural landscape. *Biological Conservation* **115**, 241–249.

14

MANAGING AN OVER-ABUNDANT NATIVE BIRD: THE NOISY MINER (*MANORINA MELANOCEPHALA*)

Michael F. Clarke and Merilyn J. Grey

1. Noisy Miners (*Manorina melanocephala*) prevent many small insectivorous birds from utilising small and degraded box-ironbark woodland remnants.
2. Noisy Miners will penetrate up to 300 m from edges into the interior of large woodland remnants.
3. Noisy Miners are more likely to colonise peninsulas and clumps of canopy vegetation protruding from the edges of large remnants, than straight edges of remnants.
4. Noisy Miners appear to dominate the more productive parts of the landscape with more fertile soils.
5. Removal of Noisy Miners results in increased bird diversity, a significant decrease in insect damage to leaves of Grey Box (*Eucalyptus microcarpa*) trees, and a measurable improvement in tree health in small box-ironbark remnants.
6. Noisy Miners spend a substantial proportion of their time foraging on the ground; therefore, promoting the growth of ground-layer vegetation has the potential to make a site less attractive to Noisy Miners.
7. Culling is the most humane, cost-effective and time-efficient method of reducing the impact of Noisy Miners in the short term.
8. Noisy Miner culls should be accompanied by revegetation and rehabilitation designed to deter Noisy Miners and provide habitat for colonising bird species and other fauna.

Introduction

We have been studying the ecological consequences of the aggressive behaviour of the native Noisy Miner (see Figure 14.1) in Victorian woodlands for over 25 years. We have worked in small remnants (<10 ha) of the box-ironbark ecosystem of north-eastern Victoria, near Benalla and Violet Town (Grey *et al.* 1997; Grey *et al.* 1998), along the edges of large remnants (10 700 to 29 000 ha) of box-ironbark forest and woodland in the Goldfields bioregion of north-central

Figure 14.1 Noisy Miner (*Manorina melanocephala*). (Photo ©Rohan Clarke/www.wildlifeimages.com.au)

Victoria, near Bendigo (Clarke and Oldland 2007; Taylor *et al.* 2008; Oldland *et al.* 2009), in large grassy woodland remnants (195 to 5400 ha) of the Goldfields bioregion, the Victorian Riverina bioregion and the Gippsland Plains bioregion (Clarke and Oldland 2007; Taylor *et al.* 2008) (see map on previous page).

The Noisy Miner is a widespread, large (about 60 g), hyper-aggressive, communally breeding honeyeater that forms large colonies and dominates grassy woodland remnants throughout eastern Australia, expelling nearly all other bird species (Dow 1977; Higgins *et al.* 2001). Clearing of woodlands and leaving small clumps of paddock trees or corridors surrounded by pasture appears to have inadvertently created an abundance of prime Noisy Miner habitat. The domination of remnants by the Noisy Miner is occurring in a region of Australia in which an estimated 80–90% of the temperate woodlands have been cleared for agriculture since European colonisation (Hobbs and Hopkins 1990). We are only now realising the full ramifications of this in terms of extinction debt (Tilman *et al.* 1994; Recher 1999; Ford *et al.* 2001). It appears that within its range, the Noisy Miner is a key threatening process in the ongoing progression of bird extinctions. Most small remnants of eucalypt woodland are dominated by Noisy Miners, and the species continues to expand its range (R. Major unpublished data) and is increasing in abundance in parts of its range (Barrett *et al.* 2002). The Noisy Miner's interspecific territorial behaviour prevents many small woodland birds (already declining because of habitat loss) from occupying many remnants. Exclusion of small insectivorous birds by the Noisy Miner may also exacerbate dieback caused by insects in small woodland remnants (Ford and Bell 1981; Loyn 1987). Substantial conservation funds are directed at large-scale revegetation projects, often with an emphasis on planting vegetation corridors to improve connectivity within the landscape. Our research attempts to inform these efforts, in the hope that we may avoid making the situation worse, by creating additional habitat for Noisy Miners.

Lessons

1. The Noisy Miner prevents many small insectivorous birds from utilising small and degraded box-ironbark woodland remnants

The Noisy Miner has a profound negative effect upon the diversity and abundance of small woodland birds. After we removed the Noisy Miner from seven small (<10 ha) box-ironbark remnants, there was a major influx of small insectivorous birds, providing the first experimental evidence that the species affects avian diversity and abundance by aggressive exclusion of small birds (Grey *et al.* 1997; Grey *et al.* 1998). Scientists had long been aware of a correlation between the presence of the Noisy Miner in a woodland remnant and low diversity and abundance of birds (especially small insectivores) (e.g. Dow 1977; Ford and Bell 1981; Loyn 1987, among many others), but importantly, our experiments also showed that when domination by the Noisy Miner is reduced, small birds will use degraded woodland remnants for nesting and feeding. Species arriving in remnants following the removal of the Noisy Miner included the endangered Regent Honeyeater (*Anthochaera phrygia*) and numerous members of the threatened woodland-dependent bird community such as the Black-chinned Honeyeater (*Melithreptus gularis*), Fuscous Honeyeater (*Lichenostomus fuscus*), and small insectivores like the Dusky Woodswallow (*Artamus cyanopterus*), Southern Whiteface (*Aphelocephala leucopsis*) and Varied Sittella (*Daphoenositta chrysoptera*) (Flora and Fauna Guarantee – Scientific Advisory Committee 2001; Department of Sustainability and Environment 2006). The Noisy Miner did not reinvade the small woodland remnants during the ensuing 12 months, and in some cases, for several years (D. Robinson, personal communication; M. Grey unpublished data). Reducing the abundance of the Noisy Miner in remnant eucalypt woodlands may be a useful measure that could assist in the recovery of threatened or endangered bird species.

2. Noisy Miners will penetrate up to 300 m from edges into the interior of large woodland remnants

Domination by the Noisy Miner even occurs into the interior of large remnants. The Noisy Miner penetrates into the interior of large remnants from 150 m to more than 300 m and this occurs in a number of habitat types (Piper and Catterall 2003; Clarke and Oldland 2007). If an aim in land management is to ensure that there is some Noisy Miner-free core habitat for other bird species, such behaviour has implications for the minimum size of a remnant, or width of a corridor. Remnants need to be over 36 ha in area and corridors over 600 m wide to avoid total domination by the Noisy Miner and to be of value as habitat for other bird species. The density of vegetation in replantings also seems to be a critical factor in providing Noisy Miner-free habitat (Lindenmayer *et al.* 2009). Narrower wildlife corridors nevertheless have value to many other taxa, apart from birds (Bennett 2003), particularly for maintaining connectivity in the landscape for less mobile taxa. Revegetation efforts for conservation of woodland birds should be targeted at increasing the size of existing remnants and widening wildlife corridors whenever possible.

3. The Noisy Miner is more likely to colonise peninsulas and clumps of canopy vegetation protruding from the edges of large remnants, than straight edges of remnants

The shape of large remnants can influence the likelihood of occupation of large remnant edges by the Noisy Miner. In our study of the edges of large remnants (>300 ha), Taylor *et al.* (2008) demonstrated that the locations of Noisy Miner colonies were most commonly associated with three features of the remnants: 1. projections of remnant vegetation from the remnant edge into the agricultural matrix (e.g. corners of remnants or peninsulas of vegetation); 2. corridors of vegetation extending into a paddock from the remnant edge; and 3. clumps of trees in the agricultural matrix within 100 m of the remnant edge (Taylor *et al.* 2008). These edge features may confer advantages for interspecific territorial defense to the Noisy Miner when attempting to detect intruders into their territory.

The Noisy Miner most commonly occupies remnants of <20 ha. They will also colonise the human-induced interface of farmland/woodland edges of larger remnants and forest blocks (e.g. Dow 1977; Ford and Bell 1981; Loyn 1987; Hobbs and Hopkins 1990; Barrett *et al.* 1994; Piper and Catterall 2003). Their occupation of the edges of large remnants is not continuous, however. This study highlights the value of considering the patterns of habitat use by the Noisy Miner when fencing and revegetating remnants in an agricultural landscape. To deter the Noisy Miner, enclose protrusions, such as peninsulas and clumps of nearby trees, and revegetate to smooth and round out the remnant edge to eliminate corners (Figure 14.2).

4. The Noisy Miner appears to dominate the more productive parts of the landscape with more fertile soils

The Noisy Miner dominates the more productive parts of the landscape. This has profound implications for other woodland birds, as the Noisy Miner is excluding them from some of the most productive and fruitful remnants in the landscape, already rare because the more fertile land has been preferentially cleared for agriculture (Reid 1999).

When examining why Noisy Miner colonies occupied some corners of remnants but not others, we discovered that sites with deeper, more fertile soils and higher proportions of Yellow Gum (*Eucalyptus leucoxylon,* a prolific, reliable nectar producer) were more likely to be colo-

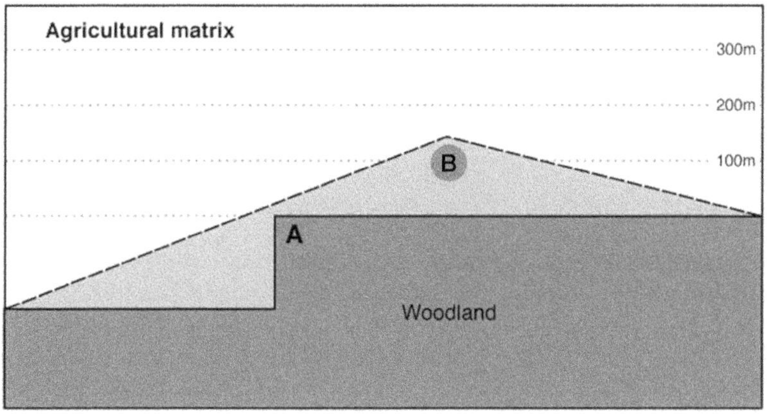

Figure 14.2 Hypothetical example of the edge of a large woodland patch with two protrusions: (A) projection (corner); and (B) clump. The broken lines (---) represent the perimeter of the proposed revegetation. To deter Noisy Miners it should extend into the agricultural matrix, enclosing both protrusions, and smoothing the patch edge. (With kind permission from SpringerScience+Business Media: Landscape Ecology, 23, 2008, 337-389, Taylor R.S., Oldland J.M., Clarke M.F. Figure 3.)

nised than less productive sites (Oldland *et al.* 2009). This 'productivity hypothesis' postulates that the high-density, sedentary colonies of the Noisy Miner require year-round food supplies of nectar and insects which are more likely to occur at more fertile sites (Recher *et al.* 1996; Keatley and Hudson 2006), particularly during droughts.

5. Removal of the Noisy Miner results in increased bird diversity, a significant decrease in insect damage to leaves of Grey Box trees, and a measurable improvement in tree health in small box-ironbark remnants

There are cascading, positive ecological consequences of removing the Noisy Miner from small remnants. A flood of small birds was able to use remnants once the Noisy Miner had been removed, including a number of species listed in the threatened Victorian temperate-woodland bird community (Grey *et al.* 1997; Grey *et al.* 1998; Department of Sustainability and Environment 2006). A disproportionate number of these invading birds were migratory. This finding suggests that the Noisy Miner may prevent migratory species from accessing vital resources which may only be intermittently available in small remnants in a highly fragmented agricultural landscape.

Many researchers have noticed an association between Noisy Miners and dieback (e.g. Ford and Bell 1981; Ford 1985; Loyn 1987, and many others). We now have experimental evidence that suggests that there is a causal link between the Noisy Miner and the exacerbation of insect-induced eucalypt dieback (Grey 2008). Removal of the Noisy Miner, and the repopulation of remnants by small birds, resulted in a significant decrease in the level of leaf damage caused by herbivorous insects in Grey Box remnants, mainly due to a reduction in damage caused by chewing and leaf-mining insects. Trees in box-ironbark remnants which were monitored following the removal of Noisy Miners also showed a steady improvement in overall 'tree health' (Grey 2008). The influx of small insectivorous birds following the removal of the Noisy Miner from woodland remnants appears to have resulted in a reduction in leaf damage and an improvement in tree health in some circumstances, showing that Noisy Miner domination of small woodland remnants has a top–down effect, which can have an adverse impact on tree health.

6. The Noisy Miner spends a substantial proportion of its time foraging on the ground. Promoting the growth of ground-layer vegetation has the potential to make a site less attractive to the Noisy Miner

Promoting the growth of ground-layer vegetation may deter the Noisy Miner by removing a foraging substrate. In contrast to most honeyeaters, which are strongly arboreal, the Noisy Miner often uses the ground as a foraging substrate (Higgins *et al.* 2001). In a study of micro-habitat preferences of the Noisy Miner (M. Grey and M. Clarke unpublished data), we found that over 12% of Noisy Miner movements result in flying to the ground. When ground foraging, the Noisy Miner shows a significant preference for grazed ground. Ground-foraging Noisy Miners exploit sites with short grass (<5 cm), which are near fallen timber (<5 m) and within 25 m of canopy cover. Obviously, canopy trees and fallen timber are vital elements of the landscape for a wide range of taxa and must be retained. We suggest that encouraging grasses or herbs to grow as long and thick as possible by fencing remnants to exclude stock would make a site less attractive to the Noisy Miner. Conversely, activities in or beside remnants that result in shortening the grass, such as mowing, burning or intensive grazing by livestock and macropods may enhance the suitability of a site for the Noisy Miner. Although allowing grasses to grow long and thick may deter some other ground-foraging species (Maron and Lill 2005), we believe that when the Noisy Miner is present, these species are very unlikely to be present anyway. While this issue requires further research, we consider overall gains in biodiversity that could be achieved by the liberation of a site from the Noisy Miner outweigh the potential disadvantage to some species. Should ground-feeding species arrive in the absence of the Noisy Miner, it is easy to reverse the process to create bare ground to accommodate them. We therefore believe manipulation of the grassy ground-layer should be considered as an option for managing the Noisy Miner.

7. Culling is the most humane, cost-effective and time-efficient method of reducing the impact of Noisy Miners in the short term

Culling is the most humane and efficient method for controlling the Noisy Miner in the short term. Translocation of the species is an ethically questionable management practice, as it places the translocated animal under prolonged acute stress and almost invariably results in its death. It is also likely to have negative effects on non-target species already living at the release site (Fulton and Ford 2001).

Translocation and culling are really the only two options if one wishes to rapidly reduce numbers of the Noisy Miner at a site. Translocation of an aggressive species like the Noisy Miner only relocates the problem. Although translocated birds may survive for a month or more, they do not assimilate into local Noisy Miner communities. Rather, they wander about the landscape, sometimes for considerable distances, and are either preyed upon, or suffer a lingering death from starvation (Clarke and Schedvin 1997; Fischer and Lindenmayer 2000).

The costs involved in capturing large numbers of the Noisy Miner using mist-nets are prohibitive, especially for such a common, widespread species. Also, mist-netting cannot guarantee complete eradication of a Noisy Miner colony and involves non-target species. From our experience, the costs of mist-netting equate to 6.5 person-hours (and a minimum of $174 [2009 costing]) per bird. This does not take into account equipment costs and the costs of translocation of the birds following mist-netting, or monitoring their fate. By contrast, culling by a professional shooter is rapid, more effective in removing whole colonies, and results in less prolonged suffering on the part of the Noisy Miners involved. Shooting equates to 11.6 person-minutes per bird and is less than 10% of the cost of mist-netting ($14 per bird

[2009 costing]), not taking into account equipment costs (M. Grey unpublished data). These costs could be reduced considerably if landholders obtained a permit to carry out their own cull on their property.

It must be remembered that the Noisy Miner is just one important component of a vast, widespread and complex problem. Habitat clearing, fragmentation, degradation and lack of regeneration are the primary causes of decline in Victoria's woodlands. The Noisy Miner is simply responding to suitable conditions. Removing the Noisy Miner is an emergency, short-term measure which can assist declining bird species to gain some breathing space. There are compelling reasons why revegetation and habitat restoration of remnants should accompany culling.

8. Noisy Miner culls should be accompanied by revegetation and rehabilitation designed to deter the Noisy Miner and provide habitat for colonising bird species and other fauna

Manipulating vegetation in existing remnants may discourage Noisy Miner activity, while at the same time providing habitat for woodland-dependent bird species. We concur with Cunningham *et al.* (2008) who recommend that conserving and enhancing existing native vegetation remnants must be the primary objective in woodland conservation, followed by planting. Our studies, however, highlight the contribution made by even the smallest, most degraded remnants, paddock trees and corridors in agricultural landscapes, if they can be freed of domination by Noisy Miners (Grey *et al.* 1998; Fischer and Lindenmayer 2002a,b,c; Cunningham *et al.* 2008). We advocate adopting a combination of approaches to diminish the destructive influence of the Noisy Miner in woodlands:

- Culling the Noisy Miner may be necessary to rapidly increase habitat available for threatened bird communities and to hasten site rehabilitation (Grey 2008).
- Aim to create remnants >36 ha in size to ensure there is some Noisy Miner-free interior habitat for other birds.
- Consider making vegetation corridors much wider (e.g. by adding to already existing roadside corridors (Robinson 2006)), again with the aim of providing Noisy Miner-free interior habitat.
- Aim for smoothing or rounding of remnant edges when revegetating to enlarge patches of vegetation and incorporate clumps of trees into the remnant when fencing.
- Consider removing some fertile areas from agricultural production, to promote biodiversity conservation, particularly areas which contain large, productive eucalypts.
- Fence remnants to re-establish all vegetation layers, including grasses. The evidence for the shrub-layer being a deterrent to the Noisy Miner is equivocal (Clarke *et al.* 1995; Hastings and Beattie 2006; Oldland *et al.* 2009); nevertheless it will provide habitat and shelter for small birds. Increased *Acacia* in the understorey of plantings may also make revegetation efforts less suitable as habitat for the Noisy Miner (Hastings and Beattie 2006; Lindenmayer *et al.* 2010).
- Avoid fencing just part of a remnant. The Noisy Miner will dominate the ungrazed fenced section from the unfenced grazed section (M. Grey unpublished data). Extend fences a minimum of 25 m from the canopy edge into the paddock to deter the Noisy Miner from ground foraging adjacent to a remnant (M. Grey unpublished data).
- Avoid management activities that shorten ground-layer vegetation, such as grazing, mowing and fire within and adjacent to remnants and corridors.

Conclusion

Revegetation and rehabilitation programs may create new Noisy Miner habitat without proper attention to habitat type (Oldland *et al.* 2009), remnant shape and dimensions (Clarke and Oldland 2007; Taylor *et al.* 2008), plant species mix (Hastings and Beattie 2006; Maron 2007; Oldland *et al.* 2009) and vegetation structural complexity (M. Grey unpublished data). The implications of not acting may be an increase in domination of woodlands by the Noisy Miner, leading to the widespread biotic impoverishment as predicted by Garrott *et al.* (1993) and the potential for irreversible loss of bird diversity from our woodlands (Recher 1999), followed by further irreparable damage to woodland remnants from rural dieback (Grey 2008).

Acknowledgements

Noisy Miner projects undertaken by the Ornithology and Avian Conservation laboratory at La Trobe University were supported with grants from the Victorian Department of Conservation and Natural Resources awarded to the Royal Australasian Ornithologists Union and M. Clarke, the Land and Water Resources Research and Development Corporation, and the Australian Research Council to M. Clarke. We thank our colleagues Joanne Oldland and Rick Taylor who conducted some of these studies, and all field assistants and landholders who assisted in these studies. Projects were conducted under permits from the La Trobe University Animal Experimentation and Ethics Committee, the Victorian Department of Conservation and Natural Resources (Permit No. RP-93-210), and the Australian Bird and Bat Banding Scheme.

Bio

Mike Clarke is Head of the Department of Zoology at La Trobe University. He is intrigued by the different ways species respond to threatening processes; in particular, inappropriate fire regimes and over-abundant native species. He serves on several recovery teams for threatened species, and was the Convener of the Scientific Advisory Committee on the Flora and Fauna Guarantee Act to the Victorian Minister for the Environment for five years. In 2007 he was awarded the D.L. Serventy Medal by the Royal Australasian Ornithologists Union for outstanding contribution to the scientific literature in ornithology.

Merilyn Grey was recently awarded a PhD from La Trobe University for her thesis 'The Impact of the Noisy Miner *Manorina melanocephala* on Avian Diversity and Insect-induced Dieback in Eucalypt Remnants' for which she received the Dean's Research Thesis Merit Citation in the Faculty of Science, Technology and Engineering. Merilyn is now an honorary research fellow in the Ornithology and Avian Conservation laboratory in the Department of Zoology at La Trobe University. She has been a resolute Noisy Miner observer for over 15 years, and her special interest is in ecological processes in small, degraded woodland remnants.

References

Barrett, G.W., Ford, H.A. and Recher, H.F. (1994). Conservation of woodland birds in a fragmented rural landscape. *Pacific Conservation Biology* **1**, 245–256.

Barrett, G.W., Silcocks, A. and Cunningham, R. (2002). Australian Bird Atlas (1998–2001). Supplementary report no. 1 – Comparison of Atlas 1 (1977–1981) and Atlas 2 (1998–2001). Environment Australia Natural Heritage Trust Fund through Birds Australia: Hawthorn East, Victoria.

Bennett, A.F. (2003). 'Linkages in the Landscape: The Role of Corridors and Connectivity in Wildlife Conservation'. IUCN: Gland, Switzerland.
Clarke, M.F., Grey, M.J., Britton, D.R. and Loyn, R.H. (1995). 'The Noisy Miner *Manorina melanocephala* and Rural Dieback in Remnant Eucalypt Woodlands'. RAOU Report No. 98. Royal Australasian Ornithologists Union: East Hawthorn, Victoria.
Clarke, M.F. and Oldland, J.M. (2007). Penetration of remnant edges by Noisy Miners *Manorina melanocephala* and implications for habitat restoration. *Wildlife Research* **34**, 253–261.
Clarke, M.F. and Schedvin, N. (1997). An experimental study of the translocation of noisy miners *Manorina melanocephala* and difficulties associated with dispersal. *Biological Conservation* **80**, 161–167.
Cunningham, R.B., Lindenmayer, D.B., Crane, M., Michael, D., MacGregor, C., Montague-Drake, R. and Fischer, J. (2008). The combined effects of remnant vegetation and tree planting on farmland birds. *Conservation Biology* **22**, 742–752.
Department of Sustainability and Environment (2006). Flora and Fauna Guarantee Act 1988 Threatened List April 2006. Taxa and communities of flora and fauna which are threatened. Department of Sustainability and Environment East Melbourne.
Dow, D.D. (1977). Indiscriminate interspecific aggression leading to almost sole occupancy of space by a single species of bird. *Emu* **77**, 115–121.
Fischer, J. and Lindenmayer, D.B. (2000). An assessment of the published results of animal relocations. *Biological Conservation* **96**, 1–11.
Fischer, J. and Lindenmayer, D.B. (2002a). The conservation value of paddock trees for birds in a variegated landscape in southern New South Wales. I. Species composition and site occupancy patterns. *Biodiversity and Conservation* **11**, 807–832.
Fischer, J. and Lindenmayer, D.B. (2002b). The conservation value of paddock trees for birds in a variegated landscape in southern New South Wales. 2. Paddock trees as stepping stones. *Biodiversity and Conservation* **11**, 832–849.
Fischer, J. and Lindenmayer, D.B. (2002c). Small patches can be valuable for biodiversity conservation: Two case studies on birds in southeastern Australia. *Biological Conservation* **106**, 129–136.
Flora and Fauna Guarantee – Scientific Advisory Committee (2001). Final recommendation on a nomination for listing. Victorian temperate-woodland bird community. Nomination 512. Flora and Fauna Guarantee Act 1988, Melbourne.
Ford, H.A. (1985). The bird community of eucalypt woodland and eucalypt dieback in the Northern Tablelands of New South Wales. In *Birds of Eucalypt Forests and Woodlands: Ecology, Conservation, Management.* (Eds A. Keast, H.F. Recher, H.A. Ford and D. Saunders) pp. 333–340. RAOU and Surrey Beatty & Sons: Chipping Norton.
Ford, H.A., Barrett, G.W., Saunders, D.A. and Recher, H.F. (2001). Why have birds in the woodlands of southern Australia declined? *Biological Conservation* **97**, 71–88.
Ford, H.A. and Bell, H. (1981). Density of birds in eucalypt woodland affected to varying degrees by dieback. *Emu* **81**, 202–208.
Fulton, G.R. and Ford, H.A. (2001). The conflict between animal welfare and conservation. *Pacific Conservation Biology* **7**, 152–153.
Garrott, R.A., White, P.J. and Vanderbilt-White, C.A. (1993). Overabundance: an issue for conservation biologists? *Conservation Biology* **7**, 946–949.
Grey, M.J. (2008). The Impact of the Noisy Miner *Manorina melanocephala* on Avian Diversity and Insect-induced Dieback in Eucalypt Remnants. PhD Thesis, School of Life Sciences, Department of Zoology, La Trobe University: Melbourne.

Grey, M.J., Clarke, M.F. and Loyn, R.H. (1997). Initial changes in the avian communities of remnant eucalypt woodlands following a reduction in the abundance of noisy miners, *Manorina melanocephala*. *Wildlife Research* **24**, 631–648.

Grey, M.J., Clarke, M.F. and Loyn, R.H. (1998). Influence of the Noisy Miner *Manorina melanocephala* on avian diversity and abundance in remnant Grey Box woodland. *Pacific Conservation Biology* **4**, 55–69.

Hastings, R.A. and Beattie, A.J. (2006). Stop bullying in the corridors: can including shrubs make your revegetation more Noisy Miner free? *Ecological Management and Restoration* **7**, 105–112.

Higgins, P.J., Peter, J.M. and Steele, W.K., eds. (2001) *Handbook of Australian, New Zealand & Antarctic Birds. Volume 5 Tyrant-flycatchers to Chats*. Oxford University Press: Melbourne.

Hobbs, R.J. and Hopkins, A.J.M. (1990). From frontier to fragments: European impact on Australia's vegetation. *Proceedings of the Ecological Society of Australia* **16**, 93–114.

Keatley, M.R. and Hudson, I.L. (2006). A comparison of long-term flowering patterns of Box-Ironbark species in Havelock and Rushworth forests. *Environmental Modelling and Assessment* **12**, 279–292.

Lindenmayer, D.B., Knight, E.J., Crane, M.J., Montague-Drake, R., Michael, D.R. and MacGregor, C.I. (2010). What makes an effective restoration planting for woodland birds? *Biological Conservation* **143**, 289–301.

Loyn, R.H. (1987). Effects of patch area and habitat on bird abundances, species numbers and tree health in fragmented Victorian forests. In *Nature Conservation: The Role of Remnants of Native Vegetation*. (Eds D.A. Saunders, G.W. Arnold, A.A. Burbidge and A.J.M. Hopkins) pp. 65–77. Surrey Beatty and Sons Pty Ltd in association with CSIRO and CALM: Chipping Norton.

Maron, M. (2007). Threshold effect of eucalypt density on an aggressive avian competitor. *Biological Conservation* **136**, 100–107.

Maron, M. and Lill, A. (2005). The influence of livestock grazing and weed invasion on habitat use by birds in grassy woodland remnants. *Biological Conservation* **124**, 439–450.

Oldland, J.M., Taylor, R.S. and Clarke, M.F. (2009). Habitat preferences of the noisy miner (*Manorina melanocephala*) – a propensity for prime real estate? *Austral Ecology* **34**, 306–316.

Piper, S.D. and Catterall, C.P. (2003). A particular case and a general pattern: hyperaggressive behaviour by one species may mediate avifaunal decreases in fragmented Australian forests. *Oikos* **101**, 602–614.

Recher, H.F. (1999). The state of Australia's avifauna: a personal opinion and prediction for the new millennium. *Australian Zoologist* **31**, 11–29.

Recher, H.F., Majer, J.D. and Ganesh, S. (1996). Eucalypts, arthropods and birds: on the relation between foliar nutrients and species richness. *Forest Ecology and Management* **85**, 177–196.

Reid, J.R. (1999). 'Threatened and declining birds in the New South Wales Sheep-Wheat belt. I. Diagnosis, characteristics and management'. Consultancy Report to New South Wales National Parks and Wildlife Service. CSIRO Sustainable Ecosystems: Canberra.

Robinson, D. (2006). Is revegetation in the Sheep Pen Creek area, Victoria, improving Grey-crowned Babbler habitat? *Ecological Management and Restoration* **7**, 93–104.

Taylor, R.S., Oldland, J.M. and Clarke, M.F. (2008). Edge geometry influences landscape-scale habitat preferences of an edge specialist in south-eastern Australia. *Landscape Ecology* **23**, 377–389.

Tilman, D., May, R.M., Lehman, C.L. and Nowak, M.A. (1994). Habitat destruction and the extinction debt. *Nature* **371**, 65–66.

15

TIPPING THE SCALES: THOUGHTS ON IMPROVING MANAGEMENT FOR WOODLAND REPTILES

Geoff W. Brown

1. Management regimes, especially where they incorporate disturbances such as livestock grazing or firewood removal, have significant impacts on reptile occurrence.
2. The distribution of reptiles is influenced by factors operating at several spatial scales.
3. The pattern and condition of remnant vegetation in the rural landscape influences the distribution of reptiles.
4. Effective management for reptiles should aim to create, retain or increase structural heterogeneity, particularly in the ground strata.
5. Care should be taken in designing reptile surveys, especially with respect to survey intensity and the techniques employed.
6. The relative scarcity of reptiles in many woodlands of south-eastern Australia is a historical artifact of land management since European settlement.

Introduction

I have investigated the biota of a range of woodland and forest environments in south-eastern Australia for the past 25 years (see map on next page). Typically, these projects have dealt with human-modified landscapes and, as such, have had an applied focus — one that explores the status of species or assemblages and the processes (natural and anthropogenic) that influence their occurrence, with a view to developing more effective management. The location of the work has varied, as often has the scale and the target taxa. Yet the principal objective has always been to better understand native terrestrial wildlife and improve the viability and conservation of populations.

My research career has been marked by several large collaborations that have investigated key management issues at the landscape scale, such as habitat fragmentation in agricultural landscapes, long-term impacts on biodiversity of ecological thinning, the drivers of reptile occurrence within the Murray-Darling Basin, and the occurrence of native fauna within grazed production systems of woodlands of the inland slopes of Victoria and New South Wales,

specifically in relation to fertiliser use, grazing management and pressure, or landscape gradients of native vegetation cover. My work has typically included plains and riverine grassy woodland complexes of the Victorian Riverina, characterised by low-density tree cover of River Red Gum (*Eucalyptus camaldulensis*) and Black Box (*E. largiflorens*), and dry box-ironbark forests of the more elevated Victorian midlands, characterised by a dense to sparse canopy of box, ironbark and gum-barked eucalypts.

These collaborations have commonly involved scientists from an array of universities and government agencies (e.g. Victorian Department of Sustainability and Environment, Parks Victoria, Victorian Department of Primary Industries, Museum Victoria, CSIRO).

Lessons

1. Management regimes, especially where they incorporate disturbances such as livestock grazing or firewood removal, have significant impacts on reptile occurrence

The dry woodlands of south-eastern Australia have experienced widespread disturbance – most of this region is privately managed and used for agriculture. Hence it has been profoundly altered by management practices embraced since European settlement, particularly livestock grazing. Recent evidence implicates grazing in the decline of reptile assemblages in a range of woodland types, primarily through a decrease in the quality of suitable habitat (Hadden and Westbrooke 1996; James 2003; Driscoll 2004; Brown *et al.* 2008). My own work in box-ironbark woodlands has revealed lower reptile abundance and fewer species on disturbed sites (generally from livestock grazing) compared with relatively undisturbed sites (Brown 2001). In a recent study of reptiles in grazed native pastures (including treed sites) in the south-eastern

Australia, overall reptile abundance was positively correlated with several habitat and management variables, including native plant richness (in the ground-layer), coarse woody debris (CWD) cover, as well as stocking rate and type (rotational grazing), and negatively correlated with available phosphorus (G. Brown unpublished data). The positive (although weak) response to stocking rate may be partly interpreted as a prior filtering of the species pool (that is, most observed reptiles were skinks favoured by grazing).

Management of reptiles on public land brings different challenges: detrimental impacts are typically associated with timber harvesting, firewood collection and, less commonly, mining. The loss or alteration of habitat structure through firewood collection is particularly acute in floodplain forests and woodlands, which are the source of most firewood. If management is to focus on individual species, then the Pink-tailed Worm-Lizard (*Aprasia parapulchella*), a highly localised, threatened reptile that is restricted in Victoria to stony woodland habitats (primarily on public land), is an example of a species with particular management requirements. Major threats include habitat loss, fragmentation and degradation, mostly associated with acute urban and agricultural pressures, including the removal or displacement of rocks, inappropriate fire regimes, predation by introduced carnivores or domestic pets, and recreational activities.

2. The distribution of reptiles is influenced by factors operating at several spatial scales

The distribution and occurrence of reptiles are influenced by processes operating at multiple spatial scales (Fischer *et al.* 2004; Brown *et al.* 2008). Reptiles respond idiosyncratically to environmental processes, and at different scales, such that no single universal factor governs their occurrence. While the primary influence at the regional scale is the environmental gradient (climate, elevation), other influences at smaller scales, such as the pattern of native vegetation (e.g. remnant size and shape, amount of proximate vegetation) and site-level characteristics (such as grass, rock or CWD cover, modified mostly through management) have been implicated as determinants of local occurrence (e.g. Fischer *et al.* 2003; Fischer *et al.* 2005b; Cunningham *et al.* 2007; Brown *et al.* 2008). To illustrate: in the Victorian Riverina, the species richness and overall abundance of reptiles at survey sites showed a general trend of increasing as temperature increased and rainfall and elevation decreased, and Boulenger's Skink (*Morethia boulengeri*) was significantly correlated with the occurrence of dead trees and Grey Box (*Eucalyptus microcarpa*).

3. The pattern and condition of remnant vegetation in the rural landscape influences the distribution of reptiles

Across fragmented, rural landscapes, most reptile species are confined to remnant vegetation and few use the intervening matrix. Hence, the size, shape and condition of these remnants have important ramifications for reptile occurrence. Tree cover *per se* is important: two recent studies in grazing landscapes have confirmed that reptile diversity is greater in landscapes with more native woodland cover, and within paddocks, tree cover is clearly important for preserving both the diversity and abundance of reptiles (Fischer *et al.* 2003; G. Brown unpublished data). The species-area relationship is a key explanation for the reduced species richness in fragmented landscapes, and has been demonstrated for reptiles in box-ironbark woodlands. As areas of habitat become smaller, species richness declines and the observed reptile assemblages are nested subsets of those found in larger areas (Mac Nally and Brown 2001). But other factors, aside from tree cover and remnant size, also influence reptile occurrence, notably remnant shape and habitat quality, the latter usually being an artifact of management (Kitchener *et al.* 1980; Hadden and Westbrooke 1996; Driscoll 2004; Brown *et al.* 2008).

Roadside vegetation is important as habitat in modified agricultural landscapes because it often provides less disturbed remnant vegetation, usually a result of management regimes with less grazing (Driscoll 2004; Brown *et al.* 2008). In the agricultural landscapes of the Victorian Riverina, roadsides were found to yield the highest values for reptile species richness, abundance and frequency of occurrence when compared with blocks of woodland. Species such as Woodland Blind Snake (*Ramphotyphlops proximus*) and Marbled Gecko (*Christinus marmoratus*) were more likely to occur at sites in roadside vegetation than in other types of remnants, and species richness was greater in roadsides and large blocks, and at lower elevation with increasing local tree cover (Brown *et al.* 2008).

4. Effective management for reptiles should aim to create, retain or increase structural heterogeneity, particularly in the ground strata

For most terrestrial woodland reptile species, habitat structure (typically in the ground-layer), is an important determinant of occurrence. Litter and CWD are used by many terrestrial reptile species for a myriad of purposes — basking, nesting, shelter, hibernation, foraging and, in the case of large logs which are able to retain moisture, potential refuge during drought or fire. CWD also may provide an increased diversity and abundance of invertebrate and vertebrate prey for reptiles. Hence the loss of, or disturbance to, this component of the habitat will significantly affect reptiles as well as other vertebrates. This is recognised by the Victorian Government in its listing of the loss of CWD from Victorian native forests and woodlands as a threatening process (Department of Sustainability and Environment 2008) and demonstrated in several recent studies in south-eastern Australian woodland environments, including box-ironbark woodland, River Red Gum floodplain forest, and Buloke (*Allocasuarina luehmannii*) remnants (Hadden and Westbrooke 1996; Mac Nally *et al.* 2001; Mac Nally *et al.* 2002; Fischer *et al.* 2003; Driscoll 2004; Brown *et al.* 2008). These studies have established a significant association between structural elements of the ground strata and the occurrence of vertebrate fauna, including reptiles. Statistical models reveal that blind snakes in remnant woodlands of the Victorian Riverina are more likely to occur on sites at lower elevations with a cover of coarse litter (Brown *et al.* 2008). Vegetation communities in the box-ironbark woodlands of central Victoria that support the greatest numbers of reptiles (whether individuals or species per site) are usually characterised by a relatively complex ground stratum (G. Brown unpublished data).

5. Care should be taken in designing reptile surveys, especially with respect to survey intensity and the techniques employed

Regrettably, the number of threatened vertebrate species, including reptiles, has increased inexorably in Australia since European settlement. Solutions to this issue require a better understanding of the biology and ecology of species and their habitats. Terrestrial fauna surveys often underpin the acquisition of such information, although basic knowledge is still lacking for most reptile species. Reptiles, like other biota, are surveyed for a number of reasons, but generally to provide information for the ongoing management of populations and habitats or to detect changes in population status that may pre-empt species loss.

Many factors influence the effectiveness of surveys (and thus the accuracy and comprehensiveness of the results), including survey effort (a factor of budget and time frame), choice of techniques (related to detectability), and sampling biases (the most common of which may be corrected). While it is rarely possible to achieve perfect detection of reptiles in a survey, their occurrence can be evaluated (even via estimation techniques) given adequate survey effort. For some (mostly rare or cryptic) species, however, substantial survey effort is necessary to have confidence that the species does not occur at a site. In my experience, reptile surveys are often inadequate (in terms of extent, intensity and the selection of survey techniques), especially if the purpose is the acquisition of comprehensive baseline data or medium to long-term monitoring data. The consequence is that management decisions are based on insufficient information. It has been consistently shown that different survey techniques yield different subsets of the total reptile fauna (e.g. Ryan *et al.* 2002; Garden *et al.* 2007; Thompson and Thompson 2007; Brown *et al.* 2008). For instance, pitfall trapping, despite its recognised shortcomings (including a variety of sampling biases), is an effective survey technique if the resources are available to apply it over a sufficient time-span – it can yield more and diverse reptile species per sampling unit than standardised searches (Mac Nally and Brown 2001) although, to strengthen survey results, it should be employed in conjunction with other trapping and observational techniques (Mac Nally and Brown 2001; Garden *et al.* 2007; Brown *et al.* 2008).

6. The relative scarcity of reptiles in many woodlands of south-eastern Australia is a historical artifact of land management since European settlement

There has been a landscape-scale decline in reptiles across some woodlands of south-eastern Australia, consistent with substantial changes to the natural environment wrought since European settlement. Recent studies in the Victorian Riverina and box-ironbark region have shown the reptilian fauna generally to be both depauperate and of low abundance, despite intensive survey effort and the employment of complementary survey techniques across a broad geographic area (Brown 2001; Mac Nally and Brown 2001; Brown *et al.* 2008; G. Brown unpublished data). In the box-ironbark woodlands, localised reptile species richness and abundance are typically very low. While the overall species complement of this region is relatively large, the distribution of most species is intermittent – even many relatively common species are patchily distributed (Brown 2001; Mac Nally and Brown 2001). This scenario is mirrored in the Victorian Riverina: a few widespread species, such as Boulenger's Skink, can persist where suitable microhabitat occurs on farmland (e.g. scattered trees, artificial shelter). Yet for most species, large paddocks of uniform cropland or grassland are an alien environment (Driscoll 2004; G. Brown, personal observations). Comparison with extensive trapping studies in other biogeographic regions of south-eastern Australia reveals the reptile fauna of the Victorian Riverina and box-ironbark woodlands to be impoverished; trapping returns in mesic forests and in semi-arid mallee vegetation consistently have greater capture rates (by three to 10-fold) (Brown *et al.* 2008). Surveys in remnant woodlands in southern New South Wales (Fischer *et al.* 2004) and in remnant Buloke woodland in the Victorian Wimmera (Hadden and Westbrooke 1996) also have revealed capture rates three times higher than those in the Victorian Riverina and box-ironbark woodlands, showing that other kinds of woodland can support higher populations of these species (Brown *et al.* 2008).

Conclusions

While the temperate woodlands of south-eastern Australia support a relatively diverse reptile fauna, this reptile assemblage is clearly in decline, primarily as a result of anthropogenic land use. Given that many local extinctions have already occurred and most species, even common 'generalists', are patchily distributed and vulnerable to changing land-use, there is an urgent need to tackle this decline. The key overarching question is whether landscape restoration or the adoption of different management practices will be most beneficial for reptiles. Either way, substantial and prompt intervention is required, and the following factors will need consideration: (1) what actions are required; (2) at what scale; (3) over what time frame; and (4) what information is available to guide changed management (since knowledge of the ecology, most importantly habitat requirements, of most reptiles is scanty). Research that disentangles the multiple influences on reptile occurrence and abundance will be invaluable.

To illustrate, some woodland reptiles may benefit from simple management actions that reduce levels of disturbance in selected areas or restore structural complexity to ground-layers, such as limiting the removal of firewood or intensity of grazing; that is, reptile species that are not constrained by the availability of suitable habitat but by ecological processes (e.g. disturbance, fire). Decisions also need to be responsive to land tenure and scope. For instance, selecting which management practices to adopt or continue on private land will have important implications, since private land tenure dominates most landscapes. Management decisions also need to consider spatial scale (Fischer *et al.* 2004; Fischer *et al.* 2005a); to encourage reptile diversity it is important to maintain high levels of habitat heterogeneity at both the landscape and microhabitat scales.

Acknowledgements

I would like to thank the many colleagues and collaborators who have contributed in myriad ways to various projects drawn upon for this chapter, and the funding agencies that made such work possible.

Bio

Geoff Brown has spent over 25 years engaged in ecological research, specialising in vertebrates (especially reptiles) in a range of south-eastern Australian environments. His work has had an applied focus aimed at improving the management of terrestrial biodiversity, mostly in rural and reserved landscapes. He has contributed to large collaborative projects that have investigated major ecosystems (e.g. Victorian Riverina, box-ironbark woodlands, Menindee Lakes) and the processes acting on them, or select management issues (e.g. ecological thinning, environmentally sensitive silvicultural systems, ecological corridors, management of native pastures for biodiversity, improving vertebrate monitoring protocols). He also has worked on threatened species, including, recently, the Southern Brown Bandicoot (*Isoodon obesulus*), Pink-tailed Worm-Lizard, Hooded Scalyfoot (*Pygopus schraderi*) and Alpine Tree Frog (*Litoria verreauxii alpina*).

References

Brown, G.W. (2001). The influence of habitat disturbance on reptiles in a Box-Ironbark eucalypt forest of south-eastern Australia. *Biodiversity and Conservation* **10**, 161–176.

Brown, G.W., Bennett, A.F. and Potts, J.M. (2008). Regional faunal decline – reptile occurrence in fragmented rural landscapes of south-eastern Australia. *Wildlife Research* **35**, 8–18.

Cunningham, R.B., Lindenmayer, D.B., Crane, M., Michael, D. and MacGregor, C. (2007). Reptile and arboreal marsupial response to replanted vegetation in agricultural landscapes. *Ecological Applications* **17**, 609–619.

Department of Sustainability and Environment (2008). Flora and Fauna Guarantee Act 1988. Threatened List. http://www.dse.vic.gov.au/DSE/nrenpa.nsf/LinkView/EADA0F1874AF9CF24A2567C1001020A388BBA5581CF9D859CA256BB300271BDB. (DSE: East Melbourne, Victoria)

Driscoll, D.A. (2004). Extinction and outbreaks accompany fragmentation of a reptile community. *Ecological Applications* **14**, 220–240.

Fischer, J., Fazey, J., Briese, R. and Lindenmayer, D.B. (2005a). Making the matrix matter: challenges in Australian grazing landscapes. *Biodiversity and Conservation* **14**, 561–578.

Fischer, J., Lindenmayer, D.B., Barry, S. and Flowers, E. (2005b). Lizard distribution patterns in the Tumut fragmentation 'Natural Experiment' in south-eastern Australia. *Biological Conservation* **123**, 301–315.

Fischer, J., Lindenmayer, D. and Cowling, A. (2003). Habitat models for the four-fingered skink (*Carlia tetradactyla*) at the microhabitat and landscape scale. *Wildlife Research* **30**, 495–504.

Fischer, J., Lindenmayer, D.B. and Cowling, A. (2004). The challenge of managing multiple species at multiple scales: reptiles in an Australian grazing landscape. *Journal of Applied Ecology* **41**, 32–44.

Garden, J.G., McAlpine, C.A., Possingham, H.P. and Jones, D.N. (2007). Habitat structure is more important than vegetation composition for local-level management of native terrestrial reptile and small mammal species living in urban remnants: a case study from Brisbane, Australia. *Austral Ecology* **32**, 669–685.

Hadden, S.A. and Westbrooke, M.E. (1996). Habitat relationships of the herpetofauna of remnant Buloke woodlands of the Wimmera Plains, Victoria. *Wildlife Research* **23**, 363–372.

James, C.D. (2003). Response of vertebrates to fenceline contrasts in grazing intensity in semi-arid woodlands of eastern Australia. *Austral Ecology* **28**, 137–151.

Kitchener, D.J., Chapman, A., Dell, J., Muir, B.G. and Palmer, M. (1980). Lizard assemblage and reserve size and structure in the Western Australian wheatbelt – some implications for conservation. *Biological Conservation* **17**, 25–62.

Mac Nally, R. and Brown, G.W. (2001). Reptiles and habitat fragmentation in the box-ironbark forests of central Victoria, Australia: predictions, compositional change and faunal nestedness. *Oecologia* **128**, 116–125.

Mac Nally, R., Horrocks, G. and Pettifer, L. (2002). Experimental evidence for potential beneficial effects of fallen timber in forests. *Ecological Applications* **12**, 1588–1594.

Mac Nally, R., Parkinson, A., Horrocks, G., Conole, L. and Tzaros, C. (2001). Relationships between terrestrial vertebrate diversity, abundance and availability of coarse woody debris on south-eastern Australian floodplains. *Biological Conservation* **99**, 191–205.

Ryan, T.J., Philippi, T., Leiden, Y.A., Dorcas, M.E., Wigley, T.B. and Gibbons, J.W. (2002). Monitoring herpetofauna in a managed forest landscape: effects of habitat types and census techniques. *Forest Ecology and Management* **167**, 83–90.

Thompson, G.G. and Thompson, S.A. (2007). Usefulness of funnel traps in catching small reptiles and mammals, with comments on the effectiveness of the alternatives. *Wildlife Research* **34**, 491–497.

16

CONSERVATION OF MICROBATS IN TEMPERATE WOODLANDS

Linda F. Lumsden and Andrew F. Bennett

1. Microbats comprise a large component of the mammal fauna in woodland ecosystems.
2. All types of woodland vegetation have value as roosting or foraging habitat for microbats.
3. Hollow-bearing trees are a critical resource for microbats.
4. Patterns of roosting ecology and roost selection are complex, even for common species of bats.
5. Conservation of habitat for bats in rural landscapes requires a large-scale, long-term perspective.
6. Microbats provide ecological services in rural environments.
7. Bats can be used successfully to promote the conservation of remnant vegetation.
8. Although microbats appear capable of persisting in rural environments, we cannot be complacent about their future.

Introduction

We have studied the ecology of microbats (Family Microchiroptera), also known as insectivorous bats, in remnant woodlands and forests in the Northern Plains of Victoria (see map on next page). This region has been extensively cleared during the last 150 years and now is used intensively for agriculture, primarily cropping and grazing. Native vegetation occurs predominantly as small remnants in farm paddocks, narrow strips along creeklines and roads, and as isolated trees scattered through the otherwise cleared landscape. These are temperate grassy woodlands dominated mainly by Grey Box (*Eucalyptus microcarpa*), Yellow Box (*E. melliodora*), Black Box (*E. largiflorens*) or River Red Gum (*E. camaldulensis*). The few large areas of native vegetation that remain occur primarily along the main rivers and associated floodplains.

Our work has involved regional surveys of microbats in different types and sizes of remnant vegetation, and ecological studies of the movements, foraging and roosting ecology of two species, the Lesser Long-eared Bat (*Nyctophilus geoffroyi*) and Gould's Wattled Bat (*Chalinolobus*

gouldii). Studies of roosting ecology were undertaken in the extensive Barmah forest on the Murray River floodplain and in small remnants in farmland (Lumsden 2004; Lumsden *et al.* 2002a,b).

Lessons

1. Microbats comprise a large component of the mammal fauna in woodland ecosystems

Bats are an important component of native mammal assemblages in all regions in Australia, with approximately 90 species known from Australia. Microbats often are an overlooked component of the mammal fauna. Their abundance and diversity are seldom recognised by land managers or landholders, due to their small size and nocturnal and cryptic behaviour. In extensively cleared landscapes, such as the Northern Plains of Victoria, the contribution of bats to local assemblages may be even greater than in areas with intact native vegetation. This is because such modified landscapes generally have experienced loss of terrestrial mammals. Thirteen species of microbats are known from the Northern Plains, which represents 39% of the extant mammal species. At least six species of terrestrial mammals have become extinct in this region since European settlement, while there is no evidence of regional extinctions for any species of bats (Bennett *et al.* 1998).

2. All types of woodland vegetation have value as roosting or foraging habitat for microbats

Microbats use all types of forest or woodland vegetation for roosting or foraging habitat, including remnants ranging from large blocks (>200 ha) to small isolated patches (<5 ha), linear strips of woodland along roads and streams, and scattered trees in farm paddocks (Lumsden *et al.* 1995). Foraging of most species is centred on trees, with little foraging undertaken in farmland devoid of trees, except by the Southern Freetail Bat (*Mormopterus* sp.) and White-striped Freetail Bat (*Tadarida australis*). The extensive use by bats of small remnants and scattered trees (Lumsden and Bennett 2005) is in marked contrast to other faunal groups, such as arboreal marsupials or birds, of which only a sub-set of species typically occurs in these habitats (Law *et al.* 2000; Fischer and Lindenmayer 2002).

Bats appear to be less severely affected than other vertebrate groups by the kinds of landscape change typical of temperate woodland regions. First, the scale of movements undertaken by bats allows them to move widely through the landscape and obtain resources from multiple landscape elements. For example, individuals of the Lesser Long-eared Bat and Gould's Wattled Bat regularly move up to 12 km between roost sites and foraging areas, which gives ready access to a range of different woodland remnants in a landscape (Lumsden *et al.* 2002a). Second, these species can fly across open farmland to access isolated habitats without requiring interconnected strips of wooded vegetation (Lumsden *et al.* 2002a; Lumsden and Bennett 2005). Third, their aerial foraging mode and roosts high in trees mean that they are less affected by habitat degradation at ground level, such as from cultivation, grazing by stock, or weed invasion. Last, their social organisation (e.g. colonial roosting, extensively overlapping foraging and roosting areas, high intra-specific tolerance) means they do not live in semi-exclusive territories or home ranges that impose a minimum habitat area (e.g. minimum patch size) for occurrence, that would result in low densities of individuals.

3. Hollow-bearing trees are a critical resource for microbats

Microbats are active at night and spend the day in roosts. Roosts provide refuge from predators, shelter from climatic extremes such as hot weather, and during the breeding season they are used for sheltering and raising young (maternity roosts). Most roosts used by the Lesser Long-eared Bat and Gould's Wattled Bat are in trees (Lumsden *et al.* 2002b; Lumsden 2004). Despite the general availability of anthropogenic structures in farmland areas (such as houses and farm sheds) in the Northern Plains, little use was made of these as roosts. The high use of natural roost sites suggests that tree hollows are preferred roosting locations, where available. When roosting in trees, males and non-breeding females of the Lesser Long-eared Bat often roosted under exfoliating bark, but most maternity roosts used by breeding females of this species, and all roosts used by Gould's Wattled Bat were in hollows. Tree hollows are also important for the other 11 species of microbats in the Northern Plains of Victoria, all of which are tree-hole roosting species (Menkhorst 1995). Tree hollows are more common in large old trees (Bennett *et al.* 1994), and so any loss of these trees from felling, dieback or natural senescence will be detrimental to the conservation of bats.

Dead trees were used extensively as roost sites by both species studied. The Lesser Long-eared Bat used dead trees at a higher frequency than their availability in the landscape: 69% of all tree roosts of this species were in dead trees. Dead trees often are undervalued by land managers and landholders, however. They continue to be removed for firewood, or for aesthetic or fire protection reasons, with little recognition of their importance to species such as bats. Greater awareness is needed of the value of dead trees, both on public and privately managed land, and increased protection is required in managing these areas.

Typical of tree-hole roosting bats throughout the world (Kunz and Lumsden 2003), the Lesser Long-eared Bat and Gould's Wattled Bat roost in small colonies and regularly move between multiple roost sites within a discrete roosting area (Lumsden 2004). Thus, a relatively high density of suitable roost trees is required. The frequency of movement between roosts was similar in the breeding and non-breeding seasons; this trait may become a limiting factor during the breeding season because specialised roost sites are required as maternity roosts.

4. Patterns of roosting ecology and roost selection are complex, even for common species of bats

The Lesser Long-eared Bat and Gould's Wattled Bat are among the most widespread bats in Australia and have been considered 'generalists' in their roosting ecology (e.g. Maddock and Tidemann 1995). We found an astonishing complexity, however, in the roosting requirements of these two species. There were significant inter- and intra-specific differences in the location of roosts in the landscape (Lumsden *et al.* 2002a), in the types of sites selected as roosts (Lumsden *et al.* 2002b), in roosting behaviour (Lumsden 2004) and in how they responded to changes in roost availability (Lumsden 2004). For example, both sexes of the Gould's Wattled Bat roosted predominantly in dead spouts in large, live trees; while the Lesser Long-eared Bat used a broader range of roost types, which varied between the sexes and between environments. Males of the Lesser Long-eared Bat typically roosted in farmland close to their foraging areas, using loose bark on trees and man-made structures. When there was extensive forest nearby, breeding females roosted in these areas, flying many kilometres to forage in farmland remnants. In the forest, they roosted in narrow fissures in large-diameter, dead trees. In farmland remote from extensive forest, where there was a lower availability of large dead trees, females used a wider range of maternity roosts, also making use of sites under bark, in buildings, and in small-diameter, live and dead trees (Lumsden 2004).

5. Conservation of habitat for bats in rural landscapes requires a large-scale, long-term perspective

The mobility of microbats and the scale at which they operate means that planning for conservation must be undertaken at large scales. Individuals regularly moved up to 12 km between foraging and roosting areas, and the area encompassed by a local population probably spans tens of thousands of hectares. This has several implications. First, conservation at a landscape scale requires integration and cooperation across all land tenures. Remnant vegetation in temperate woodland environments occurs on both public and private land. Second, the long-term status of bats and other wildlife in rural environments is influenced by region-wide changes in environmental quality, including processes such as rising saline groundwater and the paucity of tree regeneration across private land. Third, the key resource requirements of microbats, roost sites and foraging areas, are not evenly distributed across the landscape and individuals may need to select different parts of the landscape for their requirements. Many individuals that we tracked, foraged in farmland remnants but roosted in the extensive Barmah forest (Lumsden *et al.* 2002a).

A long-term approach is required to ensure the ongoing provision of roosting resources for microbats in temperate woodland environments. Regenerating trees or those planted in revegetation programs, typically will take more than 100 years to produce hollows (Gibbons and Lindenmayer 2002). Further, in many remnant woodlands where old hollow-bearing trees are still present, grazing by stock has inhibited regeneration of a new cohort of trees to replace these veterans when they die (Bennett *et al.* 1994). It is critical for bats and other hollow-using species, that remaining trees with hollows are retained in the landscape for as long as possible and that there is adequate regeneration for ongoing replacement of hollow-bearing trees into the future.

Revegetation programs are being undertaken widely throughout rural areas of Australia. Planted areas often are small and sometimes isolated from existing natural vegetation, thus limiting their potential value to many faunal species. Bats, because of their mobility, are potentially able to use revegetated areas sooner than other vertebrate groups. Such areas may provide foraging habitat long before the trees have grown sufficiently to supply roosting sites (Law and Chidel 2006). It is not clear, however, whether the provision of such new foraging habitat increases the size of bat populations, or merely redistributes existing individuals. If roost availability, not foraging habitat, is the factor limiting populations, it may be necessary to also provide artificial roosting structures before revegetated areas can play a useful role and increase overall population sizes.

6. Microbats provide ecological services in rural environments

Although quantitative estimates of population sizes have not been made, it is apparent that large numbers of microbats are present in many rural environments (Figure 16.1). Foraging bats consume large quantities of invertebrates, potentially eating up to half their body mass per night (Hill and Smith 1984). Such levels of predation are likely to influence population sizes of invertebrate prey species, with potential benefits for the health of trees, pastures, crops and revegetated areas. For example, in the Northern Plains of Victoria, the Southern Freetail Bat feeds extensively on the Rutherglen Bug (*Nysius vinitor*) (Order Hemiptera) (Lumsden and Wainer unpublished data), a serious pest in a range of agricultural crops (McDonald and Farrow 1988). Compared with insectivorous birds, however, for which the role of 'natural pest controller' is well recognised (e.g. Ford 1985), the role that bats may play in regulating insect numbers is rarely mentioned. Bats may be especially important among scattered trees in

Figure 16.1 Large numbers of bats are present in some rural environments. A total of 143 individuals were trapped in one night in a harp trap (see centre of photograph among trees) set in this roadside vegetation at Picola, Victoria. Over 600 individual bats were trapped in this 2 km strip of vegetation during a two-week period.

paddocks: these areas are used extensively for foraging by bats but less frequently by insectivorous woodland birds (Lumsden and Bennett 2000, 2005). Further work is required to fully elucidate the diets of bat species, including spatial and temporal variation in foods eaten and the extent to which bats may regulate invertebrate populations.

7. Bats can be used successfully to promote the conservation of remnant vegetation

Microbats can be used successfully as a focus for extension and education activities about the values of remnant woodlands. Bats are poorly known within the general community and, being nocturnal, are seldom seen by landholders. Our experience, however, is that most people find them fascinating when they have an opportunity to see them at first hand and to learn more about their biology. Microbats can be portrayed as having a beneficial role in rural environments because they feed on insects.

8. Although microbats appear capable of persisting in rural environments, we can not be complacent about their future

Our research suggests that insectivorous bats have adapted to rural landscapes more successfully than some other faunal groups, but it remains highly likely that they have suffered major population declines due to the loss of over 90% of the native vegetation in many districts. In addition, the full consequences of such historic clearing and subsequent landscape changes may not yet have been realised. For woodland bird species, for example, the relationship between habitat loss and species decline is not linear (e.g. Radford *et al.* 2005), and there may be a substantial time lag between habitat change occurring and the outcome for populations. Mature trees, for example, continue to be lost from rural landscapes, despite legislated controls on the clearing of native vegetation and an increased awareness of the value of remnant vegetation. Further, even if regeneration and revegetation are widely undertaken now, it is likely that there will be a 'gap' between the time when veteran trees in farm paddocks die and collapse, and when a new cohort of trees are old enough to provide hollows suitable as roost sites.

Conclusions

Knowledge of the ecology and conservation biology of microbats is increasing, but many knowledge gaps remain. New insights have been gained into processes such as foraging and roost selection for a few species, particularly at the patch scale, but further knowledge is required of how bats are affected by landscape pattern and structure. For example, knowledge is required on the density and demographic structure of bat populations, reproductive success, survival rates and social organisation. Such understanding is needed for all species of bats, not just for a few. These aspects are essential if we are to understand the processes that determine the long-term viability of microbat species in modified rural landscapes typical of temperate woodlands. We need to know, for example, whether some kinds of woodland remnants may be acting as demographic 'sinks', while others such as larger stands of vegetation (e.g. along the river systems) may be population 'sources'.

Microbats are relatively small, active at night and cryptic during the day, and so they often are overlooked as part of the wildlife of a local area. Despite this, they are a major component of the mammal fauna, have the capacity to adapt to modified landscapes and are likely to have an important ecological role in temperate woodlands in rural landscapes. They deserve a higher profile and greater recognition in conservation planning and public education.

Acknowledgements

This research was carried out under permits and ethics approval from the Department of Sustainability and Environment. A large number of people assisted with the field components of these studies over a number of years. We would especially like to mention the contribution made by John Silins. We gratefully acknowledge the willingness of the landholders in the region to give us access to their properties.

Bio

Lindy Lumsden is a Principal Research Scientist with the Department of Sustainability and Environment, and manages the Wildlife Ecology Section at the Arthur Rylah Institute for Environmental Research. She has been undertaking surveys and ecological research on bats for 30 years. For approximately 10 years, this work focused on bats in agricultural landscapes, but she has also studied bats throughout Australia and in a number of countries overseas. Lindy is passionate about reversing the often negative perceptions of bats in the general community, and conducts a wide range of extension activities, including courses, field days and talks, to educate and enthuse people about these fascinating animals.

Andrew Bennett has more than 25 years research experience in wildlife ecology and conservation biology. After 12 years with the Arthur Rylah Institute for Environmental Research, Department of Sustainability and Environment, he joined Deakin University in 1996 where he teaches ecology and landscape ecology, and leads an active research group in landscape ecology. He has carried out wildlife research in many regions of Victoria, and led teams investigating factors that influence the conservation of species and assemblages (mammals, birds, reptiles) in woodland and mallee ecosystems. A particular focus, both in Australia and internationally, is wildlife conservation in agricultural landscapes and the effect of human land-use and landscape change on nature conservation. He holds a Personal Chair in Landscape Ecology at Deakin University.

References

Bennett, A., Brown G., Lumsden, L., Hespe, D., Krasna, S. and Silins, J. (1998). *Fragments for the Future. Wildlife in the Victorian Riverina (the Northern Plains).* Department of Natural Resources and Environment: East Melbourne.

Bennett, A.F., Lumsden, L.F. and Nicholls, A.O. (1994). Tree hollows as a resource for wildlife in remnant woodlands: spatial and temporal patterns across the northern plains of Victoria, Australia. *Pacific Conservation Biology* **1**, 222–235.

Fischer, J. and Lindenmayer, D.B. (2002). The conservation value of paddock trees for birds in a variegated landscape in southern New South Wales. 1. Species composition and site occupancy patterns. *Biodiversity and Conservation* **11**, 807–832.

Ford, H.A. (1985). The bird community in eucalypt woodland and eucalypt dieback in the Northern Tablelands of New South Wales. In *Birds of Eucalypt Forests and Woodlands: Ecology, Conservation, Management.* (Eds A. Keast, H.F. Recher, H. Ford and D. Saunders) pp. 333–340. RAOU and Surrey Beatty and Sons: Chipping Norton.

Gibbons, P. and Lindenmayer, D. (2002). *Tree Hollows and Wildlife Conservation in Australia.* CSIRO Publishing: Melbourne.

Hill, J.E. and Smith, J.D. (1984). *Bats: A Natural History.* Rigby Publishers: Adelaide.

Kunz, T.H. and Lumsden, L.F. (2003). Ecology of cavity and foliage roosting bats. In *Bat Ecology*. (Eds T.H. Kunz and M.B. Fenton) pp. 3–89. The University of Chicago Press: Chicago.

Law, B.S. and Chidel, M. (2006). Eucalypt plantings on farms: use by insectivorous bats in south-eastern Australia. *Biological Conservation* **133**, 236–249.

Law, B.S., Chidel, M. and Turner, G. (2000). The use by wildlife of paddock trees in farmland. *Pacific Conservation Biology* **6**, 130–143.

Lumsden, L.F. (2004). The ecology and conservation of insectivorous bats in rural landscapes. PhD thesis. Deakin University: Burwood, Victoria.

Lumsden, L.F. and Bennett, A.F. (2000). Bats in rural landscapes: a significant but largely unknown faunal component. In *Balancing Conservation and Production in Grassy Landscapes. Proceedings of the Bushcare Grassy Landscapes Conference, Clare, South Australia, 19–21 August 1999*. (Eds T. Barlow and R. Thorburn) pp. 42–50. Environment Australia: Canberra.

Lumsden, L.F. and Bennett, A.F. (2005). Scattered trees in rural landscapes: foraging habitat for insectivorous bats in south-eastern Australia. *Biological Conservation* **122**, 205–222.

Lumsden, L.F., Bennett, A.F., Krasna, S.P. and Silins, J.E. (1995). The conservation of insectivorous bats in rural landscapes of northern Victoria. In *People and Nature Conservation. Perspectives on Private Land Use and Endangered Species Recovery*. (Eds A. Bennett, G. Backhouse and T. Clark) pp. 142–148. The Royal Zoological Society of New South Wales: Mosman.

Lumsden, L.F., Bennett, A.F. and Silins, J.E. (2002a). Location of roosts of the lesser long-eared bat *Nyctophilus geoffroyi* and Gould's wattled bat *Chalinolobus gouldii* in a fragmented landscape in south-eastern Australia. *Biological Conservation* **106**, 237–249.

Lumsden, L.F., Bennett, A.F. and Silins, J.E. (2002b). Selection of roost sites by the lesser long-eared bat (*Nyctophilus geoffroyi*) and Gould's wattled bat (*Chalinolobus gouldii*) in south-eastern Australia. *Journal of Zoology, London* **257**, 207–218.

Maddock, T.H. and Tidemann, C.R. (1995). Lesser Long-eared Bat *Nyctophilus geoffroyi*. In *The Mammals of Australia*. New revised edn. (Ed. R. Strahan) pp. 502–503. Reed Books: Chatswood, NSW.

McDonald, G. and Farrow, R.A. (1988). Migration and dispersal of the Rutherglen bug, *Nysius vinitor* Bergroth (Hemiptera: Lygaeidae), in eastern Australia. *Bulletin of Entomological Research* **78**, 493–509.

Menkhorst, P.W., ed. (1995). *Mammals of Victoria. Distribution, Ecology and Conservation*. Oxford University Press: Melbourne.

Radford, J.Q., Bennett, A.F. and Cheers, G.J. (2005). Landscape-level thresholds of habitat cover for woodland-dependent birds. *Biological Conservation* **124**, 317–337.

17

THE ROLE OF FARM PRODUCTION SYSTEMS IN DETERMINING VEGETATION PATTERNS AND OPTIONS FOR BROAD-SCALE CONSERVATION IN TEMPERATE WOODLANDS

Josh Dorrough

1. High-input livestock production systems are not compatible with maintaining native plant diversity and overstorey tree populations.
2. Sustainable grazing systems depend on maintaining diverse perennial ground-layer vegetation with associated drought tolerance and soil protection.
3. Overstorey tree populations and a moderate diversity of ground-layer plants can be maintained under extensive and/or infrequent livestock grazing systems, but only in the absence of soil nutrient enrichment.
4. Profitable alternatives to intensive high-input production systems do exist; however, there are significant barriers to widespread uptake.
5. In any particular situation there will be considerable uncertainty about whether reducing or excluding grazing and/or fertilisation will lead to a recovery in the diversity and abundance of native ground-layer plants.
6. Strategies for conserving and enhancing temperate woodland vegetation should include whole-of-farm approaches in addition to targeting remnants within farms.
7. Conservation planners need to consider different approaches across the landscape, owing to systematic variation in farm productivity and ecological processes.

Introduction

This chapter provides seven key lessons derived from collaborative research on vegetation patterns and dynamics in the temperate woodland landscapes of central and northern Victoria and the southern slopes and tablelands of New South Wales (see map on next page). These regions differ from the northern and coastal rainshadow grassy forests and woodlands of New South Wales in the lack of a summer-dominant rainfall pattern and the presence of frequent summer soil moisture deficits. The research has been directed towards: (1) understanding how management practices associated with agriculture (tree clearing, livestock grazing, nutrient

enrichment and cultivation) affect the composition, distribution and abundance of plant species (ground-layer plants and overstorey trees) (e.g. Dorrough *et al.* 2004; Dorrough *et al.* 2006; Dorrough and Scroggie 2008); and, (2) identification of options for conserving native vegetation (e.g. Dorrough *et al.* 2007; Crosthwaite *et al.* 2008; Duncan and Dorrough 2009). Understanding how biota respond to agriculture has required examination of the full gradient of land use intensities, from never-fertilised remnant woodlands through to cleared, sown and fertilised exotic pastures. While the research has all been conducted within the temperate woodland zone, it has also examined other vegetation types such as grasslands and dry grassy forests that occur in a mosaic with temperate woodland. All of these vegetation types share similarities in how they respond to agricultural management practices and so the lessons provided here draw on the full range of research.

The lessons are derived primarily from ecological experiments and surveys conducted both at fine scales (<1 m) and broader scales within and among local (paddocks and farms, approximately hundreds of metres to tens of kilometres) and regional landscapes (hundreds of kilometres). Developing options for conserving native vegetation in temperate woodlands not only requires knowledge of the relationships between management practices and biota, however, but also knowledge of the farm production system itself and the role of native vegetation within that (Crosthwaite and Malcolm 2000). Therefore a number of the lessons are the result of collaborative investigations that have combined ecological, farm management and farm business data to explore alternative conservation scenarios at paddock, farm and landscape scales (Dorrough *et al.* 2007; Crosthwaite *et al.* 2008; Dorrough *et al.* 2008a; Duncan and Dorrough 2009). Some of these ideas have been trialed through a conservation market-based incentive program in Victoria (www.gbcma.vic.gov.au/downloads/Biodiversity/Green_Graze_final_report.pdf).

Lessons

1. High-input livestock production systems are not compatible with maintaining native plant diversity and overstorey tree populations

Intensification of grazing and fertiliser practices lead to dramatic declines in the diversity of native plants and changes in the types and abundances of plant species (Kirkpatrick *et al.* 2005; Dorrough *et al.* 2006; Dorrough and Scroggie 2008; McIntyre 2008). These changes have significant implications for key ecosystem functions such as nutrient cycling, carbon storage and soil protection (McIntyre 2008). While a small number of native grasses and broad-leaved forbs can tolerate moderate soil fertility and grazing intensities, most geophytes (orchids and lilies), shrubs, and many native grasses and forbs persist only when such management practices are absent or of very low frequency and intensity (Dorrough and Scroggie 2008). Intensive grazing and nutrient enrichment (via stock camping and fertiliser) also reduce the likelihood of overstorey tree recruitment (Dorrough and Moxham 2005; Fischer *et al.* 2009b) and lead to declines in health of mature trees (Close *et al.* 2008). Therefore, in the medium to long-term, intensive grazing and nutrient enrichment lead to the loss of both overstorey and understorey native vegetation. There is a need for more public discussion on whether there should be limits to the location and extent of intensive production systems at both farm and landscape scales (e.g. McIntyre 2002).

2. Sustainable grazing systems depend on maintaining diverse perennial ground-layer vegetation with associated drought tolerance and soil protection

Diverse perennial vegetation underpins sustainable grazing systems that have good drought tolerance and soil protection (McIntyre 2008), providing an important link between production and native vegetation conservation. The loss of perennial native vegetation and replacement by a small suite of exotic annual plants, as a result of intensive grazing and soil nutrient enrichment, is a key factor in the deterioration of ecological condition, compromising the production base and increasing risk to drought. Although annual pastures can be productive, they provide poor soil protection (McIntyre 2008) and collapse rapidly under drought conditions (Dorrough *et al.* 2008b). Data from hundreds of paddocks across the southern temperate woodland zone demonstrate that in the medium to long-term, and despite the best management intentions and skills, perennial pastures eventually become dominated by annual plants when soil fertility is high (>20 ppm Colwell phosphorus) (Dorrough *et al.* 2006; McIntyre 2008). This same conclusion was reached during the earliest pasture experiments (e.g. Davies *et al.* 1934).

3. Overstorey tree populations and a moderate diversity of ground-layer plants can be maintained under extensive and/or infrequent livestock grazing systems, but only in the absence of soil nutrient enrichment

A moderate diversity of native ground-layer plants is often associated with infrequent livestock grazing, but only when soil fertility is low (Dorrough *et al.* 2006). The likelihood of tree recruitment is also greater under infrequent (Dorrough and Moxham 2005) or rotational grazing regimes (Fischer *et al.* 2009b), but also could depend on pasture productivity (Vesk and Dorrough 2006). A large group of native plant species, however, are intolerant of even low levels of grazing and nutrient enrichment (Prober *et al.* 2002; Dorrough and Scroggie 2008). Woodlands of low fertility and that are lightly grazed, typically have lower diversity than those within stock reserves, cemeteries and other localities with a long-term absence of stock grazing pressure (e.g. McIntyre *et al.* 2002). Despite this, grazed native pastures with scattered woodland trees are the most extensive native-dominated habitat element in the temperate woodland zone (Dorrough *et al.* 2007). Their management may be critical for the persistence of native biota and delivery of key ecosystem services (Cunningham *et al.* 2008; Fischer *et al.* 2009a, 2009b).

4. Profitable alternatives to intensive high-input production systems do exist; however, there are significant barriers to widespread uptake

Low-input rotational grazing systems appear to provide a profitable pathway for management of grazed temperate woodlands (Crosthwaite *et al.* 2008) and are predicted to directly benefit woody plant recruitment (Fischer *et al.* 2009b) and indirectly benefit the diversity of native perennial plants due to lower fertiliser use (Dorrough *et al.* 2008b). Profitable adoption of low-input systems across whole farms often entails significant capital investment in livestock grazing infrastructure (e.g. fencing and water) to enable more efficient use of pastures and thus compensate for production lost through ceasing fertiliser applications (Crosthwaite *et al.* 2008). In the absence of large holdings, opportunities to expand or to obtain alternative incomes, many producers will not feel able or willing to change their management philosophy. High land prices and agricultural extension and agribusiness will no doubt continue to encourage intensification as the major pathway to maintaining or improving farm profits. As a result, there is also a continuing need to identify strategies for managing woodlands that can be targeted towards high-input producers.

5. Vegetation recovery following relief from fertiliser and grazing is uncertain

There is still much uncertainty about how relief from livestock grazing and fertilisation affect the diversity, composition and abundance of native vegetation (Lunt *et al.* 2007; Dorrough and McIntyre unpublished data). Most of our current assumptions relating to possible benefits of reducing livestock pressure and nutrient concentrations are derived from studies examining the long-term historical impact of imposing these management activities (Lunt *et al.* 2007). Fewer studies have explored how plants respond as the intensity of agricultural land management declines. Ecological variation (e.g. plant composition, seed availability, landscape position, soil fertility, climate and weather) and a lack of empirical data lead to considerable uncertainty about the possible outcomes in any particular situation. More research is needed to examine these 'reverse' trajectories (Lunt *et al.* 2007). Policies and financial incentives to improve conservation outcomes in woodlands should take this uncertainty into account - maintaining low-input grazing systems in localities with little or no fertiliser use should not be considered equivalent to reducing stock pressure and ceasing fertiliser in previously fertilised woodlands.

6. Strategies for conserving and enhancing temperate woodland vegetation should include whole of farm approaches in addition to targeting remnants within farms

Strategies that target only remnant woodland patches are likely to be insufficient in ensuring persistence of all native biota across landscapes as well as providing minimal contribution to other resource management objectives (e.g. soil protection, water quality, nutrient cycling, stock shelter) (Vesk and MacNally 2006; Fischer *et al.* 2009a). Site-based incentive funding could have unintended perverse outcomes and facilitate declines in the condition of the more extensive scattered tree and native pasture habitats, despite assisting protection and management of remnants within properties. Site-based payments, that potentially include opportunity costs, can be invested towards intensification elsewhere on the property, and infrastructure established through incentive payments (e.g. fencing and water points) could facilitate more intensive management in areas with remaining native vegetation (e.g. scattered trees and native pastures).

Conservation strategies that treat the farm as the unit of management within the context of a regional management plan, rather than focusing on individual patches within a farm, may be more effective in achieving broad-scale habitat reconstruction and conservation. Whole-farm approaches, that: (1) identify and account for future trends in all native vegetation; (2) jointly develop farm business plans and native vegetation management strategies; and (3) provide advice on alternative on- or off-farm investments that avoid further vegetation degradation, could prove advantageous (Crosthwaite and Malcolm 2000). Whole-farm conservation strategies that take into account the farm business and farm management practices may also identify options that are more cost-effective than direct incentive payments (e.g. farm restructuring, extension advice and training) (Crosthwaite and Malcolm 2000). A whole-farm approach would also provide a framework through which adoption of low-input systems can be facilitated. As discussed above, profitable alternatives to intensification do exist and whole-farm approaches can be useful in identifying them and facilitating their adoption.

7. Conservation planners need to consider different approaches across the landscape, owing to systematic variation in farm productivity and ecological processes

In low productivity native pastures, medium to long-term initiatives that target lower stocking rates, rest from grazing and cessation of fertiliser inputs are likely to be most cost-effective (e.g. Dorrough *et al.* 2008a). High land prices and a shift away from wool production have already led to significant changes in livestock type and intensities and hence there is potential to facilitate substantial passive revegetation in many native pasture areas with little direct public investment. Conservation strategies for highly threatened woodlands in productive parts of the landscape will, however, require considerably more financial investment. De-intensification at any reasonable scale imposes great cost in productive landscapes and active replanting and targeted conservation management of existing remnants are likely to be the most cost-effective strategies. Broad scale replanting of scattered trees should be considered for livestock grazing systems (Fischer *et al.* 2009a) while block planting may be the only feasible option in areas under crops, or transforming back towards cropping.

Conclusions

Ecological research and conservation management strategies for temperate woodland need to be better placed within an understanding of the current agricultural management systems, practices, trends and drivers of change. There is also a need for more research tackling the implications of current trends in land use, or predicting future plausible scenarios in an effort to inform current conservation planning (e.g. Duncan and Dorrough 2009; Thomson *et al.* 2009). Conservation policies and management strategies need flexibility to enable them to adapt to often rapid changes in farm management trends and production systems. Collaborative research with other disciplines is important in this regard. Knowledge gained through developing connections with agribusiness, producer groups, agricultural research and development and extension also will be important in providing the context for research.

Ecologists working in temperate woodlands have done some important research examining historical patterns of vegetation change and describing the outcomes of past management on current vegetation patterns (Lunt *et al.* 2007). However, we still have limited data to guide active management or broad-scale restoration. There is a glaring lack of research examining the recovery of vegetation following relief from both intensive grazing (or changes in grazing strategy) and fertiliser application. More research funding and effort needs to be directed towards these problems as the current lack of knowledge seriously hampers our ability to make decisions about how and where to invest for conservation. There is funding directed towards medium-term stewardship payments (e.g. the Australian Government Box Gum Grassy Woodland Project, www.nrm.gov.au/stewardship) but often the management strategies being recommended have little empirical data to support them. The onus is on State and Federal Governments to ensure that such initiatives are adequately monitored and in such a way that they improve our current ecological knowledge and contribute to smarter and more cost-effective management.

Acknowledgements

The ideas presented here have developed over several years with significant input from many colleagues. In particular I would like to thank Jim Crosthwaite, Sue McIntyre, Peter Vesk, David Duncan, Jim Moll, Vivienne Turner, Ian Lunt and Joern Fischer.

Bio

Josh Dorrough undertook a PhD on native plant persistence in grazed grasslands through the Australian National University, completing in 2002. From 2002 to 2008 he worked as a research ecologist with the Arthur Rylah Institute for Environmental Research, Department of Sustainability and Environment, Victoria. He is currently a Visiting Research Scientist with CSIRO Sustainable Ecosystems in Canberra and living on a small farm in the far south-east of New South Wales.

References

Close, D.C., Davidson, N.J. and Watson, T. (2008). Health of remnant woodlands in fragments under distinct grazing regimes. *Biological Conservation* **141**, 2395–2402.

Crosthwaite, J. and Malcolm, B. (2000). 'Looking to the farm business: approaches to managing native grassland in south-eastern Australia. National Research and Development Program on Rehabilitation, Management and Conservation of Remnant Vegetation, Research Report 5/00'. Land and Water Resources Research and Development Corporation, Canberra

Crosthwaite, J., Malcolm, B., Moll, J. and Dorrough, J. (2008). Future investment in landscape change in southern Australia. *Landscape Research* **33**, 225–239.

Cunningham, R.B., Lindenmayer, D.B., Crane, M., MacGregor, C., Montague-Drake, R. and Fischer, J. (2008). The combined effects of remnant vegetation and tree planting on farmland birds. *Conservation Biology* **22**, 742–752.

Davies, J.G., Scott, A.E. and Fraser, K.M. (1934). Natural pastures: their response to superphosphate. *Commonwealth of Australia Council for Scientific and Industrial Research Bulletin* **83**, 1–76.

Dorrough, J., Ash, J. and McIntyre, S. (2004). Plant responses to livestock grazing frequency in an Australian temperate grassland. *Ecography* **27**, 798–810.

Dorrough, J., McIntyre, S., Stol, J., Brown, G. and Barrett, G. (2008b). Understanding the interactions between biodiversity and the management of native pastures in the Murray Darling Basin. Meat and Livestock Australia: Sydney.

Dorrough, J., Moll, J. and Crosthwaite, J. (2007). Can intensification of temperate Australian livestock production systems save land for native biodiversity? *Agriculture Ecosystems and Environment* **121**, 222–232.

Dorrough, J. and Moxham, C. (2005). Eucalypt establishment in agricultural landscapes and implications for landscape-scale restoration. *Biological Conservation* **123**, 55–66.

Dorrough, J., Moxham, C., Turner, V. and Sutter, G. (2006). Soil phosphorus and tree cover modify the effects of livestock grazing on plant species richness in Australian grassy woodland. *Biological Conservation* **130**, 394–405.

Dorrough, J. and Scroggie, M.P. (2008). Plant responses to agricultural intensification. *Journal of Applied Ecology* **45**, 1274–1283.

Dorrough, J., Vesk, P.A. and Moll, J. (2008a). Incorporating ecological uncertainty and farm-scale economics when planning restoration. *Journal of Applied Ecology* **45**, 288–295.

Duncan, D.H. and Dorrough, J. (2009). Historical and current land use shape landscape restoration options in the Australian wheat and sheep farming zone. *Landscape and Urban Planning* **91**, 124–132.

Fischer, J, Sherren, K., Stott, J., Zerger, A., Warren, G. and Stein, J. (2009a). Toward landscape-wide conservation outcomes in Australia's temperate grazing region. *Frontiers in Ecology and the Environment* in press.

Fischer, J., Stott, J., Zerger, A., Warren, G., Sherren, K. and Forrester, R.I. (2009b). Reversing a tree regeneration crisis in an endangered ecoregion. *Proceedings of the National Academy of Sciences* **106,** 10386–10391.

Kirkpatrick, J.B., Gilfedder, L., Bridle, K. and Zacharek, A. (2005). The positive and negative conservation impacts of sheep grazing and other disturbances on the vascular plant species and vegetation of lowland subhumid Tasmania. *Ecological Management and Restoration* **6**, 51–60.

Lunt, I.D., Eldridge, D.J., Morgan, J.W. and Witt, G.B. (2007). A framework to predict the effects of livestock grazing and grazing exclusion on conservation values in natural ecosystems in Australia. *Australian Journal of Botany* **55**, 401–415.

McIntyre, S. (2002). The way forward – from principles to practice. In *Managing and Conserving Grassy Woodlands.* (Eds S. McIntyre, J.G. McIvor and K.M. Heard) pp. 201–218. CSIRO Publishing: Melbourne.

McIntyre, S. (2008). The role of plant leaf attributes in linking land use to ecosystem function in temperate grassy vegetation. *Agriculture Ecosystems and Environment* **128**, 251–258.

McIntyre, S., Heard, K.M. and Martin, T.G. (2002). How grassland plants are distributed over five human-created habitats typical of eucalypt woodlands in a variegated landscape. *Pacific Conservation Biology* 7, 274–289.

Prober, S.M., Thiele, K.R. and Lunt, I. (2002). Identifying ecological barriers to restoration in temperate grassy woodlands: soil changes associated with different degradation states. *Australian Journal of Botany* **50**, 699–712.

Thomson, J.R., Moilanen, A.J., Vesk, P.A., Bennett, A.F. and Mac Nally, R. (2009). Where and when to revegetate: A quantitative method for scheduling landscape reconstruction. *Ecological Applications* **19**, 817–828.

Vesk, P.A. and Dorrough, J. (2006). Getting trees on farms the easy way? Lessons from a model of eucalypt regeneration in pastures. *Australian Journal of Botany* **54**, 509–519.

Vesk, P.A. and Mac Nally, R. (2006). The clock is ticking-Revegetation and habitat for birds and arboreal mammals in rural landscapes of Southern Australia. *Agriculture Ecosystems and Environment* **112**, 356–366.

18

THE ROLE OF LINEAR STRIPS AND SMALL PATCHES OF WOODLAND IN CONSERVING ENDANGERED MAMMAL FAUNA

Rodney van der Ree

1. Small patches of native vegetation in highly cleared landscapes are valuable for biodiversity.
2. Linear strips often support some of the highest quality remaining woodland.
3. Linear strips are irreplaceable, despite being suboptimal in shape.
4. Many animals need numerous trees with hollows, spread evenly across the landscape.
5. Existing structural elements provide critical resources for wildlife and should be the building blocks for restoration.
6. Active management of woodlands is critical.
7. Create and maintain a well-connected network of linear strips.
8. Tensions in the management of remnant woodlands need to be resolved.

Introduction

My research focuses broadly on quantifying and mitigating the impacts of human activities on wildlife and ecosystems. I have studied the response of a range of fauna (birds, reptiles, bats and small mammals) to habitat loss, fragmentation and degradation in agricultural and urban landscapes, two of the most human-dominated systems in the world. Most of my effort has focused on arboreal marsupials, however, especially the Squirrel Glider (*Petaurus norfolcensis).* Since 1995, I have studied the ecology of the Squirrel Glider in several environments in south-eastern Australia (see map on next page), but particularly in an agricultural landscape with less than 10% tree cover, where fauna must contend with extensive loss of habitat, ongoing decline in habitat quality, and most remaining habitat occurring as a network of narrow, linear strips along roads and watercourses.

I have used intensive and extensive spotlighting, hairtubing, trapping and radio-tracking to study the Squirrel Glider, Common Brushtail Possum (*Trichosurus vulpecula*), Brush-tailed Phascogale (*Phascogale tapoatafa*), Sugar Glider (*Petaurus breviceps*) and Yellow-footed

Antechinus (*Antechinus flavipes*), and in doing so, have uncovered many important aspects of their ecology. My research has an applied focus – the intensity of negative pressures facing the woodland and its wildlife is enormous, and ignoring these pressures to focus solely on theoretical questions will not result in effective conservation outcomes. My current research includes collaborations with all levels of government, conservation groups, road and property developers, and resource managers to better understand the impacts of management actions and to design and assess the effectiveness of mitigation actions.

Lessons

1. Small patches of native vegetation in highly cleared landscapes are valuable for biodiversity

Small patches of native vegetation are valuable for conservation in highly cleared landscapes because they often provide the majority of habitat for biodiversity. In my study area to the west of Euroa in the northern plains of Victoria, there is approximately 5–10% tree cover remaining, most of which occurs as linear strips along roads and streams or as scattered paddock trees (van der Ree 2000). The endangered Squirrel Glider and Brush-tailed Phascogale persist in this area only because of the existence of the linear strips and small patches of woodland.

In conservation biology, priority is usually given to conserving and managing large patches of native vegetation. There are good reasons for this, including the positive relationship between patch size and species richness, large core areas being less affected by edge effects, and the inclusion of different habitat types (Simberloff and Abele 1982). Large patches of native vegetation within the woodland zone are rare, and hence the small, thin and fragmented remnants become the priority. A useful analogy is the relationship between the cost of a product and its availability. As a commodity becomes rarer, harder to find, or more expensive to produce, its cost (and hence its value) to the consumer increases. The same principle holds for woodlands – the greater the extent of loss, the more valuable are the remaining pieces.

2. Linear strips often support some of the highest quality woodland that remains

Linear reserves, including roadsides, water frontages and travelling stock reserves, often contain the highest quality habitat in many districts. In some regions, the plant and animal communities within the linear strips may more closely resemble the pre-European conditions than large patches of woodland, which typically have a history of timber harvesting, grazing by stock and mining. At Euroa, the mean density of large diameter trees (>70 cm) in roadsides and unused road reserves ranged from 20.1 to 21.3 stems per ha, while the density in large blocks of similar woodland ranged from 2–4.5 per ha (van der Ree and Bennett 2001). Of great concern is that prescriptions for the retention of large trees in woodlands subject to timber harvesting is from two to six trees per ha of a size >50 cm diameter (van der Ree and Bennett 2001). This will result in stands of woodland with few large trees and reduced hollow abundance (large trees support more hollows in a range of sizes than smaller trees), which will affect the abundance and diversity of hollow-dependent species negatively.

The high density of large trees at Euroa plays an important role in the ecology of the threatened hollow-dependent fauna. For example, the home ranges of the Brush-tailed Phascogales occupying linear roadside remnants near Euroa (5–8 ha for females) was almost an order of magnitude smaller than reported for this species in large tracts of forest with low densities of large trees (~40 ha for females, >100 ha for males) (Soderquist 1995; Traill and Coates 1993; van der Ree *et al.* 2001). Similarly, compared with populations in larger blocks of contiguous habitat with fewer large trees, Squirrel Gliders within roadside strips at Euroa had comparatively small home ranges and lived at the highest densities recorded for this species (van der Ree 2002; van der Ree and Bennett 2003).

3. Linear strips of habitat are irreplaceable, despite being sub-optimal in shape

Despite their sub-optimal size and shape (Shafer 1995), narrow woodland strips and isolated trees in the northern plains region of Victoria support numerous threatened and endangered species, including the Squirrel Glider and Brush-tailed Phascogale. Not only do these and other species of conservation concern use linear strips and small fragments of woodland, in some cases their very survival depends on these remnants. For example, approximately 70% of all extant groups of the Grey-crowned Babbler (*Pomatostomus temporalis*) in Victoria incorporate roadside vegetation into their home range (Doug Robinson unpublished data). In the wheatbelt in Western Australia, Hobbs (1993) found that 39% of 348 known populations of particular species of rare flora occurred on roadsides.

There is a general misconception that single trees or small patches of trees in farmland (paddock trees) are of little value to wildlife. In highly cleared landscapes, paddock trees are disproportionately important resources for feeding and denning for the Squirrel Glider, Brush-tailed Phascogale, Common Brushtail Possum and Yellow-footed Antechinus (van der Ree and Bennett 2003; van der Ree *et al.* 2003). Importantly, the distance into the paddocks that animals venture is influenced by their capacity and willingness to traverse open gaps: for example, there is a threshold distance of about 70 m for the Squirrel Glider (van der Ree *et al.* 2003). Other groups of fauna also use paddock trees, including bats, insects and many bird species (Abensperg-Traun 2000; Fischer and Lindenmayer 2002; Law *et al.* 2000).

4. Many animals need numerous trees with hollows, spread evenly across the landscape

Over 300 species of Australian native animals use tree hollows (Gibbons and Lindenmayer 2002) and many switch trees on a daily or seasonal basis. Squirrel Gliders living in the roadside network at Euroa changed dens regularly (once every three to five days); used multiple dens (five to six trees per season, with one individual using 15 different trees over a 12-month period); and these dens were distributed evenly across most (~90%) of their home range (van der Ree 2000). Other species of hollow-dependent animals, such as the Brush-tailed Phascogale (van der Ree *et al.* 2006) and Common Brushtail Possum (van der Ree unpublished data) showed similar trends.

An abundant and evenly distributed supply of trees with hollows is important because animals attempt to minimise the distance they travel (and hence energy expended) to fulfill their daily food, shelter and mating requirements – a concept known as central place foraging (Covich 1976). In large patches of vegetation, the most efficient approach would be a circular home range centered on a single den site, thus minimising the distance between the den and any area within the home range. In linear strips, it is impossible for an animal with a large home range to achieve this arrangement, because the home range is restricted to the available habitat, and is usually narrow and elongated. To conserve hollow-dependent fauna in landscapes dominated by linear strips of habitat, an abundant and even distribution of large-diameter and hollow-bearing trees is required in perpetuity. Nest boxes may be suitable as a short-term measure, but in the long-term an ongoing recruitment of trees with hollows is critical (Gibbons and Lindenmayer 2002).

5. Existing structural elements provide critical resources for wildlife and should be the building blocks for restoration

Structural habitat elements (e.g. hollows, logs and leaf litter, branches for perching, peeling bark) provide numerous resources and foraging substrates for wildlife and, when present, should form the basis for habitat restoration works. Squirrel Gliders at Euroa preferentially foraged (Holland *et al.* 2007) and denned (van der Ree 2000) in the large-diameter trees, presumably because of the greater diversity and abundance of resources these large trees provided. For similar reasons, Brush-tailed Phascogales also preferentially foraged and denned in large-diameter trees (van der Ree *et al.* 2001).

Habitat quality is influenced by both floristic and structural attributes. Different species of flora provide a variety of resources for wildlife at different times of the year. For example, Squirrel Gliders foraged almost exclusively (98–100% of observations) in Grey Box trees from summer to winter, compared to 60% in Spring when it also foraged in Golden Wattle (Holland *et al.* 2007). Habitat structure is similarly important for wildlife. Restoring the structure of woodland is easier to achieve when existing elements (e.g. live and dead trees, fallen branches, logs) are retained and incorporated into the restoration project; if they are absent (due to natural decay or intentional removal), it may take many years for them to be replaced by natural processes. Revegetation or natural regeneration around existing large eucalypts (dead or alive) is likely to immediately contain tree hollows usable by vertebrates, which would normally take 120–180 years to develop (Gibbons and Lindenmayer 2002).

6. Active management of woodlands is critical

All woodland remnants require active and intentional management to maintain and enhance their quality and extent. The quality of many patches of woodland has declined since European settlement and continues to decline because they are primarily managed and used for other purposes (e.g. transport, emergency grazing for stock during drought, places to dump rubbish and collect firewood). The absence of saplings (and shrubs) in most grazed roadsides at Euroa (van der Ree and Bennett 2001) demonstrates that 'benign neglect' will result in a woodland dominated by a senescing canopy of large trees with little recruitment to replace it. This scenario has been observed in a woodland remnant in the Western Australian wheatbelt, where grazing by stock has prevented tree recruitment since 1929. Due to a high rate of mortality and zero recruitment, just 11% of the Salmon Gums (*Eucalyptus salmonophloia)* and 17% of the York Gums (*Eucalyptus loxophleba)* present in 1981 were predicted to still be alive in 2125 (Saunders *et al.* 2003).

The potentially important contribution of linear reserves to conservation has been recognised increasingly over the past 20 years. Management practices, however, do not appear to have improved at the same rate as our understanding of the importance of such reserves. The value of these woodland strips needs to be recognised by government and the community, and managed appropriately before the cost of repair becomes too great. One approach is to buy back (or confiscate) leases from landholders that do not manage linear reserves in an ecologically sustainable way (e.g. excessive grazing, failing to repair fences, rubbish dumping, weed/vermin invasion) and to provide incentives for those who achieve positive ecological outcomes. Government could provide labour or materials for fences adjacent to Crown reserves if landholders are willing to permanently set aside and manage minimum widths of land for conservation.

7. Create and maintain well-connected networks of linear strips

Connectivity, or the ability of plants and animals to move around the landscape to find food, shelter and mates, is a fundamental requirement for population survival (Bennett 1999). In landscapes where most wooded vegetation occurs as linear strips, creating and maintaining a connected 'network' with multiple routes across the landscape is critical. A population viability analysis of a population of the Squirrel Glider at Thurgoona (near Albury, NSW) predicted that the population was more likely to persist into the future as the number of links and connections in the area was increased (Stewart and van der Ree 2009).

In continuous habitat, animals can easily bypass an area that is either permanently unsuitable (e.g. a lake for a terrestrial species) or temporarily unsuitable (e.g. recently burnt) as habitat. In a linear strip that is 20 m wide, it may be impossible for some species to traverse a gap that has been created, either due to a permanent feature such as a road (e.g. Epps *et al.* 2005) or temporary disturbance such as fire or a territorial male. In these circumstances, animals need alternative routes across the landscape. Intersections of linear strips may play an important role in facilitating social interactions of wildlife. Groups of Squirrel Gliders living at intersections of linear strips occurred at higher densities and had higher rates of home range overlap than animals living just in straight sections (van der Ree 2000). Intersections are important locations for dispersing animals and likely for gene flow, as well as allowing animals to have a home range that is traversed in a more efficient way.

8. Tensions in the management of remnant woodlands need to be resolved

The conflict between the original role of most linear reserves and current conservation objectives must be reconciled to protect and enhance woodland values. My studies clearly demonstrate the very high value of the linear strips of woodland for the conservation of numerous threatened species, including the Squirrel Glider and Brush-tailed Phascogale (e.g. van der Ree 2002; van der Ree *et al.* 2001; van der Ree and Bennett 2003). The linear strips of woodland, however, were reserved in the early days of European settlement to move people and stock from farm to market, not for the conservation of biodiversity. Consequently, the design and maintenance standards of these original uses are usually in conflict with the conservation objectives for the strips.

The appropriate management regime for roadside vegetation is a contentious issue. As traffic volume and speed increases, the size of the clear-zone (distance between travel lanes and non-frangible vegetation) may also increase, resulting in the removal of vegetation and a reduction in the amount of habitat in the area. Vegetation is also removed when narrow roads are widened to allow access by larger farm machinery and trucks. These requirements for human safety and efficient access are significant, as are the requirements of threatened species inhabiting the woodlands.

Conclusions

Agricultural landscapes with networks of linear strips of woodland and scattered paddock trees are common features of many woodland areas and pose a unique set of challenges for conservation management. While small and narrow strips are generally considered suboptimal for conservation, these woodland remnants often provide the majority of habitat in highly modified landscapes, and are therefore critical for the conservation of both common and rare species of plants and animals. Their value for conservation of biodiversity is due to their high quality habitat (often exceeding that in the large patches) and their arrangement as an interconnected network of linear strips. Habitat quality of the linear strips for many animals is primarily influenced by the high density and relatively even spatial distribution of large trees with hollows.

Future management of linear strips must focus on creating and enhancing the quality, extent and connectivity of woodland. For arboreal mammals in the northern plains of Victoria, this means ensuring an ongoing recruitment of trees into each age class and providing a perpetual supply of large trees with hollows. The incremental loss of woodland and decline in patch quality due to road widening and other management actions that are in conflict with conservation goals need to be identified and resolved with the wide range of stakeholders involved.

Traditional approaches to conservation, such as the establishment of large reserves of intact vegetation, will not achieve conservation goals in production landscapes because such large areas of intact woodland do not exist. Similarly, agricultural practices that ignore ecological sustainability will ultimately result in species extinction and loss of agricultural productivity. Creative and visionary leadership based on rigorous scientific data is required to recover endangered species in temperate woodlands and to achieve the twin aims of conservation and ecologically sustainable agricultural production.

Acknowledgements

Thanks to Andrew Bennett for introducing me to the woodlands of north-eastern Victoria. Andrew Bennett and David Lindenmayer provided constructive comments that improved this manuscript. Numerous people have provided advice, discussion, financial support and time in the field during my work in the woodlands of south-eastern Australia. Rodney van der Ree is currently supported by The Baker Foundation.

Bio

Rodney van der Ree is currently the Deputy Director of the Australian Research Centre for Urban Ecology, a division of the Royal Botanic Gardens Melbourne, located within the School of Botany at the University of Melbourne. His research focuses primarily on the effects of urbanisation (and associated infrastructure required to support urban areas – such as roads, agricultural and water catchment areas to feed and water the urban residents) on fauna. Current research includes quantifying the negative effects of roads on wildlife, evaluating the use and effectiveness of structures designed to facilitate connectivity for fauna, and studying the ecology of bats in urban ecosystems.

References

Abensperg-Traun, M. (2000). In defence of small habitat islands: termites (Isoptera) in the Western Australian central wheatbelt, and the importance of dispersal power in species occurrence. *Pacific Conservation Biology* **6**, 31–39.

Bennett, A.F. (1999). *Linkages in the Landscape. The Role of Corridors and Connectivity in Wildlife Conservation.* IUCN: Gland, Switzerland and Cambridge.

Covich, A.P. (1976). Analyzing shapes of foraging areas: some ecological and economic theories. *Annual Review of Ecology and Systematics* **7**, 235–257.

Epps, C.W., Palsboll, P.J., Wehausen, J.D., Roderick, G.K., Ramey, R.R. and McCullough, D.R. (2005). Highways block gene flow and cause a rapid decline in genetic diversity of desert bighorn sheep. *Ecology Letters* **8**(10), 1029–1038.

Fischer, J. and Lindenmayer, D.B. (2002). The conservation value of paddock trees for birds in a variegated landscape in southern New South Wales. 2. Paddock trees as stepping stones. *Biodiversity and Conservation* **11**, 833–849.

Gibbons, P. and Lindenmayer, D.B. (2002). *Tree Hollows and Wildlife Conservation in Australia.* CSIRO Publishing: Melbourne.

Hobbs, R.J. (1993). Effects of landscape fragmentation on ecosystem processes in the Western Australian wheatbelt. *Biological Conservation* **64**, 193–201.

Holland, G., Bennett, A.F. and van der Ree, R. (2007). Time-budget and foraging behaviour of the Squirrel Glider, Petaurus norfolcensis, in remnant linear habitat *Wildlife Research* **34**(4), 288–295.

Law, B.S., Chidel, M. and Turner, G. (2000). The use by wildlife of paddock trees in farmland. *Pacific Conservation Biology* **6**, 130–143.

Saunders, D.A., Smith, G.T., Ingram, J.A. and Forrester, R.I. (2003). Changes in a remnant of salmon gum *Eucalyptus salmonophloia* and York gum *E. loxophleba* woodland, 1978 to 1997. Implications for woodland conservation in the wheat-sheep regions of Australia. *Biological Conservation* **110**(2), 245–256.

Shafer, C.L. (1995). Values and shortcomings of small reserves. *BioScience* (45), 80–88.

Simberloff, D. and Abele, L.G. (1982). Refuge design and island biogeographic theory: effects of fragmentation. *The American Naturalist* **120**, 41–50.

Soderquist, T.R. (1995). Spatial organization of the arboreal carnivorous marsupial *Phascogale tapoatafa*. *Journal of Zoology (London)* **237**(Part 3), 385–398.

Stewart, C.W. and van der Ree, R. (2009). *Population Viability Analysis for Squirrel Gliders in the Thurgoona and Albury Ranges Region of New South Wales.* Australian Research Centre for Urban Ecology: Melbourne.

Traill, B.J. and Coates, T.D. (1993). Field observations on the brush-tailed phascogale *Phascogale tapoatafa* (Marsupialia: Dasyuridae). *Australian Mammalogy* **16**, 61–65.

van der Ree, R. (2000). Ecology of arboreal marsupials in a network of remnant linear habitats. PhD thesis, Deakin University: Melbourne.

van der Ree, R. (2002). The population ecology of the Squirrel Glider *Petaurus norfolcensis*, within a network of remnant linear habitats. *Wildlife Research* **29**(4), 329–340.

van der Ree, R. and Bennett, A.F. (2001). Woodland remnants along roadsides – a reflection of pre-European structure in temperate woodlands? *Ecological Management and Restoration* **2**(3), 226–228.

van der Ree, R. and Bennett, A.F. (2003). Home range of the Squirrel Glider *Petaurus norfolcensis* in a network of linear habitats. *Journal of Zoology (London)* **259**, 327–336.

van der Ree, R., Bennett, A.F. and Gilmore, D.C. (2003). Gap-crossing by gliding marsupials: thresholds for use of isolated woodland patches in an agricultural landscape. *Biological Conservation* **115**(2), 241–249.

van der Ree, R., Bennett, A.F. and Soderquist, T.R. (2006). Nest tree selection by the threatened brush-tailed phascogale *Phascogale tapoatafa* (Marsupialia: Dasyuridae) in a highly fragmented agricultural landscape. *Wildlife Research* **33**, 113–119.

van der Ree, R., Soderquist, T.R. and Bennett, A.F. (2001). Home range use by the Brush-tailed Phascogale *Phascogale tapoatafa* (Marsupialia: Dasyuridae) in high-quality, spatially limited habitat. *Wildlife Research* **28**, 517–525.

19

FROM PATTERN TO PROCESS: TOWARDS UNDERSTANDING DRIVERS OF DIVERSITY IN TEMPERATE WOODLANDS

David M. Watson

1. Using entire woodland fragments as sampling units yields data of sufficient resolution at the most appropriate scale to inform meaningful inferences and guide effective management.
2. Patch quality is paramount to long-term persistence.
3. Explicit resource-based approaches can reveal the processes underlying occurrence patterns, enabling a mechanistic understanding of the determinants of diversity in woodlands.
4. Two or three years is rarely sufficient to generate representative data on occurrence patterns of animals at patch and landscape scales.
5. Some woodland-dependent species are sufficiently mobile that targeted restoration and rehabilitation can make a real difference to their distributions. Build it and they will come.
6. Other woodland-dependent species are dispersal-limited and strategic translocations should be considered as part of restoration efforts.
7. Fragmented landscapes are still in flux – many plants, animals and processes have not stabilised sufficiently to reflect the net consequences of habitat fragmentation over the last century.
8. The heterogeneity of nutrient inputs through space and time is crucial to sculpting the below-ground processes driving the food-webs that underpin woodland ecosystems.

Introduction

My research on temperate woodlands spans a 15-year period and includes empirical studies in *Eucalyptus*, *Acacia* and *Allocasuarina* dominated habitats in eastern Australia (see map on next page). I have also conducted work in various forested systems in the 'New World', and broad-based differences in the associated flora, fauna, ecological processes and history have

provided opportunities to place my emerging understanding of Australian woodlands in context and in more synthetic and generalised terms.

My current research interests fall into three broad areas: biodiversity conservation in agricultural landscapes, the biological consequences of habitat fragmentation, and ecological interactions between plants and animals. Combining community-scale descriptive work with species-specific studies, most of my work is restricted to vertebrates, although I recognise the central role that arthropod and microbial assemblages play in these systems. I have complemented this community-level descriptive approach with a resource-based experimental approach, treating mistletoe and other parasitic plants as model systems. While some of my work has been conducted in National Parks, travelling stock reserves and other public lands, most of my field sites are on private land and I work closely with natural resource agencies, regional bodies and individual landholders to convert my findings into practical on-ground outcomes.

Lessons

1. Using entire woodland fragments as sampling units yields data of sufficient resolution at the most appropriate scale to inform meaningful inferences and guide effective management

While many of the ecological processes underpinning woodland ecosystems vary at the landscape scale, the scale that defines many of the patterns within fragmented landscapes is the patch itself (Andrén 1994; Watson *et al.* 2000; Watson 2003a). Grazing regimes and other aspects of land use; fire frequency and other disturbances vary primarily at the patch scale and, while moderated by within-patch heterogeneity, it is often patch-scale factors that ultimately determine whether or not particular species occur within a habitat fragment (Ewers and Didham 2006). Yet the way most patchy landscapes are studied is using fixed-effort methods applied within patches – quadrats, transects and plots of fixed areas arranged in woodland fragments and the surrounding matrix (e.g. Mac Nally and Watson 2007). While useful from a statistical standpoint and the basis of rigorous comparisons, resultant data can be misleading – even nonsensical – and can obfuscate the underlying patterns. Thus using a fixed area transect to measure tree density in a woodland patch misses the point – organisms occurring within this woodland are not influenced by any single value: they are responding to the variation in structure and composition of trees within the entire patch. Likewise, a particular species of lizard is not influenced by the presence or absence of a particular substrate within a randomly placed quadrat, but responds to the density of this substrate within the entire patch (Watson and Peterson 1999; Watson 2003b).

This approach also extends to animal sampling: if the key question hinges on species-richness estimates in habitat patches, it is counter-intuitive to exclude those species seen within the patch that happened to be detected outside the transect boundaries during sampling. Various approaches can be used to design sampling regimes at the patch scale, including the use of results-based stopping rules to ensure samples from patches of differing size and quality are of equivalent completeness (Watson 2003a). Most importantly, patch-scale data yields information at the most relevant scale for land managers. Grazing regimes and fire frequencies are not applied at the quadrat or transect scale, but to the entire patch. Hence, not only do patch-scale data represent reality better, they give managers the information essential to set meaningful targets and achieve real on-ground improvements. When interpreting the findings of studies conducted using fixed-effort sampling methods at smaller scales, the reader should be careful not to extrapolate the data up to the patch scale: even though a species was not recorded within the sampling area, it still may reside in the patch.

2. Patch quality is paramount to long-term persistence

'Location' is frequently touted as all-important, whether by estate agents selling a house, conservation agencies justifying their actions or landscape ecologists interpreting their data. Position in the landscape, proximity to putative corridors, distance from presumed source areas are all key criteria driving the way we consider and manage populations of plants and animals in woodland remnants and other fragmented systems. Studies from older systems – hundreds and thousands of generations after fragmentation – paint a very different picture (Watson 2002a; Ewers and Didham 2006; Watson 2009). Isolation, connectivity and other landscape-scale measures are invariably eclipsed by measures of patch quality: the bigger fragments containing the better habitats and the wider range of microhabitats retain more species. Thus if we are concerned about long-term persistence of local populations and averting extinctions of species now restricted to fragmented systems, we must apply these lessons from older systems and strive to retain and restore the largest and most intact habitats remaining. Bigger patches necessarily contain a wider range of landforms and associated soil types and habitats, yet even the largest fragments may not be large enough – thresholds for persistence in some systems extend into the thousands of hectares (Watson 2003a). As well as validating the large-scale connectivity and restoration initiatives presently underway in Australian woodlands, this realisation calls into question most of the investment of regional and local agencies in small-scale revegetation and rehabilitation.

3. Explicit resource-based approaches can reveal the processes underlying occurrence patterns, enabling a mechanistic understanding of the determinants of diversity in woodlands

One shortcoming with patch and landscape-based approaches is that the measured variables rarely influence individual organisms directly – ravens (*Corvus spp.*) do not selectively predate nests in smaller fragments, nor do Greater Gliders (*Petauroides volans*) prefer trees growing in larger patches. Rather, individual animals respond directly to particular resources – food, shelter, cover from predators. These responses are typically directional, rapid, and predictable (i.e. nest predators prefer sites with higher nest densities, arboreal folivores avoid sites with insufficient hollows; Andren 1994; Lindenmayer 1997; Gibbons and Lindenmayer 2002). By studying distribution patterns of organisms relative to availability of carefully selected resources, not only can direct patterns of dependence be revealed, but indirect interactions with patch and landscape-scale factors can be discerned (i.e. larger patches typically support a greater size range of trees, thereby explaining why gliders occur frequently in larger patches). In addition to yielding more meaningful inferences, resource-based approaches provide testable predictions and concrete targets ideally suited to adaptive management techniques (Watson 2001; Watson and Herring 2006).

4. Two or three years is rarely sufficient to generate representative data on occurrence patterns of animals at patch and landscape scales

Two of the climatic factors that define temperate woodlands – inherently low rainfall and high variation in annual rainfall – mean that any one season is not representative of actual occurrence patterns (Maron *et al.* 2005). The summary statistics typically used to characterise ecosystems (long term averages of temperature, rainfall, number of frosts, etc.) have little meaning. Rather it is the variation in these parameters and the occasional extreme event that matter more: defining range limits, constraining distributions and explaining timing of reproduction (Stevens and Watson 2006; Barea and Watson 2007). The best way to quantify this variation and discern signal from noise is to move beyond the standard two or three-year study – periods which are simply not long enough to gain a representative sample of many patterns and processes (Watson 2009). This approach need not involve continuous monitoring. Single samples every spring for five years may be far more informative than three continuous years of sampling for many questions in many regions, despite requiring a fraction of the resources. Researchers need to be creative about designing studies to maximise their temporal relevance, and funding bodies and policy makers need to be far more realistic in the level of investment directed towards empirical research. Short-term studies of landscapes yield many answers and most of them are wrong (Strayer *et al* 1986, Stohlgren *et al.* 1995).

5. Some woodland-dependent species are sufficiently mobile that targeted restoration and rehabilitation can make a real difference to their distributions. Build it and they will come

Some woodland-dependent plants and animals are highly mobile, able to move long distances to find suitable conditions or required resources (Watson *et al.* 2000). Given inherent variation in resource availability (whether water, nectar, insects or rodents), many woodland species are readily able to travel long distances – often through relatively inhospitable habitat – buffering populations against local shortages and short-terms periods of scarcity. This is most clearly demonstrated by birds and bats, two diverse groups that contain many woodland-dependent species. Studies from isolated remnants and longer-term studies measuring inter-annual variation across multiple sites have documented high levels of species turnover in these groups (e.g.

Maron *et al.* 2005). This dynamism is driven by the comings and goings of species, prospecting across the landscape looking for particular resources. While this has previously been considered in terms of demographic and genetic consequences for species persistence in relictual landscapes, the inherent mobility of many woodland species also has important implications for revegetation and restoration initiatives. Although adding to existing remnants will always be preferable, revegetation plots in greenfield sites are visited by a large number of woodland species. If the new patch is big enough and contains sufficient resources, these visitors will stay for longer and may even breed. As well as providing recruits to arrest declining populations, the status of revegetated plots as habitat represents a much-needed success story– just the sort of thing needed to inspire landholders and management agencies to maintain their motivation and continue their efforts.

6. Other woodland-dependent species are dispersal-limited and strategic translocations should be considered as part of restoration efforts

Although many woodland species are sufficiently mobile to colonise new or improved patches, many species are not and are effectively marooned in habitat remnants. Whether unable to move long distances or subject to predation outside woodland habitats, many reptiles, amphibians and small mammals can become stranded. This has two consequences. Firstly, those woodland remnants where a species persists soon contain more individuals than can be supported, leading to increased levels of density-dependent mortality and greater likelihood of local extinction. Secondly, once the population of a particular species has gone extinct within a particular patch, the woodland is unlikely to be recolonised in the future, even though it may contain all the resources needed by that species. So, for a particular species in a given landscape, there may be too many individuals in some patches of decreasing quality and no individuals in patches of increasing quality. A short-term solution to this situation is targeted translocations - moving individual animals to unoccupied patches. Rather than willy-nilly, these actions would need to be carried out in an informed and carefully monitored manner (Vitt *et al.* 2009). The Landcare movement has demonstrated the great willingness of landholders to repopulate their production-dominated properties with the plants that used to occur there - why not give them the opportunity to put some of the animals back as well?

7. Fragmented landscapes are still in flux – many plants, animals and processes have not stabilised sufficiently to reflect the net consequences of habitat fragmentation over the last century

Most fragmented woodlands in southern Australia are the result of widespread clearing for agriculture that occurred 80–120 years ago. While this may seem ample time for various groups of organisms to have adjusted, consider this time period relative to the taxa that comprise this habitat. The various tree species that define these woodlands typically take decades to attain reproductive maturity and a century or more to acquire the hollows necessary for many woodland-dependent animals. As such, the time since fragmentation in most temperate woodlands represents less than one cohort of these plants and the effects of initial clearing and subsequent perturbation on the regeneration and overall structure of the habitat are yet to be fully realised (Watson 2003b). Recognising the age of the landscapes being studied relative to the organisms and processes being studied is critical (Watson 2002b) – the woodland fragment that has supported two generations of trees and 10 generations of cicadas feeding on their sap has also yielded more than 50 generations of *Antechinus* and hundreds of generations of the isopods they eat. Rather than choose the organism before the question, it is far more meaningful to select the question of interest, consider the appropriate time frame and then study the organism best suited to revealing changes over that duration.

8. The heterogeneity of nutrient inputs through space and time is crucial to sculpting the below-ground processes driving the food-webs that underpin woodland ecosystems

All organisms ultimately depend on water and nutrients, yet the importance of nutrient availability in explaining occurrence patterns of plants, animals and fungi in fragmented landscapes has only recently emerged and has yet to be fully appreciated in temperate woodlands. The nitrogen-fixing trees and shrubs that originally provided most of the soil-borne nitrates have undergone dramatic changes in abundance. Some were cleared, others actively promoted for commercial purposes (e.g. wattle bark for tanning leather) and most are preferentially consumed by domestic stock. The soils that have some of the lowest available phosphorous on earth require large and frequent inputs to grow food and fibre, with surface run-off and spray drift dramatically elevating levels in adjacent woodlands. The mistletoes and parasitic shrubs that concentrate nitrogen, phosphorous and various micronutrients have undergone marked changes in abundance, becoming super-abundant in some regions, locally extinct in others (March and Watson 2007). Finally, fire regimes in these woodlands have drastically changed, typically becoming less frequent and more intense, increasing or decreasing availability of some elements by orders of magnitude. Applying findings of similar patterns described in other ecosystems (Harris 2009), these changes to nutrient inputs have likely driven dramatic changes in the below-ground microbial community, altering their composition and thereby fundamentally modifying the foundation of woodland food-webs. Understanding these changes, following them through food webs and investigating the potential to restore the structure and function of microbial assemblages are key goals for woodland biologists, and new collaborations with other groups of biologists are needed to apply this bottom–up approach.

Conclusions

When considering woodlands, the starting position for most managers, policy makers and landholders is the status quo: current habitats provide a reliable indication of their original composition and represent templates for restoration and long-term preservation. This steady-state mentality is at odds with a growing understanding of how woodlands function and how ecological processes within them have changed over the last century. Species come and go, disturbances and other rare events have a disproportionate effect on community composition and changes in surrounding habitats entrain dramatic changes within remnants. Acknowledging these sources of variation and the extended time frames within which they operate is essential, both to interpret existing studies of these systems, but also to design new studies to fill critical knowledge gaps.

Most of our woodlands are gone, and most of the remainder have been greatly modified, yet we have lost surprisingly few species. While noting that the full repercussions of our actions are yet to run their course, this demonstrates the great resilience of these ecosystems and suggests that populations and processes may rebound if we make the right decisions. Applying the lessons learned, investing in ongoing research, rewarding sound land management and keeping our eyes on longer term targets, I am confident we can safeguard the long-term functionality of our temperate woodlands.

Acknowledgements

My research group has been supported by grants from Charles Sturt University, the Australian Research Council (DP0342549, DP0449417, LP0882952), the Winifred Violet Scott Trust, and the Holsworth Wildlife Research Endowment.

Bio

Dave Watson completed a Bachelor of Science at Monash University before conducting his Honours research on fragmented Buloke woodlands with Ralph Mac Nally and Andrew Bennett. His PhD at The University of Kansas examined the long-term effects of habitat fragmentation using case studies of highland birds in Mesoamerica. He returned to Australia in 1999, taking up a position at Charles Sturt University where he is now Associate Professor in Ecology. He lives with his family on a rural property near Albury, which is slowly being transformed from degraded farmland to grassy-box woodland.

References

Andrén, H. (1994). Effects of habitat fragmentation on birds and mammals in landscapes with different proportions of suitable habitat: a review. *Oikos* **71**, 355–366.

Barea, L. and Watson, D.M. (2007). Temporal variation in food resources determines onset of breeding in an Australian mistletoe specialist. *Emu* **107**, 203–209.

Ewers, R.M. and Didham, R.K. (2006). Confounding factors in the detection of species responses to habitat fragmentation. *Biological Reviews* **81**, 117–142.

Gibbons, P. and Lindenmayer D. (2002). *Tree Hollows and Wildlife Conservation in Australia.* CSIRO Publishing: Melbourne.

Harris, J. (2009). Soil microbial communities and restoration ecology: facilitators or followers? *Science* **325**, 573–576.

Lindenmayer, D.B. (1997). Differences in the biology and ecology of arboreal marsupials in forests of Southeastern Australia. *Journal of Mammalogy* **78**, 1117–1127.

Mac Nally, R.C. and Watson, D.M. (1997). How does habitat-island area affect species richness? Birds in Victorian Buloke remnants. *Austral Ecology* **22**, 227–232.

March, W.A. and Watson, D.M. (2007). Parasites boost productivity: effects of mistletoe on litter dynamics in a temperate Australian forest. *Oecologia* **154**, 339–347.

Maron, M., Lill, A., Watson, D.M., Mac Nally, R. (2005). Temporal variation in bird assemblages: how representative is a one-year snapshot? *Austral Ecology* **30**, 383–394.

Stevens, H.C. and Watson, D.M. (2006). Effect of rainfall on breeding of grey shrike-thrushes *Colluricincla harmonica*. *Corella* **30**, 16–20.

Stohlgren, T.J., Binkley, D., Veblen, T.T. and Baker, W.L. (1995). Attributes of reliable long-term landscape-scale studies: malpractice insurance for landscape ecologists. *Environmental Monitoring and Assessment* **36**, 1–25.

Strayer, D, Glitzenstein, J.S., Jones, C.G., Kolasa, J., Likens, G.E., McDonnell, M.J., Parker, G.G., Pickett, S.T.A. (1986). Long-term ecological studies: an illustrated account of their design, operation, and importance to ecology. *Occasional Publication of the Institute of Ecosystem Studies* **2**, 1–38.

Vitt, P., Havens, K. and Hoegh-Guldberg, O. (2009). Assisted migration: part of an integrated conservation strategy. *Trends in Ecology and Evolution* **24**, 473–474.

Watson, D.M. (2009). Continental Islands. In *Encyclopaedia of Islands*. (Eds R. Gillespie and D. Clague). University of California Press: Berkeley.
Watson, D.M. (2003a). The 'standardized search': an improved way to conduct bird surveys. *Austral Ecology* **28**, 515–525.
Watson, D.M. (2003b). Long-term consequences of habitat fragmentation: highland birds in Oaxaca, Mexico. *Biological Conservation* **111**, 283–303.
Watson, D.M. (2002a). A conceptual framework for the study of species composition in islands, fragments and other patchy habitats. *Journal of Biogeography* **29**, 823–834.
Watson, D.M. (2001) Mistletoe – a keystone resource in forests and woodlands worldwide. *Annual Review of Ecology and Systematics* **32**, 219–249.
Watson, D.M. and Herring, M. (2006). Of mistletoe and mechanisms – drivers of declining biodiversity in remnant woodklands. *Proceedings of Veg Futures Conference*: 1–7.
Watson, D.M., Mac Nally, R.C. and Bennett, A.F. (2000). The avifauna of remnant Buloke (*Allocasuarina luehmanni*) woodlands in western Victoria. *Pacific Conservation Biology* **6**, 46–60.
Watson, D.M. and Peterson, A.T. (1999). Determinants of diversity in a naturally fragmented landscape: bird faunas of humid montane forests in Mesoamerica. *Ecography* **22**, 582–589.

20

EIGHT STEPS TO CONSERVE DIVERSE UNDERSTOREYS IN GRASSY WOODLANDS

Ian D. Lunt, Suzanne M. Prober and Kevin R. Thiele

1. Prevent further losses: so little temperate woodland remains in most regions that all remnants must be conserved and degrading processes avoided.
2. Promote diverse solutions: diverse policy and management approaches are needed to capture the wide mix of woodland conditions, land tenures and management objectives.
3. Protect small patches: all else being equal, bigger patches are better, but small remnants are critical, as remnants containing high quality understoreys are typically small.
4. Adopt ecological management: many woodlands on productive soils require regular burning, crash grazing or slashing to promote plant diversity.
5. Restore degraded woodlands: degraded woodlands need to be revegetated and restored to promote plant diversity, control weeds and achieve sustainable woodland landscapes.
6. Reduce soil nutrients: a continuing restoration challenge is to develop techniques to reduce elevated soil nutrient levels, as weeds out-compete native plants on disturbed and fertilised soils.
7. Manage for climate change: conserve networks of diverse patches, especially interconnected, little-disturbed remnants on unfertilised soils, as these have the greatest potential to support native plants and resist exotic invasion.
8. Coordinate and support individual efforts: this vision must be coordinated and supported through a national system of management networks and stewardship schemes.

Introduction

For over 20 years we have been fortunate to work in some of Australia's most beautiful ecosystems: remnant grassy woodlands and grasslands in the wheat–sheep belt. Our research has focused on one of the most sensitive components of woodland biodiversity – the native

understorey plants, especially the grasses and vibrant wildflowers that make woodlands so spectacular. In the most diverse woodlands, one can find up to 45 different plant species in a single square metre (Lunt 1990). Our research includes woodlands in southern Queensland, New South Wales, the ACT and Victoria (see map above), particularly threatened woodlands dominated by White Box (*Eucalyptus albens*), Yellow Box (*Eucalyptus melliodora*), Grey Box (*Eucalyptus microcarpa*) and Red Gums (*E. camaldulensis* and *E. tereticornis*).

We have learnt how woodlands have changed since European settlement, how they have been affected by activities such as burning and livestock grazing, and we are beginning to learn how to restore functional, diverse understoreys (e.g. Prober and Thiele 1995; Lunt 1997; Prober *et al.* 2002, 2005, 2007; Lunt *et al.* 2006; Prober and Lunt 2009). These insights have been gained from complementary research approaches, including historical studies, regional and cross-regional surveys (of species, vegetation structure, genetics and soils), and many short and long-term experiments, the longest of which has now run for 16 years. The following eight points represent the major lessons we have distilled from our work.

Lessons

1. Prevent further losses: so little temperate woodland remains in most regions that all remnants must be conserved and degrading processes avoided

Many temperate woodlands have been cleared for agriculture and most remnants are small, isolated and degraded. For example, less than 1% of endangered White Box–Yellow Box–Red Gum woodlands remain in good condition, and most surviving remnants are heavily degraded. Considerable progress has been made over the past 20 years in listing endangered ecosystems, creating new reserves, promoting off-reserve conservation networks and raising public awareness. Woodlands continue to be lost, however, through a 'death by a thousand cuts', as paddock trees, native pastures, roadside strips and remnants are destroyed and degraded for other land uses (Prober *et al.* 2001). Activities such as intensification of fertilisation and livestock grazing, tree clearing and firewood collection continue to degrade many areas. As in other systems, woodlands are easy to degrade but hard to restore. Our threatened woodlands cannot survive this attrition; all woodland remnants must be conserved and degrading processes avoided.

2. Promote diverse solutions: diverse policy and management approaches are needed to capture the wide mix of woodland conditions, land tenures and management objectives

Every patch of woodland in agricultural landscapes has been altered by past management. Remnants with different management histories support different species and processes, and require different management goals and interventions; for example, some patches are valuable for understorey diversity and others for hollow-bearing trees (Lunt and Spooner 2005). It is rare for all biological attributes to be well represented within a single patch. These patterns are reflected in different land tenures (cemeteries, roadsides, travelling stock reserves, scenic reserves, grazed pastures, private woodlots, home and back paddocks), which have been managed for different purposes by many people with differing outcomes (Prober and Thiele 1995; Prober *et al.* 2001). All of these land tenures and histories are needed to make an effective conservation network. Woodlands cannot be conserved on public land alone, and cannot be conserved by a single policy or management approach. Diverse policy and management approaches are needed to capture the wide mix of woodland conditions and management objectives.

3. Protect small patches: all else being equal, bigger patches are better, but small remnants are critical, as remnants containing high quality understoreys are typically small

Big, interconnected remnants are essential for sustainable conservation of woodland landscapes. But not all remnants are the same, and for small understorey plants, past disturbance and management are stronger drivers of species' occurrences and population sizes than are spatial attributes such as patch size or connectivity (Prober and Thiele 1995). Typically, only small remnants (such as those in some cemeteries) have escaped degradation through livestock grazing and nutrient enrichment. These can support large, genetically diverse populations of declining species (Lunt 1994; Prober *et al.* 1998) and thus make a critical contribution to woodland conservation. Small *patches* are not the same as small *populations*. We need to protect small patches and develop ways to boost small populations.

4. Adopt ecological management: many woodlands on productive soils require regular burning, crash grazing or slashing to promote plant diversity

Some (but not all) woodlands require regular management such as burning, crash grazing or slashing to maintain plant diversity and vigorous growth of native grasses. This is most relevant in productive sites with few trees, where large dominant grasses such as Kangaroo Grass (*Themeda australis*) can rapidly out-compete smaller plants (Lunt and Morgan 2002; Prober *et al.* 2007). Many native woodland plants do not form persistent seed banks in the soil, so if existing plants die after mismanagement, the species may disappear completely from a site (Lunt and Morgan 2002). Fire is a preferred management tool in sites that have historically been burnt (e.g. many cemetery, road or rail easements, Prober *et al.* 2007), but other methods such as short-duration livestock grazing and slashing are useful in many places (Lunt *et al.*

2007). Management goals should be clearly articulated, and effects monitored to ensure that goals are being met.

5. Restore degraded woodlands: degraded woodlands need to be revegetated and restored to promote plant diversity, control weeds and achieve sustainable woodland landscapes

The enormous scale of woodland destruction and degradation means that merely conserving the few surviving remnants is not good enough. Restoration and revegetation are critical for promoting plant diversity, controlling weeds and achieving sustainable woodland landscapes (Prober and Thiele 2005). Since woodlands contain few tree species but many ground-layer plants, effort must be given to restoring diverse ground-layer species, not just planting trees and shrubs. Trees and shrubs benefit many fauna but tree planting has little (if any) benefit for ground-layer plants (Munro *et al.* 2009). Progress has been made in restoring dominant grasses through weed and seedbank management (e.g. Prober *et al.* 2005), but new strategies are needed to restore diverse understoreys over large areas (e.g. Gibson-Roy 2007).

6. Reduce soil nutrients: a continuing restoration challenge is to develop techniques to reduce elevated soil nutrient levels, as weeds out-compete native plants on disturbed and fertilised soils

A key constraint on restoration outcomes is elevated soil nutrient levels, often arising from fertilisation, livestock grazing or soil disturbance. High levels of soil nitrate or phosphorus promote ubiquitous fast-growing exotic plants which out-compete native plants, preventing sustainable restoration outcomes (Prober *et al.* 2005; Dorrough *et al.* 2006). Recently we have found that restored swards of *Themeda australis* can effectively reduce soil nitrate and control exotic annuals, hence re-establishing this natural grassland dominant is an important way to restore degraded understoreys (Prober and Lunt 2009). It is not known, however, whether *Themeda* or other native species can fulfill this function in sites with extremely high nutrient levels. Given the enormous areas across which fertilisers have been added to woodland landscapes (either deliberately or through drift from adjacent paddocks), a continuing restoration challenge is to develop practical techniques to reduce soil nutrient levels over large areas.

7. Manage for climate change: conserve networks of diverse patches, especially interconnected, little-disturbed remnants on unfertilised soils, as these have the greatest potential to support native plants and resist exotic invasion

Woodland structure, composition and function will change in unforeseen ways as climate continues to change. The aim of woodland conservation should not be to preserve current ecological conditions, let alone relics of pre-European ecosystems. Instead, it should be to actively conserve diverse, resilient, dynamic ecosystems (Dunlop and Brown 2008). The high degree of fragmentation and degradation in woodland ecosystems makes this goal challenging. The most practical way to achieve this may be to conserve (and where possible create) networks of little-disturbed sites on unfertilised soils, as these areas have the greatest potential to support diverse native understoreys and resist exotic invasion (Prober *et al.* 2002). Travelling stock reserve networks are critical in this regard.

8. Coordinate and support individual efforts: this vision must be coordinated and supported through a national system of management networks and stewardship schemes

Woodland conservation depends on diverse efforts by many individuals. Every woodland remnant and every on-ground action to restore woodland landscapes contributes to a greater conservation vision. Some new 'off-reserve' conservation models, such as Conservation Management Networks, Natural Resource Management councils and the box-gum woodland stewardship program have already been established to coordinate and support the plethora of on-ground actions being undertaken. These programs require ongoing development and integration, to support landholders at the local level and to ensure we achieve the best possible outcomes from the collective effort at bioregional and national scales (e.g. Prober *et al.* 2001; Fitzsimons and Wescott 2005).

Conclusions

When we began our research, conservation attention in Australia was focused on ecosystems such as rainforests and tall wet forests; few people were interested in declining species in fragmented woodlands. Over the last two decades, huge progress has been made in understanding woodland ecology, developing management options and policies, and raising community awareness. We now often have a good understanding of what we need to do, but we don't always know *how* to do it, especially at the large scales required. Just as woodland decline can be characterised as a 'death by a thousand cuts', woodland restoration will be achieved by a thousand small actions, not one grand symbolic gesture or research breakthrough. This may seem frustrating, but the scale of the issue means that progress will be slow even if we can multiply our efforts 10-fold. Consequently, we must highlight successes, however small they may appear, rather than focus on declines; otherwise we risk disenchanting the communities (and funding and government agencies) who have the greatest potential to enact widespread change. At the same time, we must remain vigilant that ground is not lost. In this light, proposals to sell parts of the travelling stock reserve network are destined to be condemned by future generations. Such shortsighted agendas must be vigorously opposed.

Acknowledgements

We are most grateful to the funding and land management agencies that have supported our research over many years, including the Australian Research Council, National Heritage Trust, NSW Department of Environment, Climate Change and Water, State Forests NSW, National Parks and Wildlife Service NSW, the NSW Environment Trust, Department of Sustainability and Environment Victoria, Parks Victoria, ACT Department of Environment, Climate Change, Energy and Water, and many local government and catchment management authorities, as well as community management networks and many private landholders, all of whom have aided our understanding in incalculable ways.

Bios

Ian Lunt lectures in vegetation ecology and ecosystem management at Charles Sturt University in Albury, NSW. His research documents how disturbances such as grazing and burning regimes affect biodiversity in fragmented native grasslands and grassy woodlands in south-east Australia. His work has helped managers to: (a) understand *historical* influences on natural ecosystems ('how did we get here?'), (b) refine *present* management regimes ('how do we stay afloat?'), and (c) develop restoration methods for the *future* ('how can we get ahead?'). Ian provides expert advice to many government committees in Victoria and New South Wales, and regularly writes extension articles for the broader community.

Suzanne Prober is a plant ecologist at CSIRO Sustainable Ecosystems in Perth, and has worked for many years in temperate eucalypt woodlands of southern Australia. Her research is centred on understanding and restoring ecological function and diversity in woodlands and other ecological communities of fragmented agricultural landscapes, with a focus on plant-soil interactions, weed invasion, fire ecology and native grasses. She has been closely involved in the development of novel policy initiatives for conserving and restoring fragmented and degraded temperate woodlands. Current projects focus on understanding and restoring ecological resistance and resilience in woodlands and shrublands under climate change.

Kevin Thiele is a plant taxonomist and systematist and is the Curator of the Western Australian Herbarium, part of the Department of Environment and Conservation in Perth. He has worked on the systematics, taxonomy and phylogenetics of diverse groups including *Banksia*, *Viola* and the family Rhamnaceae. With Suzanne Prober, he has contributed to ecological research seeking to understand the conservation needs of grassy woodland communities and to develop a policy framework that can meet those needs. In his role at the Western Australian Herbarium, he is involved in conservation assessment and planning for Western Australian plants, algae and fungi.

References

Dorrough, J., Moxham, C., Turner, V. and Sutter, G. (2006). Soil phosphorus and tree cover modify the effects of livestock grazing on plant species richness in Australian grassy woodland. *Biological Conservation* **130**, 394–405.

Dunlop, M. and Brown, P.R. (2008). Implications of Climate Change for Australia's National Reserve System: A Preliminary Assessment. Department of Climate Change: Canberra.

Fitzsimons, J.A. and Wescott, G. (2005). History and attributes of selected Australian multi-tenure reserve networks. *Australian Geographer* **36**, 75–93.

Gibson-Roy, P., Delpratt, J. and Moore, G. (2007). Restoring Western (Basalt) Plains grassland. 2. Field emergence, establishment and recruitment following direct seeding. *Ecological Management & Restoration* **8**, 123–132.

Lunt, I.D. (1990). Species-area curves and growth-form spectra for some herb-rich woodlands in western Victoria, Australia. *Australian Journal of Ecology* **15**, 155–161.

Lunt, I.D. (1994). Variation in flower production of nine grassland species with time since fire, and implications for grassland management and restoration. *Pacific Conservation Biology* **1**, 359–366.

Lunt, I.D. (1997). Effects of long-term vegetation management on remnant grassy forests and anthropogenic native grasslands in south-eastern Australia. *Biological Conservation* **81**, 287–297.

Lunt, I.D. and Morgan, J.W. (2002). The role of fire regimes in temperate lowland grasslands of south-eastern Australia. In *Flammable Australia: The Fire Regimes and Biodiversity of a Continent*. (Eds R. Bradstock, J. Williams and A.M. Gill) pp. 177–196. Cambridge University Press: Cambridge.

Lunt, I.D. and Spooner, P.G. (2005). Using historical ecology to understand patterns of biodiversity in fragmented agricultural landscapes. *Journal of Biogeography* **32**, 1859–1873.

Lunt, I.D., Jones, N., Spooner, P.G. and Petrow, M. (2006). Effects of European colonization on indigenous ecosystems: post-settlement changes in tree stand structures in *Eucalyptus-Callitris* woodlands in central New South Wales, Australia. *Journal of Biogeography* **33**, 1102–1115.

Lunt, I.D., Eldridge, D.J., Morgan, J.W. and Witt, G.B. (2007). Turner Review No. 13. A framework to predict the effects of livestock grazing and grazing exclusion on

conservation values in natural ecosystems in Australia. *Australian Journal of Botany* **55**, 401–415.
Munro, N.T., Fischer, J., Wood, J. and Lindenmayer, D.B. (2009). Revegetation in agricultural areas: the development of structural complexity and floristic diversity. *Ecological Applications* **19**, 1197–1210.
Prober, S.M. and Lunt, I.D. (2009). Restoration of *Themeda australis* swards suppresses soil nitrate and enhances ecological resistance to invasion by exotic annuals. *Biological Invasions* **11**, 171–181.
Prober, S.M. and Thiele, K.R. (1995). Conservation of the grassy white box woodlands: relative contributions of size and disturbance to floristic composition and diversity of remnants. *Australian Journal of Botany* **43**, 349–366.
Prober, S.M. and Thiele, K.R. (2005). Restoring Australia's temperate grasslands and grassy woodlands: integrating function and diversity. *Ecological Management and Restoration* **6**, 16–27.
Prober, S.M., Spindler, L.H. and Brown, A.H.D. (1998). Conservation of the grassy white box woodlands: effects of remnant population size on genetic diversity in the allotetraploid herb *Microseris lanceolata. Conservation Biology* **12**, 1279–1290.
Prober, S.M., Thiele, K.R. and Higginson, E. (2001). The Grassy Box Woodlands Conservation Management Network: Picking up the pieces in fragmented woodlands. *Ecological Management and Restoration* **2**, 179–188.
Prober, S.M., Thiele, K.R. and Lunt, I.D. (2002). Identifying ecological barriers to restoration in temperate grassy woodlands: soil changes associated with different degradation states. *Australian Journal of Botany* **50**, 699–712.
Prober, S.M., Thiele, K.R., Lunt, I.D. and Koen, T.B. (2005). Restoring ecological function in temperate grassy woodlands: manipulating soil nutrients, exotic annuals and native perennial grasses through carbon supplements and spring burns. *Journal of Applied Ecology* **42**, 1073–1085.
Prober, S.M., Thiele, K.R. and Lunt, I.D. (2007). Fire frequency regulates tussock grass composition, structure and resilience in endangered temperate woodlands. *Austral Ecology* **32**, 808–824.

21

CONSERVATION OF WOODLAND VERTEBRATE BIOTA IN THE TEMPERATE WOODLANDS OF SOUTHERN NEW SOUTH WALES

David Lindenmayer, Mason Crane, Damian Michael, Rebecca Montague-Drake and Christopher MacGregor

1. Remnant vegetation and plantings are perceived as markedly different kinds of habitat by some elements of the biota and should be treated as different components of a portfolio of vegetation assets on a farm.
2. There are cumulative effects of remnant vegetation and plantings on vertebrate biota at several spatial scales.
3. Large patches are important for biodiversity conservation but small patches also can make an important conservation contribution.
4. Forgotten environments, particularly rocky outcrops, are critical for the conservation of species-rich groups like reptiles.
5. The content, configuration and context of plantings all make an important contribution to the value of these areas for bird biodiversity.
6. Landscape context is a key driver of the composition of vertebrate assemblages occupying temperate woodland remnants.
7. There are few, if any, robust shortcuts to biodiversity conservation in temperate woodland landscapes.

Introduction

The temperate woodlands of south-eastern Australia are some of the most heavily cleared and degraded vegetation types in Australia (Benson 2008). There also have been significant changes in biodiversity in these vegetation types, including increases in exotic plant species (Burrows 1999) and reductions in many native plant species (McIntyre *et al.*1993; McIntyre and Lavorel 1994; Prober *et al.* 2005), as well as changes in populations of many species of mammals and birds (Ford *et al.* 2001; Barrett *et al.* 2003).

Over the past decade, we have conducted an extensive series of multi-scaled studies of the vertebrate biota of remnant temperate woodland and planted native vegetation in the South-west Slopes bioregion of southern New South Wales (see map on next page), (e.g. Cunningham

et al. 2007; 2008). The seven lessons for temperate woodland vertebrate conservation that we briefly summarise in this chapter are drawn from this ongoing work.

The South-west Slopes bioregion

The South-west Slopes bioregion is the most extensively and intensively disturbed of the 13 botanical regions of New South Wales. An estimated 85% of the original cover of native vegetation has been removed in the past 220 years (Benson 2008) with most remaining temperate woodland stands occurring as small remnant patches, often less than 1 ha in size (Gibbons and Boak 2002). Woodland clearing to grow agricultural crops or to improve pasture growth for grazing has removed some plant species and modified or destroyed habitat for other woodland-dependent organisms. Where temperate woodlands have not been completely cleared, ecosystems can be under substantial pressure from traditional forms of grazing (Spooner *et al.* 2002). Furthermore, many areas of temperate woodland in the South-west Slopes bioregion are potentially susceptible to salinity (Stirzaker *et al.* 2002). In an effort to counter some of these environmental problems, major planting programs have taken place in the South-west Slopes bioregion over the past 15–20 years (Freudenberger and Harvey 2004; Lindenmayer *et al.* 2007; Barrett *et al.* 2008). In addition, a Woodland Stewardship Program has recently commenced to improve the protection of existing areas of high quality, remnant native woodland on private land (Department of Environment 2009).

Lessons

1. Remnant vegetation and plantings are perceived as markedly different kinds of habitat by some elements of the biota and should be treated as different components of a portfolio of vegetation assets on a farm

Results from several of our field studies suggest that old-growth woodland, regrowth woodland, and plantings are quite different habitats for different elements of the biota

(Cunningham *et al.* 2007). For example, in the case of reptiles, we found that the Ragged Snake-eyed Skink (*Cryptoblepharus pannosus*) was significantly more likely to be found in regrowth whereas old-growth woodlands were the vegetation types most likely to support the Marbled Gecko (*Christinus marmoratus*). Similar results were obtained for birds, for which we found that plantings and remnant woodlands supported markedly different assemblages (Cunningham *et al.* 2008). In terms of the relative contribution to the number of bird species on a farm, remnant native vegetation was found to be more important than plantings in a ratio of ~3:1. Some species, however, including a number of conservation concern (e.g. Rufous Whistler *Pachycephala rufiventris* and White-browed Woodswallow *Artamus superciliosus*) are found predominantly in plantings.

Such vegetation type differences in biotic responses are perhaps not surprising given differences in the structural attributes and patterns of plant species composition among different kinds of vegetation. They have significant ramifications for vegetation management policies, however. Informed woodland conservation will require the maintenance and continued recruitment of a range of vegetation types on a given farm and landscape. Our on-ground experience suggests that, for example, regrowth woodland stands are regarded as 'rubbish' country by many landholders and are often those areas targeted for clearing and/or repeated burning, although new stands of old-growth woodland cannot develop without first passing through a regrowth phase. Clearly then, there is a need for scientists to improve their communication about the different habitat roles that different native vegetation can play on farms and for policy-makers to ensure that land management practices recognise these differences.

2. There are cumulative effects of remnant vegetation and plantings on vertebrate biota at several spatial scales

Our work has demonstrated the combined effects on biodiversity (particularly birds) of many components of vegetation cover at the farm level – woodland remnants, scattered paddock trees, and native pastures as well as the complementary role that planted areas play with existing remnant native vegetation cover on farms (Cunningham *et al.* 2008). Higher farm-level species richness of groups such as birds is associated with increased levels of heterogeneity in the types of vegetation cover. On this basis, a key management recommendation is to maintain or increase remnant native vegetation at the farm level so that farms support an array of vegetation elements – remnant woodland patches, fallen timber, native pastures (Cunningham *et al.* 2008) as well as plantings (particularly large block plantings) (Lindenmayer *et al.* 2007). Based on the relative contribution to bird species gained on a farm (see Lesson #1), we argue that for most farms, restoration efforts should focus first on conserving and enhancing existing areas of native vegetation (including native pastures) and second on replanting (Cunningham *et al.* 2008).

3. Large patches are important for biodiversity conservation but small patches can also make an important conservation contribution

The conservation biology literature highlights the importance of species-area relationships (reviewed by Rosenzweig 1995) and, in turn, the importance of large patches for biodiversity conservation. Our studies generally support this principle and have indicated that large remnant patches of temperate woodland are important for groups such as birds, arboreal marsupials and reptiles (Cunningham *et al.* 2007; Montague-Drake *et al.* 2009). Large patches are also important for many species of conservation concern such as several species of declining woodland birds (Montague-Drake *et al.* 2009) and the Carpet Python (*Morelia spilota metcalfei*) (Michael and Lindenmayer 2008). Our work, however has also shown that small patches of woodland (such as woodland remnants less than 1–2 ha in size) also can be important places for: (1) providing habitat for many elements of native vertebrate biota (Montague-Drake *et al.* 2009); (2) making a cumulative contribution to farm-level biodiversity conservation (Cunningham *et al.* 2008); and (3) providing habitat and landscape connectivity by acting as stepping stones that facilitate animal movements between larger patches (Fischer and Lindenmayer 2002).

Given these insights about patch-size relationships, an important management lesson has been that while the maintenance of large patches is important in temperate woodland landscapes, the conservation of small patches should not be overlooked and they should not be allowed to become degraded or to be cleared simply because they are small. This reinforces Lessons #1 and 2 about the combined contribution for biodiversity conservation of different vegetation assets on a farm and the value of heterogeneity in these assets.

4. Forgotten environments, particularly rocky outcrops, are critical for the conservation of species-rich groups like reptiles

Vegetation cover has (rightly) been a major focus of temperate woodland conservation efforts, but we have found that environments such as denuded rocky outcrops that have largely been overlooked in the past can be important habitats for many species of reptiles (Michael *et al.* 2008; Michael and Lindenmayer 2008).

The way rocky outcrops are managed can make a substantial difference to their conservation value for reptiles. The density of tree plantings and regrowth vegetation around rocky outcrops needs to be considered so that thermal environments are not modified in ways that are detrimental for reptiles.

Other practices need to be carefully managed to avoid negative impacts on the integrity of rocky outcrops. These include the control of wildfires to limit damage to vegetation cover as habitat for reptiles (Michael *et al.* 2008), especially those of conservation concern, e.g. the Carpet Python (Michael and Lindenmayer 2008).

5. The content, configuration and context of plantings all make an important contribution to the value of these areas for bird biodiversity

How plantings are done can have a significant effect on their use by biota. A major recent study on the South-west Slopes (Lindenmayer *et al.* 2010) has demonstrated that maximising the value of planted areas for birds requires consideration of: (i) vegetation cover in the neighbourhood of a planting; (ii) configuration of a planting (that is, where a planting is located on a farm and its geometry, particularly its size and shape); and (iii) the content of planting or the vegetation structure and plant species composition of a revegetated area. That research has indicated that it is best to establish plantings in places with pre-existing areas of remnant native vegetation or other planted areas. There was also increased value for bird species richness of locating plantings in gullies and riparian areas. In addition, block-shaped plantings rather than strip-shaped ones were those most likely to support several individual species of native birds (e.g. Red-capped Robin, *Petroica goodenovii*). Planting content attributes like the number of logs, the density of understorey vegetation and the prevalence of mistletoe are also important, particularly for some individual bird species (Lindenmayer *et al.* 2010). The spatial configuration of plantings also can be important, with the intersections of linear strip plantings and block-shaped plantings supporting more species than isolated linear strip plantings (Lindenmayer *et al.* 2007). Our data also suggest that plantings add more taxa to overall species richness when a farm has low rather than high levels of remnant native vegetation (Cunningham *et al.* 2008). Thus, an important lesson has been that even highly degraded farms should not be exempt from planting efforts because of the relatively high contributions that planted areas might make to bird diversity on such farms.

6. Landscape context is a key driver of the vertebrate assemblages occupying temperate woodland remnants

Studies from widely different kinds of environments around the world have demonstrated that the landscape context of a patch (and hence the attributes of the matrix surrounding a patch) can significantly influence the number of species as well as the occurrence of individual species found in that patch. We have demonstrated that temperate woodland patches are no different in this regard, both from our experimental studies (Lindenmayer *et al.* 2008), as well as our observational investigations (e.g. Michael *et al.* 2008; Montague-Drake *et al.* 2009). For example, an experimental study we established in the late 1990s has shown that the vertebrate biota inhabiting woodland patches surrounded by maturing stands of Radiata Pine plantation contrasts markedly with the biota found in matched woodland remnants where the landscape context is grazing paddocks (Lindenmayer *et al.* 2008). Over the 12 years of the study to date, the bird assemblage within the woodland patches where the landscape context is dominated by a softwood plantation has transformed from being one characteristic of woodland environments to one that is a novel combination of woodland and forest birds (Lindenmayer *et al.* 2008). This has implications for biodiversity conservation in semi-cleared agricultural areas which currently support patches of (often threatened) woodland vegetation communities but which are now being targeted for plantation development or may become the target of such land use transformations in the future.

A key lesson from our studies has been that informed management of temperate woodlands must take into consideration not only patch-level factors but also the landscape context of patches and hence the condition of the matrix in which patches are located.

7. There appear to be few, if any, robust 'shortcuts' to biodiversity conservation in temperate woodland landscapes

Many kinds of biodiversity shortcuts have been advocated in conservation biology to reduce the complexity of managing large numbers of species with varying habitat and other requirements. To date, however, although our work has encompassed four vertebrate groups (arboreal marsupials, birds, reptiles and terrestrial mammals), we have found that no individual species or set of species within or among any of these groups would be a robust 'biodiversity shortcut' such as an indicator species, flagship species, or focal species. This outcome is most likely to arise because of marked differences between groups, and species within groups, to different kinds of environments, vegetation types and disturbance regimes. For example, although many reptile taxa exhibit strong associations with rocky outcrops (Michael *et al.* 2008), these environments are not extensively used by birds or mammals. Similarly, plantings are rarely inhabited by arboreal marsupials (Cunningham *et al.* 2007), but a subset of the bird assemblage use them in preference to woodland remnants (Cunningham *et al.* 2008).

An important lesson from our work has been that landholders and policy makers need to be aware of the limitations of an uncritical application of biodiversity surrogate schemes. Instead, they should explicitly articulate the objectives of woodland management and, in turn, carefully consider which elements of biodiversity they are aiming to conserve.

Conclusions

More than a decade of intensive field-based empirical research has led to some important new lessons and insights about the conservation and management of temperate woodland vegetation and replanted vegetation on the South-west Slopes of New South Wales. Some findings were anticipated prior to the start of our work, such as the importance of large patches of remnant native vegetation. Others were not, like the value of small patches of remnant native vegetation and the identification of previously overlooked but species-rich environments such as rocky outcrops. Additional discoveries to underpin improved woodland management are likely to be made. There are also key knowledge gaps which need to be filled to improve future management of woodland biota. One major one is that there is no current understanding of longitudinal change in biotic responses to initiatives like planting and enhanced management of woodland remnants. For example, how does the value of planted native vegetation change over time? Most work on planting-age effects has been cross-sectional and based on space-for-time contrasts in the use by biota of plantings of different ages. These kinds of studies are valuable but also have substantial limitations because of factors like the effects of site history and past land use (Pickett 1989). Answers to questions about longitudinal changes in the values of plantings will be important for temporal planning of vegetation assets on farms and in landscapes.

Another knowledge gap revolves around the current limited understanding of relationships between vegetation condition and biodiversity. Condition metrics are now widely used in many parts of Australia, including in areas dominated by temperate woodland, but how different components of woodland biota (birds, reptiles, etc.) respond to woodland in different condition is poorly understood. This lack of understanding is problematic because assessments of vegetation condition underpin initiatives like the Federal Government's Woodland Stewardship Program (Department of Environment 2009), and are the basis of efforts to improve condition through financial incentives to private landholders. A better understanding of relationships between vegetation condition and biotic response is fundamental to assessing patch size and condition trade-offs in woodland management. For example, is a large patch in poor condition a higher priority for conservation than a small one in good condition? Our hope is that we will close these and other knowledge gaps in the coming years and contribute to improved conservation outcomes in the temperate woodland ecosystems of south-eastern Australia.

Acknowledgements

Our temperate woodlands research over the past decade has been possible only because of outstanding statistical collaboration from Ross Cunningham, Emma Knight, Jeff Wood, Alan Welsh and Christine Donnelly. Many organisations and funding schemes have provided financial assistance that has facilitated our ongoing work. These include: Land & Water Australia, the Australian Research Council, Rural Industries Research and Development Corporation, the Natural Heritage Trust, Caring for Our Country, the Kendall Foundation, the Thomas Foundation, the Herman Slade Foundation and the Murray Catchment Management Authority.

Bios

David Lindenmayer is a Research Professor in the Fenner School of Environment and Society at the Australian National University. He has worked in the temperate woodlands of southern New South Wales for more than a decade and published many books and scientific articles on their conservation and management.

Mason Crane started working on temperate woodland research with David Lindenmayer's team in 1999. Growing up in Gundagai, NSW he has developed a passion for biodiversity conservation in rural landscapes and an intimate understanding of rural communities. He has special expertise in woodland ecology and management, and in recent years has focused his research on possum and glider conservation in agricultural landscapes.

Damian Michael is a Senior Research Officer at the Fenner School of Environment and Society. He has been part of David Lindenmayer's research team since 2000, and has special expertise in botany, ecology and herpetology. Damian completed a PhD on the ecology and conservation of reptiles in agricultural landscapes with a focus on managing rocky outcrops.

Rebecca Montague-Drake is a Senior Research Officer at the Fenner School of Environment and Society. She joined David Lindenmayer's research team in 2005, and is based in Gundagai, NSW. She is passionate about working to improve biodiversity conservation in woodland environments. Her particular areas of interest include small mammals and birds.

Christopher MacGregor is a Senior Research Officer at the Fenner School of Environment and Society. He has been working as a wildlife ecologist with David Lindenmayer's research team since 1998. A keen ornithologist and botanist, he has experience in wildlife research on the south coast and South-west Slopes of New South Wales and the central highlands of Victoria.

References

Barrett, G.W., Freudenberger, D., Drew, A., Stol, J., Nicholls, A.O. and Cawsey, E.M. (2008). Colonisation of native tree and shrub plantings by woodland birds in an agricultural landscape. *Wildlife Research* **35**, 19–32.

Barrett, G.W., Silcocks, A., Barry, S., Cunningham, R. and Poulter, R. (2003). *The New Atlas of Australian Birds*. Birds Australia: Melbourne.

Benson, J.S. (2008). New South Wales vegetation classification and assessment: Part 2 Plant communities of the NSW South-western Slopes bioregion and update of the NSW Western Plains plant communities, version 2 of the NSWVCA database. *Cunninghamia* **10**, 599–673.

Burrows, G.E. (1999). A survey of 25 remnant vegetation sites in the south west slopes, New South Wales. *Cunninghamia* **6**, 283–299.

Cunningham, R.B., Lindenmayer, D.B., Crane, M., Michael, D. and McGregor, C. (2007). Reptile and arboreal marsupial response to replanted vegetation in agricultural landscapes. *Ecological Applications* **17**, 609–619.

Cunningham, R.B., Lindenmayer, D.B., McGregor, C., Crane, M. and Michael, D. (2008). The combined effects of remnant vegetation and replanted vegetation on farmland birds. *Conservation Biology* **22**, 742–752.

Department of Environment, Water, Heritage and the Arts (2009). 'Environmental Stewardship Program. Box Gum Grassy Woodland Project.' Department of Environment, Water, Heritage, and the Arts: Canberra.

Fischer, J. and Lindenmayer, D.B. (2002). The conservation value of paddock trees for birds in a variegated landscape in southern New South Wales. 2. Paddock trees as stepping stones. *Biodiversity and Conservation* **11**, 833–849.

Ford, H.A., Barrett, G.W., Saunders, D.A. and Recher, H.F. (2001). Why have birds in the woodlands of southern Australia declined? *Biological Conservation* **97**, 71–88.

Freudenberger, D. and Harvey, J. (2004). Predicting biodiversity benefits of the Saltshaker Project, Boorowa, NSW. *Ecological Management and Restoration* **5**, 5–14.

Gibbons, P. and Boak, M. (2002). The value of paddock trees for regional conservation in an agricultural landscape. *Ecological Management and Restoration* **3**, 205–210.

Lindenmayer, D.B., Cunningham, R., Crane, M., Michael, D. and Montague-Drake, R. (2007). Farmland bird responses to intersecting replanted areas. *Landscape Ecology* **22**, 1555–1562.

Lindenmayer, D.B., Cunningham, R.B., McGregor, C., Crane, M., Michael, D., Montague-Drake, R., Fischer, J., Felton, A. and Manning, A. (2008). The changing nature of bird populations in woodland remnants as a pine plantation emerges: results from a large-scale 'natural experiment' of landscape context effects. *Ecological Monographs* **78**, 567–590.

Lindenmayer, D.B., Knight, E.J., Crane, M.J., Montague-Drake, R., Michael, D.R. and MacGregor, C.I. (2010). What makes an effective restoration planting for woodland birds? *Biological Conservation* **143**, 289–301.

McIntyre, S., Huang, Z. and Smith, A.P. (1993). Patterns of abundance in grassy vegetation of the New England Tablelands: identifying regional rarity in a threatened vegetation type. *Australian Journal of Botany* **41**, 49–64.

McIntyre, S. and Lavorel, S. (1994). How environmental and disturbance influence composition in temperate Australian grasslands. *Journal of Vegetation Science* **5**, 373–384.

Michael, D.R., Cunningham, R. and Lindenmayer, D.B. (2008). A forgotten habitat? Insular granite inselbergs conserve reptile diversity in fragmented landscape in south-eastern Australia. *Journal of Applied Ecology* **45**, 1742–1752.

Michael, D.R. and Lindenmayer, D.B. (2008). Records of the Inland Carpet Python *Morelia spilota metcalfei* (Serpentes: Pythoniade) from the South West Slopes of New South Wales. *Proceedings of the Linnean Society of New South Wales* **129**, 253–261.

Montague-Drake, R., Lindenmayer, D.B. and Cunningham, R.B. (in press). Habitat determinants of site occupancy for woodland bird species of conservation concern. *Biological Conservation.*

Pickett, S.T.A. (1989). Space-for-time substitution as an alternative to long-term studies. In *Longterm Studies in Ecology: Approaches and Alternatives.* (Ed. G.E. Likens) pp. 110–135. Springer-Verlag, New York.

Prober, S.M., Thiele, K.R., Lunt, I.D. and Koen, T.B. (2005). Restoring ecological function in temperate grassy woodlands: manipulating soil nutrients, exotic annuals and native perennial grasses through carbon supplements and spring burns. *Journal of Applied Ecology* **42**, 1073–1085.

Rosenzweig, M.L. (1995). *Species Diversity in Space and Time.* Cambridge University Press: Cambridge.

Spooner, P., Lunt, I. and Robinson, W. (2002). Is fencing enough? The short-term effects of stock exclusion in remnant grassy woodlands in southern NSW. *Ecological Management and Restoration* **3**, 117–126.

Stirzaker, R., Vertessey, R. and Sarre, A., eds. (2002). *Trees, Water and Salt. An Australian Guide to Using Trees for Healthy Catchments and Productive Farms.* Joint Venture Agroforestry Program: Canberra.

22

THE ROAD TO CONSERVING WOODLANDS: RE-AFFIRMING SOME PAST PRESCRIPTIONS AND DEVELOPING THE NEW

Peter G. Spooner

1. Protect the best patches, by minimising current threats to key assets.
2. Expand woodland patches, by exploiting site resilience and promoting natural regeneration.
3. Develop suitable woodland disturbance regimes, such as fire, grazing, thinning, or other management interventions.
4. 'Plug the holes' by restoring degraded patches and then monitoring the outcomes.
5. Implement conservation programs that are congruent with farm operations, economics and infrastructure.
6. Ensure conservation programs are tied to long-term binding covenants.
7. Improve and maintain communication strategies in regards to the 'so what' question about woodlands.

Introduction

For over 10 years I have worked principally on threatened and endangered woodlands in southern NSW (see map on next page), dominated by Yellow Box (*Eucalyptus melliodora*), Grey Box (*E. microcarpa*), White Box (*E. albens*), and co-dominates of White Cypress-pine (*Callitris glaucophylla*), Blakelyi's and River Red-Gums (*E. blakelyi and E. camaldulensis).* My research has focused on (1) assessing the outcomes of fencing grassy woodlands on private land – a popular intervention to arrest woodland decline; and (2) investigating the impacts of past and present disturbance regimes on roadside vegetation, and influences on plant recruitment processes. From this research, I have learnt how past land-use history influences present-day woodland vegetation structures and compositions (Spooner and Lunt 2004; Spooner 2005), and how management actions such as grazing can result in both positive and negative outcomes (Spooner *et al.* 2002; Spooner and Briggs 2008), due to often irreversible changes to ecosystem conditions (Lunt and Spooner 2005; Spooner and Allcock 2006). The following seven points encompass the major lessons I have learnt from my work and experiences in southern NSW.

Lessons

1. Protect the best patches by minimising current threats to key assets

Current threats to key woodland assets need to be minimised. Identification of key assets should be determined based on available data (e.g. remote sensing, roadside vegetation surveys, and condition assessments). But where this data does not exist, local land manager experience is an equally important and vital source of information. Most land managers know where the 'best' woodland patches are: in travelling stock routes and reserves (TSRs), in wide roadside reserves (>40 m; Spooner and Lunt 2004), in cemeteries (Prober and Thiele 1995), or as hidden gems in the 'back paddock' on private lands. Given the state of woodlands in general, it is vital that **we don't stuff these up!** This means preventing further incremental clearing, minimising the removal of fallen timber, reducing fertiliser inputs, and ensuring that management inputs such as grazing are appropriate to the conservation values of the site.

Woodlands on public lands face additional challenges, such as road maintenance and inappropriate weed control activities. To address this problem, programs are urgently needed to: (1) identify, monitor and literally signpost high conservation areas; and (2) educate local workers on the environmental values of key assets. It is also imperative that no further clearing or sale of public lands occur where high conservation patches exist. For example, the present NSW Government's plan to sell off TSRs is nonsensical, given that: (1) many TSRs contain woodlands (e.g. Yellow, White and Grey Box communities) of high conservation status (Hibberd 1978); and (2) the responsibility entrusted to governments to protect listed environmental assets such as woodlands. Such a move is akin to throwing out the crown jewels in a garage sale (Spooner and Smallbone 2008). As these issues highlight, many of the best woodland patches which exist on public lands are still under threat, and so there is ample opportunity for relevant government agencies to implement policies and programs to improve the protection of the condition of woodland remnants in perpetuity.

2. Expand woodland patches, by exploiting site resilience and promoting natural regeneration

A simple and cost-effective way to conserve and manage patches is to facilitate a natural increase in their size. In many areas, woodlands often exist as remnants in small or long, linear road reserves of varying widths which are usually surrounded by a hostile farming matrix. Many such woodland remnants serve as 'time capsules' of conditions from over a century ago (Spooner 2005; Spooner and Smallbone 2009), as they often possess strong site resilience, and can serve as a seed source for encouraging natural regeneration (Gunderson 2000). Therefore, this aim could be easily achieved by placing fences further away from a roadside verge or remnant patch (Venning 1998; Spooner *et al.* 2002), and as necessary, provide inputs such as soil scarification and weed control, to facilitate natural regeneration processes to occur. Instead of engineering costly 'designer' solutions elsewhere (i.e. revegetation programs; Aronson and Andel 2006), this approach has enormous potential for novel, broad-scale programs to enlarge existing areas, to conserve and enhance vital genetic stock of woodland species, and buffer populations from external degrading threats (Hobbs and Saunders 1994).

3. Develop suitable woodland management (disturbance) regimes, such as fire, grazing, thinning, or other management interventions

Should a manager 'lock it up' and leave a woodland patch, or implement a management intervention (e.g. spraying or grazing)? Perhaps burning a site is a consideration? This requires consideration of another question – why do you want to do something? Perhaps there is an obvious problem (such as weed invasion), or perhaps this (need to intervene) is more a human desire to manipulate the trajectory of the site; to fast-forward its improvement (Hilderbrand *et al.* 2005). Historically, woodlands were perturbed by fire, grazing, humans and soil disturbances by native animals (Spooner *et al.* 2004; Eldridge and Kwok 2008). Most present-day woodland remnants on private property, however, now exist as isolated, modified patches, which have different structures and compositions, possess varying levels of site resilience (i.e. state of seed banks), and consequently exist in various altered states (Lunt and Spooner 2005; Spooner and Allcock 2008). As a result, application of 'cookbook' prescriptions (Hilderbrand *et al.* 2005) such as burning can be problematic, and may cause unnecessary stress to an already degraded woodland patch.

Given the changes that have occurred to woodlands and the broader landscape, frequent mowing, slashing, weed control, or some form of stock grazing may be the only practical methods available to the landholder to manage woodlands (Lunt 1991). To develop suitable conservation management prescriptions for woodlands requires understanding of a sites' disturbance history, ecological condition, scale and landscape context. The framing of suitable goals and performance criteria to assess and monitor any proposed intervention(s) is also essential (Lake 2001). For example, grazing by livestock is often considered a threat to woodlands. Grazing by stock may be useful, however, to control biomass and weeds, and promote native plant structure and diversity, if suitable objectives and monitoring criteria are selected for the given site (Spooner *et al.* 2002; Lunt *et al.* 2007).

4. 'Plug the holes', by restoring degraded patches and then monitoring the outcomes

Where woodland conditions are degraded, or connectivity is an issue, it is often necessary to implement restoration treatments to 'plug the holes'. Woodlands support a diverse array of distinctive fauna, and therefore it is critical that species habitat and resource requirements are maintained or even enhanced. For example, past land management activities have often resulted in the clearing of understorey shrubs (historically considered as cover for rabbits), or grazing of tree regeneration, resulting in degraded woodland stands with a sparse native understorey (Spooner *et al.* 2002). The need to restore woodland understoreys (Lunt 1995; Prober and Thiele 2005), and the technical expertise required to implement restoration works, is now well established.

Formal monitoring of restoration project outcomes, however, is often lacking (Spooner and Briggs 2008). This is a real issue, as timeframes for recovery are often in the order of decades, and stretch well beyond those of a project (<3 years). Seasonal differences such as drought and climate change can strongly influence recovery processes (Spooner and Briggs 2008). The establishment and ongoing support of long-term monitoring protocols is therefore critical to ensuring that ecological outcomes are quantified (e.g. regeneration from plantings; changes in flora and fauna attributes), and future management actions refined, as part of an adaptive learning framework.

5. Implement conservation programs that are congruent with farm operations, economics and infrastructure

To manage and conserve remnant woodlands on private lands better, a suite of fencing, revegetation, restoration and other incentive programs have been implemented throughout Australia. In implementing programs, however, whole farm systems need to be considered. Normally uptake of such programs requires landholders to meet some form of selection criteria to obtain funding (e.g. patch condition, location, vegetation type or area, agreed management inputs). Recent programs (e.g. Box Gum Grassy Woodland Project) tend to place greater priority on ungrazed woodland patches – the 'best' areas, where landholders are required to enter long-term covenant agreements to exclude stock from the target site (Caring for our Country 2009).

In a situation akin to 'not seeing the farm for the trees', however, problems can arise when too little thought is given to the impacts of the program intervention on the whole farm system, which in turn, can put the programs success at risk. For example, fencing to exclude stock can result in a desired improvement of ecological conditions in the target woodland patch, but deterioration elsewhere on the farm, due to an associated increase in stocking levels to maintain farming income (Spooner and Briggs 2008). The indirect effects of applying woodland conservation programs can also lead to unforeseen problems with farm management cycles (e.g. crop and grazing rotation), leading to difficulties in feeding stock, and then potential inability to meet the grazing aims for the woodland site. For better outcomes of woodland conservation agreements, project managers need to consider available infrastructure (i.e. farm-scale paddock and fencing design, water points), prevailing economics (i.e. stock and cropping type, rotation cycles), and most importantly, the motivations of the individual landholder in seeking funding (Pannell *et al.* 2006).

6. Ensure conservation programs are tied to long-term binding covenants

Much has been achieved in woodland management and restoration in recent years, which should be acknowledged. To date, various conservation programs aimed at private lands have been developed, which do not require a binding covenant agreement (Binning 1997). Although this approach has merit in terms of uptake and education, a main concern is the risk posed to this investment when farms are sold, and management subsequently changes (Driver and Davidson 2002; Spooner *et al.* 2002). For example, over recent years I have witnessed a number of properties where much revegetation work has been carried out, only to then be sold, and the revegetation area cleared or grazed by the new owner. This risk may increase as a growing number of farmers seek to retire from the land. To ensure that taxpayers' money is being well spent, it is imperative that most, if not all, future programs be tied to some form of legal binding agreement (however, see Lesson #5 above); that is, held on title for a considerable time, or in perpetuity.

7. Improve and maintain communication strategies in regards to the 'so what' question about woodlands

Why conserve woodlands? As evidenced by this book, the reasons are many; however, evidence suggests that these messages are not necessarily getting through to all land managers – such as the farmers who don't involve themselves in conservation programs. Of concern is that less money is being spent on direct forms of conservation management communication with landholders, and more on program outputs (e.g. area fenced, number of plants established), presumably on the basis that communication goals have already been met. A website does not suffice for a good, passionate on-ground conservation or Landcare officer (Mendham *et al.* 2007; Kelly *et al.* 2009). Many farmers and townspeople still do not understand why we as conservation managers or ecologists, advocate for the retention of native animals, fallen timber or vegetation on farms. This is difficult, as conservation management is still viewed by some as a threat to farming livelihoods (Moore and Renton 2002).

For landholders to adopt and apply woodland conservation practices on farms, there is a critical need to communicate clear and demonstrative advantage (i.e. economically or socially) over current approaches (Pannell *et al.* 2006). Therefore, we need to refocus our messages in terms of protecting remnant woodlands to safeguard farming systems for future generations, particularly in response to more recent threats such as climate change. Research and communication towards quantifying these values in terms of ecosystem services (e.g. pollination services, pest control) may be the key.

Conclusions

Woodlands are threatened or endangered because the soils and climate zones in which they occur are ideal for agriculture or human settlement. Early settlers knew that the presence of Yellow Box trees was a good indicator of healthy, fertile soils, and promptly set about clearing the land for cropping and grazing. As a result, remaining woodland vegetation often exists on hilltops with poorer soils considered inferior for farming, hard to access areas, or in small TSR reserve and roadside remnants. Throughout the remainder of the landscape, small remnants and isolated paddock trees are a stark reminder of once extensive tracts of woodlands, whose declining health conveys a pressing message of landscape decline (Gibbons *et al.* 2008).

Much knowledge and understanding has now been accumulated on woodland ecology and restoration practice. We need to continue our efforts to improve the quality, extent and connectivity of woodland remnants, however, and underpin on-ground works with credible long-term monitoring to evaluate their outcomes. The conservation value of vital assets such as roadsides, TSRs and other woodland patches should be identified and their condition assessed, as part of a regular process to set priorities for conservation and management. Further investment is required to develop woodland management regimes (e.g. grazing) that are congruent with whole farm planning by the individual landholder, but able to be generalised and transferred elsewhere. Here we must work with the agricultural industry, not separate from it, to learn more about conservation farming practices, and develop novel management solutions.

Acknowledgements

My work has been supported by various agencies such as the NSW Environment Trust, NSW Department of Environment and Climate Change (DECC), CSIRO, Greening Australia, and more recently the Australian Research Council. I am indebted to Ian Lunt and many other colleagues for their guidance and support. Most importantly, a hearty thank-you to the many landholders in southern NSW who have assisted in my research.

Bio

Peter Spooner is a Senior Lecturer in Ecology at Charles Sturt University. He grew up in rural areas of New South Wales, and recalls fond childhood memories of long country drives, admiring the gorgeous colours and native species of woodlands in western NSW. As a result, he has a strong desire to conduct research aimed at improving our understanding, and conservation management of woodland remnants. His research interests include the disturbance and restoration ecology of temperate ecosystems, and road ecology, with a specific interest in research of TSR and roadside vegetation patterns and processes in Australia.

References

Aronson, J. and van Andel, J. (2006). Challenges for ecological theory. In *Restoration Ecology: The New Frontier.* (Eds J. van Andel and J. Aronson) pp. 223–233. Blackwell Publishing: Oxford.

Binning, C.E. (1997). Beyond reserves: options for achieving nature conservation objectives in rural landscapes. In *Frontiers in Ecology: Building the Links.* (Eds N. Klomp and I. D. Lunt) pp. 155–168. Elsevier Science: Oxford.

Caring for our Country (2009). 'Environmental Stewardship – Box Gum Grassy Woodland Project'. Australian Government NRM Team: Canberra. http://www.nrm.gov.au/stewardship/box-gum/lachlan-murrumbidgee-cw.html#strategy

Driver, M. and Davidson, I. (2002). Deposit for the future: remnant vegetation management through community cost-sharing. *Ecological Management & Restoration* **3**, 81–89.

Eldridge, D.J. and Kwok, A.B.C. (2008). Soil disturbance by animals at varying spatial scales in a semi-arid Australian woodland. *Rangeland Journal* **30,** 327–337.

Gibbons, P., Lindenmayer, D.B., Fischer, J., Manning, A.D., Weinberg, A., Seddon, J., Ryan, P. and Barrett, G. (2008). The future of scattered trees in agricultural landscapes. *Conservation Biology* **22,** 1309–1319.

Gunderson, L.H. (2000). Ecological resilience – in theory and application. *Annual Review of Ecology and Systematics* **31**, 425–440.

Hibberd, J.K. (1978). 'The future of the long paddock: a study of travelling stock reserves, routes and roadside verges in the Southern Tablelands of NSW'. Nature Conservation Council of NSW: Sydney.

Hilderbrand, R.H., Watts, A.C. and Rondle, A.M. (2005). The myths of restoration ecology. *Ecology and Society* **10**, 46–56.

Hobbs, R.J. and Saunders, D.A. (1994). Effects of landscape fragmentation in agricultural areas. In *Conservation Biology in Australia and Oceania Moritz.* (Eds C. Moritz and J. Kikkawa.) pp. 77–95. Surrey Beatty & Sons: Chipping Norton, NSW.

Kelly, B., Allan, C. and Wilson, B.P. (2009). Soil indicators and their use by farmers in the Billabong Catchment, southern New South Wales. *Australian Journal of Soil Research* **47**, 1–9.

Lake, P.S. (2001). On the maturing of restoration: linking ecological research and restoration. *Ecological Management and Restoration* **2**, 110–115.
Lunt, I.D. (1991). Management of remnant lowland grasslands and grassy woodlands for nature conservation: a review. *Victorian Naturalist* **108**, 56–66.
Lunt, I.D. (1995). European management of remnant grassy forests and woodlands in south-eastern Australia – past, present and future? *Victorian Naturalist* **112**, 239–249.
Lunt, I.D. and Spooner, P.G. (2005). Using historical ecology to understand patterns of biodiversity in fragmented agricultural landscapes. *Journal of Biogeography* **32**, 1859–1873.
Lunt, I.D., Eldridge, D.J., Morgan, J.W. and Witt, G.B. (2007). A framework to predict the effects of livestock grazing and grazing exclusion on conservation values in natural ecosystems in Australia. *Australian Journal of Botany* **55**, 401–415.
Mendham, E., Millar, J. and Curtis, A. (2007). Landholder participation in native vegetation management in irrigation areas. *Ecological Management & Restoration* **8**, 42–48.
Moore, S.A. and Renton, S. (2002). Remnant vegetation, landholders' values and information needs: An exploratory study in the West Australian wheatbelt. *Ecological Management & Restoration* **3**, 179–187.
Pannell, D.J., Marshall, G.R., Barr, N., Curtis, A., Vanclay, F. and Wilkinson, R. (2006). Understanding and promoting adoption of conservation practices by rural landholders. *Australian Journal of Experimental Agriculture* **46**, 1407–1424.
Prober, S.M. and Thiele, K.R. (1995). Conservation of the grassy white box woodlands: relative contributions of size and disturbance to floristic composition and diversity of remnants. *Australian Journal of Botany* **43**, 349–366.
Prober, S.M. and Thiele, K.R. (2005). Restoring Australia's temperate grasslands and grassy woodlands: integrating function and diversity. *Ecological Management and Restoration* **6**, 16–27.
Spooner, P.G. (2005). On squatters, settlers and early surveyors: historical development of road reserves in southern New South Wales. *Australian Geographer* **36**, 55–73.
Spooner, P.G. and Allcock, K.G. (2006). Using a state-and-transition approach to manage endangered *Eucalyptus albens* (White Box) woodlands. *Environmental Management* **38**, 771–783.
Spooner P.G. and Briggs, S.V. (2008). Woodlands on farms in southern NSW: a longer-term assessment of vegetation changes after fencing. *Ecological Management and Restoration* **9**, 33–41.
Spooner, P.G. and Lunt, I.D. (2004). The influence of land-use history on roadside conservation values in an Australian agricultural landscape. *Australian Journal of Botany* **52**, 445–458.
Spooner, P.G. and Smallbone, L. (2008). Effects of road age on the structure of roadside vegetation in south-eastern Australia. *Agriculture, Ecosystems & Environment* **129**, 57–64.
Spooner, P.G., Lunt, I.D. and Robinson, W. (2002). Is fencing enough? The short-term effects of stock exclusion in remnant grassy woodlands in southern NSW. *Ecological Management and Restoration* **3**, 117–126.
Spooner, P.G., Lunt, I.D., Briggs, S.V. and Freudenberger, D. (2004). Effects of soil disturbance from roadworks on roadside shrubs in a fragmented agricultural landscape. *Biological Conservation* **117**, 393–406.
Venning, J. (1988). *Growing Trees for Farms, Parks and Roadsides*. Lothian Publishing: Port Melbourne.

23

WOODLAND BIODIVERSITY CONSERVATION: BASKET-CASE OR BATTLEGROUND? INSIGHTS FROM THE MALLEE

Don A. Driscoll

1. Farming in extensively cleared landscapes is not ecologically sustainable because one-fifth to one-third of animal species is on a declining trajectory. Keeping all remnant vegetation and restoring substantial areas of native vegetation is needed to halt declining trajectories and stem the loss of biodiversity.
2. Narrow linear remnants provide important habitat for reptiles and beetles, but their value is altered by livestock grazing and when adjacent to roads.
3. Large, regularly spaced reserves are needed for effective connectivity, to account for species turnover and to accommodate intraspecific genetic variation. Biodiversity conservation depends on large reserves.
4. Some species specialise in using the same habitat that is targeted for land clearing. Restoration of under-represented communities would help rebuild the balance.
5. Fire exclusion from remnant mallee woodland vegetation has the potential to drive species towards extinction, but so could frequent fire.
6. Reptile and beetle species with limited dispersal may be those most vulnerable to decline in fragmented landscapes, but a combination of traits is likely to be important.
7. Beware of assumptions about how populations of species interact with the spatial arrangement of habitat. Metapopulations may be rare.

Introduction

Mallee woodlands continue to provide a substantial focus of my research into habitat loss, fragmentation, and fire ecology. Mallee straddles the boundary of the Australian wheatbelt and the semi-arid rangelands, and so occurs in some large intact regions as well as small remnants in agricultural landscapes (see map on next page). This combination provides the opportunity to compare intact and fragmented communities; a rare opportunity among Australian temperate woodland types.

In central New South Wales, I examined the impacts of the size, shape and condition of mallee woodland remnants on reptiles and beetles (Driscoll 2004; Driscoll and Hardy 2005; Driscoll and Weir 2005). On the Eyre Peninsula in South Australia, in collaboration with many colleagues, I have examined the responses of a range of taxa to fire, and to habitat loss and fragmentation (Driscoll and Henderson 2008; Schutz and Driscoll 2008). Reptiles and beetles are an important focus of this work. Mallee woodland harbours a diverse reptile fauna with approximately 40 species found within each of the two main study regions. Beetles are diverse in almost every habitat and can be trapped at the same time as reptiles. With a large number of species available for study, these groups provide excellent opportunities to examine the relationship between species' traits and species' responses to disturbance and for exploring the importance of theoretical models of survival in fragmented landscapes, particularly metapopulations. Many lessons (and many more questions) are emerging from our mallee research, and some of the lessons will resonate with researchers and land managers working throughout Australia's temperate woodlands.

Lessons

1. Farming in extensively cleared landscapes is not ecologically sustainable because one-fifth to one-third of animal species is on a declining trajectory. Keeping all remnant vegetation and restoring substantial areas of native vegetation are needed to halt declining trajectories and stem the loss of biodiversity

The extent of the biodiversity crisis is one of the most striking lessons emerging from my experience with temperate woodland fauna. Many mallee communities have been widely cleared for agriculture, and commonly around 10% of the original woodland remains. In extensively cleared woodlands, usually 20–30% of animal species can be classified as declining, with the trend evident in birds, mammals, reptiles and beetles (Driscoll and Weir 2005). Although feral predators have caused or contributed to declines of mammals, some birds and possibly reptiles (Driscoll 2004), the overriding cause of species declines is habitat loss and degradation. Declines are ongoing as we repay the extinction debt associated with over clearing (Tilman *et al.* 1994). We might avoid paying back the extinction debt by stopping all land clearing and by rebuilding habitat; however, as long as our response to the plight of biodiversity in remnant woodlands is piecemeal land clearing and limited revegetation, we have effectively chosen the option of triage, with 20–30% of woodland species falling into the 'as good as dead' category. 'Business as usual' will not suffice to bring about the restoration that is needed in mallee woodlands (Duncan and Dorrough 2009).

2. Narrow linear remnants provide important habitat for reptiles and beetles, but their value is altered by livestock grazing and when adjacent to roads

Most agricultural regions carved from former mallee communities now support many linear remnants. Most of these remnants range in width from 10 m to 200 m and extend around paddocks, along roadsides, and in some regions, along linear sand dunes. Some widespread reptile species and many beetles that cannot use cleared land, use linear remnants. Linear mallee remnants therefore have substantial conservation value. Remnant condition is important, however. I found that some reptile species were excluded from remnants that were grazed or beside roads and others had reduced abundance. Interestingly, the impact of grazing was lower for some reptiles when the prickly clumping grass *Triodia* was present. *Triodia* is unpalatable to sheep and probably provides important ungrazed and untrampled refuges for reptiles. Condition also influenced landscape-scale connectivity. Grazing and altered vegetation on roadsides meant that fragmentation was increased for some reptile species by rendering remnants uninhabitable (Driscoll 2004). Although linear remnants are likely to be an important part of large-scale connectivity programs, carefully managing habitat quality within linear remnants will enable a broader range of species to benefit (Recher 2003; Schutz and Driscoll 2008).

3. Large, regularly spaced reserves are needed for effective connectivity, to account for species turnover and to accommodate intraspecific genetic variation. Biodiversity conservation depends on large reserves

Unfortunately, connectivity goals may not be achieved using linear remnants alone because some reptile and beetle species cannot use them. Some species have survived for 30 years since clearing, in small reserves of a few square kilometres. Two reptile species (Painted Dragon *Ctenophorus pictus*, Hooded Scaly-foot *Pygopus nigriceps*), however, may have declined from the entire remnant landscape in NSW (Driscoll 2004). The presence of these two species in large uncleared mallee regions suggest that reserves larger than a few square kilometres are needed. There are substantial reptile community differences between locations separated by just tens of kilometres. These differences in the species present or their relative abundance are probably associated with historic patterns of range changes and evolution, and responses to local environments (Driscoll and Henderson 2008). This spatial diversity emphasises the importance of having many reserves spaced through the landscape. Regularly spaced reserves are also needed to conserve the genetic variation of species. Genetic analysis of the Nobbi Dragon *Amphibolurus nobbi* suggests that substantial genetic differentiation occurs at a scale of a few tens of kilometres (Driscoll and Hardy 2005).

Proximity to a large reserve may also influence the ability of species to occupy linear remnants. On the Eyre Peninsula, we have found that common reptiles can occupy remnants within a kilometre or two of large reserves even when the remnants are not directly connected to the reserve (Schutz and Driscoll 2008). Recent research (Williams *et al.* unpublished data), however, suggests that some common species are absent from linear remnants that were between six and ten kilometres from large reserves. It seems possible that the linear remnants along the crests of sand dunes could act as stepping stones within a few kilometres of a large reserve. Blocks of habitat of at least a few square kilometres may therefore need to be separated by less than a few kilometres for the dune-top remnants to act as stepping stones, and not just population sinks.

4. Some species specialise in using the same habitat that is targeted for land clearing. Restoration of under-represented communities would help rebuild the balance

Throughout Australian woodlands, the least productive areas are those most likely to remain uncleared (Yates and Hobbs 1997). In mallee regions with linear sand dunes, the common pattern of clearing has been to retain vegetation on dune-tops but to clear the agriculturally productive swale habitat. This pattern of clearing can substantially affect reptile communities because many reptiles are specialised to use either swales or dunes (Pianka 1969). We suspect that movement of swale-preferring reptiles onto adjacent dunes increased the number of occasionally trapped species on dunes in large reserves. In the isolated dune-tops, with the adjacent swale cleared, there were fewer occasionally trapped species (Schutz and Driscoll 2008). Preferential clearing of swales is likely to influence swale-specialist species from a range of taxa (Leys and Eldridge 1998; Pate and Verboom 2009). Retaining or restoring swale areas may therefore have disproportionate biodiversity benefits, although the cost of foregone production may also be high.

5. Fire exclusion from remnant mallee woodland vegetation has the potential to drive species towards extinction, but so could frequent fire

Although we have found little evidence that ground-dwelling beetles are much affected by fire, fire has a strong influence on most other groups in mallee woodlands. A single fire provides open habitat that is favoured by several reptile and bird species and stimulates germination of plant species that may have been present in the community only as soil-stored seed (unpublished data, Driscoll and Henderson 2008). The same fire may remove leaf litter, shrubs and old trees with hollows to the disadvantage of some reptile, mammal and bird species (e.g. Brown *et al.* 2009). Some plant species that are killed by fire and must regenerate from seed, such as *Callitris* species, are particularly vulnerable to decline if the interval between fires is too short (less than 15 years for *Callitris*, Bradstock and Cohn 2002). On the other hand, this same group of species can decline if the fire interval is too long because plants senesce and canopy-stored seed declines (Bradstock *et al.* 1996). We suspect there could be an interaction of fire and habitat fragmentation for species that need to recolonise areas at some time during the fire cycle, but evidence is scant (Driscoll and Henderson 2008). Understanding the risk of extinction under different fire intervals, and the capacity of species to recolonise after extinction through cleared land and intact mallee remain critical knowledge gaps.

6. Reptile and beetle species with limited dispersal may be those most vulnerable to decline in fragmented landscapes, but a combination of traits is likely to be important

Discovering the characteristics of species that are vulnerable to decline after disturbance has important applications in conservation. It can help identify species before they decline, allowing the most appropriate species to be monitored. It can also help quantify the importance of the disturbance in terms of numbers of species that are affected. Knowledge of reptile and beetle ecology is limited, making it difficult to relate species traits to risk of decline. Nevertheless, in mallee woodlands, there is some evidence that reptiles with limited dispersal capacity may be most vulnerable to decline (Driscoll 2004). Dispersal capacity was also a key trait predicting success of beetle species (Driscoll and Weir 2005). Poorly dispersing reptiles and beetles were more likely to be confined to the largest sites in best condition, implying decline from other remnants. For beetles, the stratum of the landscape used by adults was also important, with species living underground more likely to decline compared with those that primarily live on or above ground. The risk of decline of individual species, however, was more complex than just dispersal or the stratum occupied. Combinations of traits seemed to influence how beetles responded to remnant size, shape or condition. There remains substantial uncertainty as to how important particular traits are for determining extinction risk in modified landscapes. We are not yet at a stage of narrowing the conservation focus to poorly dispersing or underground species and more focus on the mechanisms linking species' traits to disturbance impacts is needed (Norris 2008).

7. Beware of assumptions about how populations of species interact with the spatial arrangement of habitat. Metapopulations may be rare

Although patches of habitat embedded in an apparently hostile matrix fit the basic assumption of metapopulation theory (Hanski 1998), there are several conditions that must be satisfied for a metapopulation to occur. Importantly, the matrix must be hostile, and not harbour established populations. This seems to be true for most reptiles, although deep sands may ameliorate the affect (Driscoll 2004; Schutz and Driscoll 2008). In contrast, approximately one-quarter of beetle species were most abundant in the paddocks in western NSW. An intermediate dispersal capacity is also required for a metapopulation, but dispersal capacity is very poorly understood. Winged beetle species can be extremely widely dispersing and so may be able to select and occupy all favourable habitat patches (patchy metapopulation, Harrison 1991). The other extreme seems to be exhibited by several flightless species and some reptiles, with patterns of decline associated with poor dispersal and not self-sustaining metapopulations. Lessons about the influence of remnant condition on species occurrence patterns (see Lesson#2 above) provide an important warning that distinguishing habitat from non-habitat requires a species-level analysis of the use of landscape elements. Measuring dispersal and habitat occupancy remains an important first step in assessing how a species uses a landscape. Assumptions based on ecological theory may apply to very few species in the landscape (Driscoll *et al.* 2010).

Conclusions

My strongest impression from working in Australia's woodlands is that a vast amount of biodiversity is currently being lost, and that large-scale emergency intervention is needed, particularly revegetation. Linear vegetation is valuable if carefully managed, but our research hints at the importance of large patches of habitat spaced at regular intervals across the landscape. Given the imperative to reconstruct habitat, there are two critical implications for public policy.

1. Have the triage debate. Should we regard conservation in remnant woodlands as a basket-case, or should society, or perhaps the industries that benefit from land clearing, fund the

battle to halt species declines? Current funding of revegetation implies a basket-case decision in most agricultural regions, despite policy brochures proclaiming the opposite. The rate of biodiversity loss and likely ultimate extent of loss need to be widely understood, and weighed up against the cost of large-scale revegetation, and the possible benefits of alternative use of those funds.
2. If the policy decision is to fight to conserve woodland biodiversity, then investment in social, policy and ecological research becomes a priority so that land can be provided for revegetation, methods of revegetation are developed, and the human community is broadly supportive of biodiversity recovery.

In addition to developing revegetation methods, another important research goal is to link the characteristics of species with their response to habitat loss and fragmentation, and to fire. A stronger understanding of dispersal may be helpful, and is currently a major knowledge gap. If we can discover the mechanisms underlying species' responses, it may be possible to predict the outcome of new scenarios, including revegetation and climate change scenarios. A capacity to compare scenarios will be an important tool in making management decisions because field-based revegetation experiments will take several decades to complete. Defining a range of best-bet actions using sound mechanistic models would provide a basis on which to implement large-scale and long-term revegetation experiments.

Acknowledgements

Thanks to CSIRO Sustainable Ecosystems, Flinders University, Department of Environment and Heritage SA and Australia and Pacific Science Foundation for funding my mallee projects. Great thanks to the many volunteers who helped collect the reptile and beetle data, particularly Ian Sellar, Annabel Smith and Samantha Blight.

Bio

After an undergraduate science degree at Melbourne University and Honours at the Australian National University, Don Driscoll studied frog conservation and genetics at the University of Western Australia for his PhD. He worked as a threatened species officer in Brisbane, and then was granted a post-doctoral research position with CSIRO in Canberra. He undertook an ARC Post-doctorate at the University of Tasmania, examining beetles in rainforest patches, then led a Masters in Biodiversity course at Flinders University, Adelaide. Currently at ANU, he collaborates with David Lindenmayer as they challenge ecological theory with actual data, in addition to ongoing woodland research.

Mallee Fire Website: http://fennerschool-research.anu.edu.au/malleefire/

References

Bradstock, R.A., Bedward, M., Scott, J. and Keith, D.A. (1996). Simulation of the effect of spatial and temporal variation in fire regimes on the population viability of a Banksia species. *Conservation Biology* **10**, 776–784.

Bradstock, R.A. and Cohn, J.S. (2002). Demographic characteristics of mallee pine (*Callitris verrucosa*) in fire-prone mallee communities of central New South Wales. *Australian Journal of Botany* **50**, 653–665.

Brown, S., Clarke, M. and Clarke, R. (2009). Fire is a key element in the landscape-scale habitat requirements and global population status of a threatened bird: the Mallee Emu-wren (*Stipiturus mallee*). *Biological Conservation* **142**, 432–445.
Driscoll, D.A. (2004). Extinction and outbreaks accompany fragmentation of a reptile community. *Ecological Applications* **14**, 220–240.
Driscoll, D.A. and Hardy, C.M. (2005). Dispersal and phylogeography of the agamid lizard *Amphibolurus nobbi* in fragmented and continuous habitat. *Molecular Ecology* **14**, 1613–1629.
Driscoll, D.A. and Henderson, M.K. (2008). How many common reptile species are fire specialists? A replicated natural experiment highlights the predictive weakness of a fire succession model. *Biological Conservation* **141**, 460–471.
Driscoll, D.A., Kirkpatrick, J.B., McQuillan, P.B. and Bonham, K. (2010). Classic metapopulations are rare among beetle species from a naturally fragmented landscape. *Journal of Animal Ecology* **79**, 294–303.
Driscoll, D.A. and Weir, T. (2005). Beetle responses to habitat fragmentation depend on ecological traits, remnant condition and shape. *Conservation Biology* **19**, 182–194.
Duncan, D.H. and Dorrough, J.W. (2009). Historical and current land use shape landscape restoration options in the Australian wheat and sheep farming zone. *Landscape and Urban Planning* **91**, 124–132.
Hanski, I. (1998). Metapopulation dynamics. *Nature* **396**, 41–49.
Harrison, S. (1991). Local extinction in a metapopulation context: an empirical evaluation. *Biological Journal of the Linnean Society* **42**, 73–88.
Leys, J.F. and Eldridge, D.J. (1998). Influence of cryptogamic crust disturbance to wind erosion on sand and loam rangeland soils. *Earth Surface Processes and Landforms* **23**, 963–974.
Norris, K. (2008). Agriculture and biodiversity conservation: opportunity knocks. *Conservation Letters* **1**, 2–11.
Pate, J.S. and Verboom, W.H. (2009). Contemporary biogenic formation of clay pavements by eucalypts: further support for the phytotarium concept. *Annals of Botany* **103**, 673–685.
Pianka, E.R. (1969). Habitat specificity, speciation, and species density in Australian desert lizards. *Ecology* **50**, 498.
Recher, H.R. (2003). WildCountry. *Pacific Conservation Biology* **8**, 221–222.
Schutz, A.J. and Driscoll, D.A. (2008). Common reptiles unaffected by connectivity or condition in a fragmented farming landscape. *Austral Ecology* **33**, 641–652.
Tilman, D., May, R.M., Lehman, C.L. and Nowak, M.A. (1994). Habitat destruction and the extinction debt. *Nature* **371**, 65–66.
Yates, C.J. and Hobbs, R.J. (1997). Temperate eucalypt woodlands: a review of their status, processes threatening their persistence and techniques for restoration. *Australian Journal of Botany* **45**, 949–973.

24

FRAGMENTATION RESPONSES OF BIRDS, INSECTS, SPIDERS AND GENES: DIVERSE LESSONS FOR WOODLAND CONSERVATION

Richard E. Major

1. The Noisy Miner controls communities. In terms of bird diversity, anthropogenically induced increases in Noisy Miner activity provide the simplest explanation for differences in the composition of bird communities.
2. All animals are unequal. Vertebrates and invertebrates have very different responses to anthropogenic habitat fragmentation such that small remnants with depauperate bird communities can have very rich invertebrate faunas.
3. Vertebrates have teeth. Despite representing a tiny fraction of biodiversity, vertebrates operate at scales compatible with human comprehension and will continue to dominate public policy.
4. Habitat defies mapping. Variation in fine-scale habitat characteristics means that the number of hectares of different mapped vegetation types provides a poor surrogate for the distribution of biodiversity.
5. Corridors lead to sinks. Few 'corridors' link viable habitats but they increase the amount of over-represented edge habitat. Increasing the viability of remnants through fine-scale habitat improvement reduces the extinction risk and the associated need for linkages.
6. History happens. Much of the present-day distribution of animals in the landscape is a consequence of historical accidents: chance dispersal, habitat loss and extinction.
7. Understand process to protect pattern. Most research in fragmented woodlands has described snapshots of the distribution of biodiversity which does not necessarily reflect the processes that promote long-term survival.

Introduction

White Cypress Pine woodlands (*Callitris glaucophylla*) in combination with box woodlands (*Eucalyptus* spp.) are prominent in landscapes of central-western New South Wales (see map on next page) in the vicinity of the towns of Condobolin, West Wyalong, Grenfell and Forbes (Sivertsen and Metcalfe 1995). This area is one of the oldest and most heavily modified

agricultural regions in Australia (Goldney and Bowie 1990), chiefly producing sheep, wheat and more recently, canola. Intensive clearing beginning in the 1860s has left the remaining 16% of native vegetation in a highly fragmented state. Remnants differ greatly in size, shape and habitat quality, with most of the native vegetation occurring in State Forests, road reserves, travelling stock routes and rocky ridges on private land.

My research has concentrated on the effects of fragmentation at a landscape scale, considering community responses (birds, insects and spiders), species responses (the Red-capped Robin *Petroica goodenovii*), and gene flow (ants and wolf spiders) (Table 24.1). A high proportion of remnants (in terms of number, not area) in these landscapes are linear in shape, and linear additions have predominated in conservation work via tree-planting programs. Accordingly, understanding the use of linear habitats by a range of fauna has been a main research focus. The different responses of vertebrates and invertebrates means that no unitary solutions can be prescribed to the problems associated with habitat fragmentation, but seven messages that are not necessarily intuitive emerge from this research.

Lessons

1. The Noisy Miner controls communities. In terms of bird diversity, anthropogenically induced increases in Noisy Miner activity provide the simplest explanation for differences in the composition of bird communities

The presence of the aggressive, colonial Noisy Miner (*Manorina melanocephala*) provides the simplest explanation for the patterns of bird distribution observed in our studies (Major *et al.* 2001). Remnants that were home to the Noisy Miner generally lacked small insectivores and honeyeaters, the species that are responsible for much of woodland bird diversity but which tend to be declining. The Noisy Miner preferred vegetation types with an open eucalypt canopy

Table 24.1 Summary of research results in the box/cypress pine woodlands of central-western New South Wales.

Study organism	Attribute measured	Response in small/linear remnants	Reference
Terrestrial birds	Community composition	Fewer species of small birds Noisy Miners more likely	Major *et al.* 2001
Red-capped Robins	Density Age structure	Lower Younger	Major *et al.* 1999b
	Nest-predation	Higher	Major *et al.* 1999a
	Adult survival	No difference but uniformly low with little statistical power	Major and Gowing 2001
Wolf spiders	Community composition	No difference Related to microhabitat	Major *et al.* 2006
	Edge response	No difference despite community differences between woodland and matrix	Martin and Major 2001
	Population genetics	No loss of genetic variability due to recent fragmentation in both large and small/linear remnants Genetic structuring over small spatial scales in both large and small/linear remnants	Colgan *et al.* 2002
Insects	Community composition	Higher abundance and species richness	Major *et al.* 2003; Major *et al.* 1999c
	Population genetics	Three populations from either large or small/linear remnants necessary to capture 75% of genetic variability in small area	Macaranas *et al.* 2000

and little understory (e.g. open box woodlands), particularly small and/or linear remnants, which tend to be more degraded. They were less common in remnants that were greater than 100 ha in area. Retention of large remnants with a diversity of vegetation layers is essential for protecting bird diversity.

2. All animals are unequal. Vertebrates and invertebrates have very different responses to anthropogenic habitat fragmentation such that small remnants with depauperate bird communities can have very rich invertebrate faunas

No generality could be found in the responses of birds and invertebrates to fragmented habitat. Remnant size was a key determinant of bird species composition, with declining woodland birds mostly absent from remnants smaller than 200 ha, which tended to lack interior habitat (Major *et al.* 2001). In contrast, both ground insects (Major *et al.* 1999c) and foliage insects (Major *et al.* 2003) were more abundant and diverse in the edge habitats provided by small and linear remnants. Similarly, wolf spiders (Lycosidae – Figure 24.1) showed very little edge response (Martin and Major 2001) such that there was no difference in wolf spider communities associated with remnant size, even though there were highly distinctive spider communities in woodlands compared with the agricultural matrix (Major *et al.* 2006). Despite surviving in small and linear remnants, and therefore exhibiting tolerance to fragmentation, one species of

Figure 24.1 Wolf spiders (Lycosidae) are active ground predators that exhibited significant genetic structuring over small spatial scales.

woodland-dependent wolf spider showed significant genetic structuring of populations across the landscape (Colgan *et al.* 2002) indicating turnover in biodiversity over small distances. Fine-scale variation in genetic diversity was also observed among ants (Macaranas *et al.* 2000), with a minimum of three habitat patches required to represent three-quarters of the genetic diversity present in a 20 × 40 km area. To conserve all biodiversity, it is therefore necessary to conserve multiple patches over smaller spatial scales to conserve ground-living invertebrates and large patches to conserve birds. The conservation value of small and linear remnants should not be dismissed, even if they tend to have lower value for vertebrate conservation.

3. Vertebrates have teeth. Despite representing a tiny fraction of biodiversity, vertebrates operate at scales compatible with human comprehension and will continue to dominate public policy

One curious farmer who allowed us to work in his small, remote remnant, later told us that following our trip he had gone out one night to look for wolf-spider eye-shines to check whether we had been having him on. The experience inspired him to retain the remnant that he had intended to 'tidy up'. This was an exception; with the bird component of our research generating far more interest from landholders, funding bodies, conservation groups and government authorities. We began by thinking that researching the neglected fauna (invertebrates) should be a priority, but finished with the realisation that they are neglected for two reasons. Their variability at fine spatial scales makes for labour-intensive study when trying to draw conclusions at the scale of human intervention. More importantly, they appear to be handicapped when competing for public attention. If the ultimate aim of research in conservation biology is to enhance conservation actions, the research results must connect with people. At present, birds make more cost-effective vehicles for research than invertebrates.

4. Habitat defies mapping. Variation in fine-scale habitat characteristics means that the number of hectares of different mapped vegetation types provides a poor surrogate for the distribution of biodiversity

Vegetation maps based on aerial photographs were used to stratify our survey designs in terms of vegetation type and remnant size (Major *et al.* 2001); however, measurement of on-ground habitat characteristics demonstrated that mapped vegetation could not be used as a surrogate for biodiversity. There was considerable variation in habitat features such as rocks, logs, leaf litter and coarse woody debris, which was subject to the idiosyncrasies of the local land manager (Major *et al.* 2006). In general, remnants on private land were characterised by canopy and grassy understory, with low densities of shrubs, logs, coarse woody debris and leaf litter. Particular animal species (both spiders and birds) were associated with these features (Seddon *et al.* 2003; Major *et al.* 2006) which explained their distribution among particular remnants. In general, remnant attributes (e.g. vegetation type, remnant area) were more important than landscape attributes (e.g. distance to nearest large patch, area of vegetation in surrounding landscape) in determining the composition of bird communities (Major *et al.* 2001). There is considerable opportunity to improve the habitat values of remnants through well-considered management, and such action has the potential to deliver greater conservation benefit than revegetation of areas already used for agricultural production.

5. Corridors lead to sinks. Few 'corridors' link viable habitats but they increase the amount of over-represented edge habitat. Increasing the viability of remnants through fine-scale habitat improvement, reduces the extinction risk and the associated need for linkages

Much of the remnant habitat in central-western NSW is already in a linear configuration and most habitat construction programs contribute to the linear estate. Linkages between patches are only desirable if the remnant patches they link have high conservation value, notwithstanding their isolation. In highly cleared and degraded landscapes, enhancing the habitat of source patches to reduce the probability of local extinction should receive equal attention to land management actions that attempt to increase the probability of recolonisation through connectivity. There is evidence that linear strips of habitat are poor in quality, potentially acting as sinks for the limited sources: that is, they collect individuals produced in high quality habitat, causing an overall drain on the wider population (Pulliam and Danielson 1991). The age structure of the Red-capped Robin (Major *et al.* 1999b) was younger in linear strips of habitat and there was evidence of elevated activity of nest predators (Figure 22.2) in these areas (Major *et al.* 1999a). Furthermore, corridors provide the edge habitat that, among other things, allows the Noisy Miner to flourish. Not only does this limit the value of the habitat at the patch level, but it enables the Noisy Miner to permeate the landscape with the potential of controlling bird diversity at a large spatial scale (Major *et al.* 2001). While many species may conceivably prefer to disperse through vegetated corridors than across open paddocks, it is likely that the presence of the Noisy Miner will compromise use of corridors by small birds and potentially make dispersal a risky activity.

6. History happens. Much of the present-day distribution of animals in the landscape is a consequence of historical accidents: chance dispersal, habitat loss and extinction

We identified genetic structuring of wolf spider populations over the landscape, but not in a pattern that could be explained by anthropogenic habitat fragmentation or current geographical barriers (Colgan *et al.* 2002). Stochastic extinctions of local populations and unusual

Figure 24.2 Grey Butcherbird (*Cracticus torquatus*) attacking an artificial nest. Avian nest predators showed greater activity in linear strips of habitat.

dispersal events have evidently been important historically. Historical accidents are also evident in the pattern of vegetation remaining on the land. The location of roads, State Forests, stock routes and the idiosyncrasies of individual landowners are responsible for the woodlands we see today. Despite conflicting with public policy, three of five roadside strips in which we marked individual Red-capped Robins experienced partial clearing associated with road maintenance during a three-year study (Major and Gowing 2001). Incremental habitat loss will increase the extinction probability of the already small populations that provide the genetic links across the landscape. Just as historical accidents have shaped the past, stochastic events will frustrate planning for the future. Redundancy is therefore required in the reserve system to cope with stochastic events.

7. Understand process to protect pattern. Most research in fragmented woodlands has described snapshots of the distribution of biodiversity which does not necessarily reflect the processes that promote long-term survival

Our current understanding of animal responses to woodland fragmentation is largely based on patterns obtained from biodiversity surveys. Continuing declines in areas where land clearing has already ceased indicate that ecological systems have not yet reached a new equilibrium. Many patches identified from surveys as containing suitable habitat for animals on the basis of presence data are likely to contain relictual populations destined to local extinction. Accordingly, estimates of minimum patch size requirements for birds are likely to be underestimates. Demographic data are desperately needed to understand differential mortality rates, reproductive rates, dispersal and immigration in heavily fragmented landscapes (Major and Gowing 2001). It is foolhardy to attempt to conserve biodiversity by facilitating dispersal without an understanding of productivity and the relative roles of sources and sinks in determining population trajectories.

Conclusions

Clearing of woodlands has historically been the main cause of decline of biodiversity in central-western New South Wales. Despite legislation prohibiting broadscale land clearing, woodland in New South Wales continues to disappear at an alarming rate (Keith *et al.* 2009). Limiting the destruction of intact habitats remains a conservation priority that still cannot be taken for granted. We are also witnessing further declines due to the ongoing effects of habitat fragmentation, and understanding the mechanism of these declines is a research priority. These declines are presumably caused by demographic changes in reproductive output, mortality and immigration/emigration amongst vastly altered proportions of source and sink habitats. There is some evidence from work on small birds that it is the chronic shortage of individuals in the landscape that is the critical concern, rather than a lack of corridors for movement. Conservation management has so far given disproportionate recognition to the immigration/emigration side of the equation and there is an urgent need to understand the dynamic relationship between source and sink habitats. Given the high levels of degradation of remaining woodland habitats, it is likely that enhancing the quality of source habitats to lower the risk of extinction of local populations will be more cost-effective than enhancing connectivity.

Research on birds has had a disproportionate impact on our understanding of animal responses to fragmentation. Identifying consistent patterns is a far greater challenge for invertebrate populations, due to their extreme diversity and the fine spatial scales at which they exhibit variation. Although our research shows that birds cannot be considered surrogates for all biodiversity, there are good reasons for them to remain at the forefront of research activities. Their charisma means that research findings are readily incorporated into public understanding and policy. Moreover, broadly consistent patterns have emerged from studies in different parts of Australia (e.g. see Chapters 14, 25 and 29), building an irrefutable knowledge base to underpin conservation actions. Now that the patterns of bird distributions in woodlands are reasonably well known, it would be desirable to steer our research priorities towards developing an understanding of the demographic processes that drive the patterns.

Acknowledgements

I wish to thank all my Australian Museum colleagues who were responsible for undertaking the research that provided the foundation for the opinions that I present in this paper: Sarah Brown, Gerry Cassis, Fiona Christie, Don Colgan, Greg Gowing, Mike Gray, Tessa Ivison, Julie Macaranas, Tim Martin, Chris Reid and Derek Smith. I am also grateful to Forests New South Wales, the Rural Lands Protection Board and the many private landowners who gave us enthusiastic approval to work on their land.

Bio

Richard Major is a Senior Research Scientist at the Australian Museum having studied zoology at the University of Sydney prior to completing his PhD at Monash University. His two main research interests are the breeding ecology of birds, and the effect of habitat modification on animal populations and communities. His current research focuses on the ecology of birds in urban environments, with projects on species that are becoming more common (e.g. Australian White Ibis) and species that are in decline (e.g. White-fronted Chat). He had a leading role in the development of the Eureka Prize-winning *Birds in Backyards* project, with responsibility for much of the web content including the popular backyard bird surveys. He is Chairperson of the New South Wales Scientific Committee, which has responsibility for listing key threatening processes and threatened species and communities in New South Wales.

References

Colgan, D.J., Brown, S., Major, R.E., Christie, F.J., Gray, M. and Cassis, G. (2002). Population genetics of wolf spiders of fragmented habitat in the wheat belt of New South Wales. *Molecular Ecology* **11,** 2295–2305.

Goldney, D.C. and Bowie, I.J.S. (1990). Some management implications for the conservation of vegetation remnants and associated fauna in the central western region of New South Wales. *Proceedings of the Ecological Society of Australia* **16,** 427–440.

Keith, D.A., Orscheg, C., Simpson, C.C., Clarke, P.J., Hughes, L., Kennelly, S.J., Major, R.E., Soderquist, T.R., Wilson, A.L. and Bedward, M. (2009). A new approach and case study for estimating extent and rates of habitat loss for ecological communities. *Biological Conservation* **142,** 1469–1479.

Macaranas, J.M., Colgan, D.J., Major, R.E., Cassis, G. and Gray, M. (2000). Species discrimination and population differentiation in ants using microsatellites. *Biochemical Systematics and Ecology* **29,** 125–136.

Major, R.E., Christie, F.J. and Gowing, G. (2001). Influence of remnant and landscape attributes on Australian woodland bird communities. *Biological Conservation* **102,** 47–66.

Major, R.E., Christie, F.J., Gowing, G. and Ivison, T.J. (1999a). Elevated rates of predation on artificial nests in linear strips of habitat. *Journal of Field Ornithology* **70,** 351–364.

Major, R.E, Christie, F.J., Gowing, G. and Ivison, T.J. (1999b). Age structure and density of red-capped robin populations vary with habitat size and shape. *Journal of Applied Ecology* **36,** 901–908.

Major, R.E., Christie, F.J., Gowing, G., Cassis, G. and Reid, C.A.M. (2003). The effect of habitat configuration on arboreal insects in fragmented woodlands of south-eastern Australia. *Biological Conservation* **113,** 35–48.

Major, R.E. and Gowing, G. (2001). Survival of red-capped robins (*Petroica goodenovii*) in woodland remnants of central western New South Wales, Australia. *Wildlife Research* **28,** 565–571.

Major, R.E., Gowing, G., Christie, F.J., Gray, M. and Colgan, D. (2006). Variation in wolf spider (Araneae: Lycosidae) distribution and abundance in response to the size and shape of woodland fragments. *Biological Conservation* **132,** 98–108.

Major, R.E., Smith, D., Cassis, G., Gray, M.R. and Colgan, D.J. (1999c). Are roadside strips important reservoirs of invertebrate diversity? A comparison of the ant and beetle faunas of roadside strips and large remnant woodlands. *Australian Journal of Zoology* **47,** 611–624.

Martin, T.J. and Major, R.E. (2001). Changes in wolf spider assemblages across woodland/pasture boundaries in the central wheatbelt of New South Wales, Australia. *Austral Ecology* **26,** 264–274.

Pulliam, H. and Danielson, B. (1991). Sources, sinks and habitat selection: a landscape perspective on population dynamics. *American Naturalist* **137,** 60–66.

Seddon, J.A., Briggs, S.V. and Doyle, S.J. (2003). Relationships between bird species and characteristics of woodland remnants in Central New South Wales. *Pacific Conservation Biology* **9**, 95–119.

Sivertsen, D. and Metcalfe, L. (1995). Natural vegetation of the southern wheat-belt (Forbes and Cargelligo 1:250 000 map sheets). *Cunninghamia* **4,** 103–128.

25

HOW CAN WE REVERSE THE LOSS OF OUR WOODLAND BIRDS?

Hugh A. Ford

1. Bird species will continue to be lost from remnants, and decline in regions, even well after vegetation clearing has ceased.
2. We know the basic causes of these losses (habitat loss and fragmentation), but we know less about the ecological processes involved.
3. Different species of birds occupy different ecological niches – they forage in different ways, in different locations, on different plant species.
4. Species often compete for food, space and nest sites, and simplifying their habitat will increase this competition, allowing fewer species to survive.
5. Every change in management will favour some species of birds, but put others at a disadvantage – there will be winners and losers.
6. We need diverse management of remnant woodland within a region to sustain a full range of bird species.
7. Almost every patch of woodland has value for some species of native birds.
8. Our biggest challenge is to provide suitable habitat for the birds that live in open, grassy woodland, without the Noisy Miner dominating it.

Introduction

I have been studying birds in eucalypt woodlands and open forests around Armidale, northern New South Wales (see map on next page) for the past 30 years, along with a succession of students and colleagues. My initial interest was to learn more about their natural history: where do they feed, what do they eat and what is their breeding biology? In the 1970s, eucalypt dieback was a major issue in New England, with fears that we would lose most of our native trees in farmland as well as the birds associated with them. The trees have recovered, however, and so have some of the birds. Other birds, though, have declined, although not always the ones that we might have expected to suffer. From my background in basic ecology, I have tried to understand the ecological processes behind these declines (Ford *et al.* 2001). The environment has changed, in some cases for the better, for instance, due to extensive tree planting and

protection from over-grazing. My views have changed also, as I have learnt more about the ecological processes operating in woodlands. There are no simple answers to the question of how we can reverse the decline of so many woodland birds. More people are studying them now, however. There is sympathy in most rural communities for replacing some of the vegetation that has been lost, as well as support for well-designed management.

Lessons

1. Bird species will continue to be lost from remnants, and decline in regions, even well after vegetation clearing has ceased

One of the most universal rules of ecology is that the larger an island or patch of forest, the greater the number of species of plants, birds or insects that will live there. This means that the extensive clearing of native vegetation will result in small remnants, which will lose species. Ultimately, extensively cleared landscapes will lose some species entirely. What is less well appreciated is that these losses may take time. The number of species 'relaxes' to a lower diversity of species, in a process that has been called an 'extinction debt' (Possingham and Field 2001). A dramatic example of this was shown in remnant patches of forest in Victoria that were surveyed in the 1980s and again 22 years later. An average of nine bird species had gone from each patch (MacHunter *et al.* 2006). In New England we have seen bird species repeatedly disappear from remnants of native woodland and forest. This happens whether they are nature reserves, State Forests, in travelling stock routes or on private land. Eastwood State Forest, which is now Imbota Nature Reserve (a 350 ha remnant 10 km south-east of Armidale), has lost about six bird species since the late 1970s. Over time, these local losses are compounded until we see decline across the whole region.

2. We know the basic causes of these losses (habitat loss and fragmentation), but we know less about the ecological processes involved

We have been following and trying to understand the progressive decline and retreat of two bird species in New England for nearly 30 years (Ford *et al.* 2009). The Brown Treecreeper (*Climacteris picumnus*) was common in remnant woodlands around Armidale when Richard Noske studied it from 1976 to 1981 (Noske 1979, 1991). Jeff Walters found that the Brown Treecreeper was absent from several sites where it originally occurred and also from other sites of apparently suitable habitat (Walters *et al.* 1999). Even where the species was present in isolated remnants, often groups contained only male birds. Caren Cooper discovered why (Cooper and Walters 2002). The Brown Treecreeper lives in groups, and young females leave the group in search of a new group, which they can join. Small vegetation patches may contain few or no other groups. When they are isolated from other patches the dispersing females cannot find occupied patches, and conversely new females cannot reach the source remnants. One after another, local populations of one or two groups lose females and finally disappear altogether. We have now recorded 14 such local extinctions and the species has retreated from 40 km east of Armidale to about 30 km west of Armidale (Ford *et al.* 2009).

What is especially worrying is that it is not only small and isolated patches that are losing treecreepers. About 10–15 groups occurred in Imbota Nature Reserve in 1980, but these dwindled until the last individuals were seen in 2001. Most of our populations west of Armidale occurred in TSRs, where the woodland is well connected. Dispersing females should be able to find other groups by crossing only small gaps. Yet even here local populations are disappearing and in the two best sites nine out of the 14 territories occupied in 1998 are now vacant. So, we know part of the reason why the Brown Treecreeper is disappearing from the landscape around Armidale, but there are probably other causes, such as subtle changes in their habitat.

We have found a very similar pattern of disappearances in the Hooded Robin (*Melanodryas cucullata*), which also has gone from 14 previously occupied sites around Armidale over 30 years (Ford *et al.* 2009). Although we understand the causes less well, we do not believe that it is poor dispersal that is the problem. Instead, chronically low breeding success means that there are too few young robins to disperse to other patches (Fitri and Ford 2003). Again, subtle changes to the species' habitat may be responsible for this. Either the numbers of nest predators have increased or the number of safe nesting sites has declined.

3. Different species of birds occupy different ecological niches – they forage in different ways, in different locations, on different plant species

In the middle of the last century, the Reverend Peter Hartley in England and Robert MacArthur in the north-eastern USA collected information on different species of titmice and warblers, respectively, living in the same place (Hartley 1953; MacArthur 1958). They found that different species feed at different levels, on different substrates, such as bark or foliage, using different methods, such as gleaning or hawking, and on different species of trees, such as oaks or pines. In the 1980s, Harry Recher in southern New South Wales, and I, working in New England, with our colleagues, compared the foraging behaviour of Australian birds in eucalypt woodlands and forests (Recher *et al.* 1985; Ford *et al.* 1986). We, too, found that all species differed in some aspect of their foraging behaviour and ecology. Honeyeaters gleaned leaves or probed into flowers. Treecreepers probed into bark. Fantails hawked for flying insects. Robins pounced on prey on the ground. Rosellas took seeds from the trees and finches foraged on seeds from the grass or ground. The comparisons between closely related species are particularly interesting, because all showed some differences. Many of these studies are described in a book devoted to birds in eucalypt woodlands and forests (Keast *et al.* 1985).

The reason why we need to appreciate that different species forage in different ways in different places is that subtle changes to their habitat may reduce the availability of their food or foraging sites. We were primarily interested in testing ecological theory and also driven by curiosity. To manage a species, however, we need to know its ecological niche – where it belongs in its ecosystem. For instance, the Brown Treecreeper has gone from Imbota Nature Reserve, but the White-throated Treecreeper (*Cormobates leucophaea*) still thrives there. There are still Eastern Yellow or Scarlet Robins (*Eopsaltria australis* and *Petroica boodang*) in most of the remnants that have lost Hooded Robins. Brown Treecreepers spend half of their time feeding on open ground, which White-throated Treecreepers never do (Noske 1979). Hooded Robins also prefer open woodlands with plenty of bare ground and avoid areas with a lot of grasses and herbs, whereas Eastern Yellow Robins prefer sites with more trees (Templeton 2001). Perhaps changes to the amount of prey or access to this prey on the ground have also contributed to the decline of these two species.

4. Species often compete for food, space and nest sites, and simplifying their habitat will increase this competition, allowing fewer species to survive

One reason for the numerous studies comparing the foraging behaviour of different species of forest birds was the belief that species competed with each other for resources such as food, and so needed to occupy different ecological niches to be able to co-exist. Different species may also interact aggressively with each other. For instance, the Brown Treecreeper dominates the White-throated Treecreeper (Noske 1979) and the Hooded Robin dominates other robins (Fitri and Ford 1998). So direct competition from relatives would not appear to be a reason why these two species are declining. Competition may be indirect, however, via depletion of common food supplies. Different species of honeyeaters often feed on nectar from the same species of plant, and may deplete the amount of nectar (Ford 1979). When nectar is scarce, it is the smaller, subordinate species that may compete best, as they can exploit the meagre amounts of food better than the larger, dominant species, which require more energy.

Because all species of bird in a patch of woodland occupy a different ecological niche, any changes that simplify the structure or floristic diversity of the woodland will make some of these niches inadequate. The Noisy Miner (*Manorina melanocephala*) is an aggressive honeyeater which is common in degraded eucalypt woodlands. It tends to exclude almost all other medium-sized and small birds. In many studies, the presence of Noisy Miners in a site was found to have a far greater impact on how many birds and how many species occur, than any other factor. This was especially the case where the vegetation had been simplified by thinning of trees and grazing. Removing the Noisy Miner allows other bird species to return to a site (Grey *et al.* 1998). Several studies have shown that the Noisy Miner is less likely to dominate sites where there are trees and shrubs, such as wattles and bulokes, in addition to eucalypts.

5. Every change in management will favour some species of birds, but put others at a disadvantage – there will be winners and losers

Different management can change the structure and composition of the vegetation and alter the suitability of a site for different species. Gillian Dunkerley studied the Fuscous Honeyeater (*Lichenostomus fuscus*) at Eastwood State Forest in the late 1980s (Dunkerley 1989). Twenty years later she has returned to conduct a census of her study site to find that the species, previously the most abundant bird, was missing from many points. My long-term, though more sporadic, censuses also indicate a decline in the Fuscous Honeyeater. It is still common, but now the previously rare Yellow-faced Honeyeater (*L. chrysops*) is almost equally abundant. Whereas the Fuscous Honeyeater prefers grassy woodland, the Yellow-faced Honeyeater prefers open forest with a shrubby understorey. We have noticed a thickening of the vegetation at Eastwood/Imbota over the last 20 years.

There is little doubt that reducing the pressure of livestock grazing, and the addition of trees and shrubs, will provide a favourable environment for many bird species. It is the birds of shrubby forest that benefit most, however. Birds of grassy woodlands appear to be losing ground, not only Fuscous Honeyeaters, Brown Treecreepers and Hooded Robins, but also the Diamond Firetail (*Stagonopleura guttata*), Dusky Woodswallow (*Artamus cyanopterus*), and Restless Flycatcher (*Myiagra inquieta*). All were common in the 1970s, but are now scarce and local in our region. Over time there are winners and losers, and these depend on the type of management deployed.

6. We need diverse management of remnant woodland within a region to sustain a full range of bird species

We have learnt a lot about the natural history of woodland birds over the last few decades. We have also learnt something of the ecological processes that have caused some species to decline – interrupted dispersal, competition from Noisy Miners, predation on nests, and probably loss of suitable foraging areas.

There are a number of clear messages to managers from this research. Larger and well-connected patches are better, so plant native trees and shrubs around existing patches to enlarge them and plant corridors to join up neighbouring patches. These recommendations have been made for a long time and are well accepted, but they may not be enough. We believe, but cannot confirm, that there have been important changes in the habitat that have been detrimental to many bird species. Any management approach will alter the structure of the habitat and there will be winners and losers. This means that we need a diverse approach as to how we manage a region. Many forms of management, such as grazing by livestock, will already be widespread. Some patches of vegetation may be protected from grazing. But equally important, we need to experiment with periodic grazing – enough to prevent extensive growth of grasses and weeds, but not too much, which might prevent some regeneration of trees and shrubs. Often the sites that allow the best opportunity for this type of management are TSRs. It is no surprise that they are the best places to see many of the declining woodland birds today. Travelling stock routes often have the best display of spring flowers: moderate grazing retards aggressive weeds, but allows the rare natives a chance to bloom. But they are under threat of being sold or leased to local landowners, and hence experiencing the same type of management that is common in the region. Alternatively, they might be reserved and protected from grazing. Ironically, this too could also be detrimental to some woodland birds.

7. Almost every patch of woodland has value for some species of native birds

A more positive viewpoint is that, even if by default, there is always a range of management approaches in a region. This means that almost any patch of woodland provides the right conditions for some woodland birds. The history of biological conservation is replete with examples of once common species that have declined, often through subtle changes in management. Patches of woodland on private land, even if they only harbour common species, have a place on the palette for building the whole picture of regional conservation.

8. Our biggest challenge is to provide suitable habitat for the birds that live in open, grassy woodland, without the Noisy Miner dominating it

Open eucalypt woodlands, with little cover, favour the Noisy Miner. Yet these are the preferred habitat of many of the declining woodland birds. We have a major dilemma. If we manage woodland in the hope that it will become suitable for the Brown Treecreeper, Hooded Robin and Diamond Firetail, perhaps we shall simply provide more habitat for the Noisy Miner? We need to find sites that have these desirable species, but not the Noisy Miner. What is special about such sites, and the way in which they are being managed? We need to experiment with management to see if we can deter Miners, while retaining or attracting the species we want.

Conclusions

We have many of the tools to arrest and reverse the decline of many woodland birds. As we start to understand the ecological processes that cause a species to decline, this gives us the clues of how we can alter management to encourage the species. We have also made great strides in rehabilitating and reconstructing woodlands. By appreciating that different species occupy different ecological niches we can see how one type of management would not suit all of them. This means that we require a regional diversity of management to maintain the full range of species. Therefore, collaboration among all land-managers is needed to identify which types of management are being neglected. Although we can learn much by copying best practices, we also have much to gain from taking a novel approach to management.

Acknowledgements

I am most grateful to a long succession of Honours and postgraduate students, and colleagues, who have worked on woodland birds around Armidale over the past 30 years. I also thank the Worldwide Fund for Nature (Australia), the Environmental Fund of New South Wales, the National Science Foundation, and the University of New England for support.

Bio

Hugh Ford was born in England and received his university education in Scotland, before moving to the University of Adelaide as a Post-doctoral Fellow. He then became a lecturer at the University of New England, where he has stayed for over 30 years and is now an Emeritus Professor. His main research interests have been Australian honeyeaters, the ecological basis of cooperative breeding, and bird migration, as well as, of course, the ecology and conservation of birds of eucalypt woodlands. He wrote a book, *Ecology of Birds: an Australian perspective*, has edited two others and written over 100 journal articles and book chapters. In 1993 he was awarded the Serventy Medal for his contributions to ornithology.

References

Cooper, C.B. and Walters, J.R. (2002). Experimental evidence of disrupted dispersal causing decline of an Australian passerine in fragmented habitat. *Conservation Biology* **16**, 471–478.

Dunkerley, G.M. (1989). The Fuscous Honeyeater: food resources and the bird community. PhD thesis, University of New England: Armidale, NSW.

Fitri, L.L. and Ford, H.A. (1998). Aggression among Hooded Robins *Melanodryas cucullata* and other birds. *Corella* **22**, 24–29.

Fitri, L.L. and Ford, H.A. (2003). Breeding biology of Hooded Robins *Melanodryas cucullata* in New England, New South Wales. *Corella* **27**, 68–77.

Ford, H.A. (1979). Interspecific competition in Australian honeyeaters – depletion of common resources. *Australian Journal of Ecology* **4**, 145–164.
Ford, H.A., Barrett, G.W., Saunders, D.A. and Recher, H.F. (2001). Why have birds in the woodlands of southern Australia declined? *Biological Conservation* **97**, 71–88.
Ford, H.A., Noske, S. and Bridges, L. (1986). Foraging of birds in eucalypt woodland in north-eastern New South Wales. *Emu* **86**, 169–179.
Ford, H.A., Walters, J.R., Cooper, C.B., Debus, S.J.S. and Doerr, V.A.J. (2009). Extinction debt or habitat change? Ongoing losses of woodland birds in north-eastern New South Wales, Australia. *Biological Conservation* **142**, 3182–3190.
Grey M.J., Clarke, M.F. and Loyn, R.H. (1998). Influence of the noisy miner *Manorina melanocephala* on avian diversity and abundance in remnant grey box woodland. *Pacific Conservation Biology* **4**, 55–69.
Hartley, P.H.T. (1953). An ecological study of the feeding habits of English titmice. *Journal of Animal Ecology* **22**, 261–288.
Keast, A., Recher, H.F., Ford, H.A. and Saunders, D.A. (Eds) (1985). *Birds of Eucalypt Forests and Woodlands: Ecology, Conservation and Management.* Surrey Beatty and Sons: Chipping Norton, NSW.
MacHunter, J., Wright, W., Loyn, R.H. and Rayment, D. (2006). Bird declines over 22 years in forest remnants in south-eastern Australia: evidence of faunal relaxation? *Canadian Journal of Forest Research* **36**, 2756–2768.
McArthur, R.H. (1958). Population ecology of some warblers of northeastern coniferous forests. *Ecology* **39**, 599–619.
Noske, R.A. (1979). Co-existence of three species of treecreepers in north-eastern New South Wales. *Emu* **79**, 120–128.
Noske, R.A. (1991). A demographic comparison of cooperatively breeding and non-cooperative treecreepers (Climacteridae). *Emu* **91**, 73–86.
Possingham, H. and Field, S.A. (2001). Regional bird extinctions and their implications for vegetation clearance policy. *LifeLines* **7**, 15–16.
Recher, H.F., Holmes, R.T., Schultz, M., Shields, J. and Kavanagh, R. (1985). Foraging patterns of birds in eucalypt forest and woodland on the tablelands of south-eastern Australia. *Australian Journal of Ecology* **10**, 399–419.
Templeton, A.E. (2001). Distribution and behaviour of declining woodland birds in a fragmented and degraded rural landscape. B.Sc. (Hons) thesis, University of New England: Armidale, NSW.
Walters, J.R., Ford, H.A. and Cooper, C.B. (1999). The ecological basis of sensitivity of brown treecreepers to habitat fragmentation: a preliminary assessment. *Biological Conservation* **90**, 13–20.

26

THE CONSERVATION OF REPTILES IN THE TEMPERATE WOODLANDS OF SOUTHERN NEW SOUTH WALES

Damian Michael and David Lindenmayer

1. Nature reserves, travelling stock reserves and remnant vegetation provide a network for conserving regional reptile diversity and represent nodal points for concentrating restoration efforts.
2. Regrowth vegetation is important for declining woodland birds but represents a conservation challenge for reptiles in changing agricultural landscapes.
3. A subset of reptiles use currently restored habitat, but a different planting design is required to provide habitat for other species of reptiles and prevent existing habitat from being compromised for them.
4. Rocky outcrops and surface rocks play an important role in structuring reptile assemblages in patches of remnant native vegetation and represent reptile hotspots.
5. Grazing landscapes can provide habitat for reptiles and matrix conditions can influence patterns of aggregation in the rock-dwelling Crevice Skink (*Egernia striolata*).

Introduction

Reptiles are a diverse group of animals in terrestrial ecosystems worldwide. In similar fashion to amphibians (Beebee and Griffiths 2005), however, reptiles have experienced unprecedented declines on a global scale (Whitfield Gibbons *et al.* 2000). Major threats to reptile populations include habitat loss and degradation, introduced invasive species, environmental pollution, disease and climate change (Whitfield Gibbons *et al.* 2000). Extensive habitat loss and degradation of grassy woodland ecosystems are causes of region-wide declines in reptile fauna in south-eastern Australia (Brown *et al.* 2008; Michael and Lindenmayer 2010). For this reason, the conservation of reptile fauna needs to be a fundamental part of natural resource management in agricultural landscapes (Michael and Lindenmayer 2010).

Over the past decade, we have conducted a series of long-term and cross-sectional studies of reptiles of remnant temperate woodlands, planted native vegetation and rocky outcrops in the South-western Slopes bioregion of southern New South Wales (see map on next page). The lessons summarised below are based on that work.

Lessons

1. Nature reserves, travelling stock reserves and remnant vegetation provide a network for conserving regional reptile diversity and represent nodal points for concentrating restoration efforts

Habitat restoration is a fundamental part of wildlife conservation and should be guided by sound science to achieve maximum biodiversity benefits. We identify three areas where habitat restoration efforts may enhance reptile conservation in the South-western Slopes bioregion.

Conservation reserves in the South-western Slopes have considerable value for reptiles. For example, Nail Can Hill Flora and Fauna Reserve (~1000 ha), Benambra Nature Reserve (1400 ha) and Ulandra Nature Reserve (3960 ha) collectively support >30 species of reptiles (Caughley and Gall 1985; Michael 2004). These and other nature reserves in the South-western Slopes support species that are absent or rare in lowland agricultural landscapes, including the Nobbi Dragon (*Amphibolurus nobbi*), Burton's Snake Lizard (*Lialis burtonis*) (Cunningham *et al.* 2007), Pink-tailed Worm-lizard (*Aprasia parapulchella*) (Michael and Herring 2005) and the Inland Carpet Python (*Morelia spilota metcalfei*) (Michael and Lindenmayer 2008).

Travelling stock reserves (TSRs) also support a diverse range of regionally rare reptile species. For example, 26 species have been recorded in Mates Gully TSR (approximately 100 ha) near Tarcutta (Sass 2003; Michael *et al.* 2009b). This TSR is one of the few places in the South-western Slopes where small snakes such as the Southern Death Adder (*Acanthophis antarcticus*), Eastern Bandy Bandy (*Vermicella annulata*) and Prong-snouted Blind Snake (*Ramphotyphlops bituberculatus*) have been recorded (Sass 2003; Michael *et al.* 2009b). The large size of nature reserves (and many TSRs) and comparative lack of human disturbance are key reasons why these areas are critical to the conservation of reptiles in southern NSW.

Patches of remnant vegetation on private property can also provide important habitat for a diverse range of reptiles (Michael and Lindenmayer 2010). Over the past eight years, we have recorded more than 25 reptile species from small patches of remnant vegetation ranging from <3 ha to 20 ha in size (Cunningham *et al.* 2007; Michael *et al.* 2008b).

The three different types of land tenure mentioned above provide an important network of habitats from where habitat restoration efforts should be focused. Creating buffer zones around nature reserves, TSRs and remnant vegetation, and linking these habitats may assist the recovery of reptile populations in agricultural landscapes.

2. Regrowth vegetation is important for declining woodland birds but represents a conservation challenge for reptiles in changing agricultural landscapes

Managing regrowth vegetation presents a conservation challenge to land managers concerned with enhancing biodiversity in changing agricultural landscapes. Regrowth (natural seedling regeneration and coppice regrowth) is critical in the recovery of declining woodland birds, many of which are dependant on this type of vegetation structure (Montague-Drake *et al.* 2009). Reptile populations, however, may be disadvantaged by shade effects caused by dense regrowth. Two fundamental differences between regrowth vegetation and old-growth remnant vegetation are tree density and the size of overstorey trees. These attributes influence canopy cover and the amount of sunlight that reaches the ground – an important factor in the physiological requirements of reptiles (most temperate zone reptiles require access to sunlight before metabolic and daily activities can be conducted).

We found reptile diversity was lower in granite outcrops supporting regrowth vegetation than outcrops supporting widely spaced, old-growth vegetation (Michael *et al.* 2008b). We also found certain types of reptiles were less abundant in patches of regrowth vegetation in areas without rocks (Michael *et al.* in review). Many of the species that responded negatively to regrowth were terrestrial generalists, a group of reptiles that account for a large proportion of the reptile fauna in agricultural landscapes. Dense regrowth reduces the suitability of basking and shelter sites by changing thermal gradients. This finding presents a fundamental challenge to the maintenance and recovery of reptiles in landscapes undergoing changes in vegetation succession, as it may mean novel approaches are required to manage regrowth vegetation and create open-canopy environments for heliothermic organisms. Two management recommendations are to increase the internal structural heterogeneity of regrowth remnants by partially thinning dense stands, or to manage natural regeneration by introducing ecologically appropriate grazing and fire regimes.

3. A subset of reptiles use currently restored habitat, but a different planting design is required to provide habitat for other species of reptiles and prevent existing habitat from being compromised for them

Tree plantings have the potential to provide new habitat for species in heavily cleared landscapes. The value of plantings in the recovery of biodiversity, however, varies among faunal groups (Cunningham *et al.* 2007; Lindenmayer *et al.* 2007). Our work in the South-western Slopes has shown that a small sub-set of reptiles use tree plantings (Cunningham *et al.* 2007; Michael *et al.* 2008a). Several species were not recorded in plantings (e.g. Inland Carpet Python and Lace Monitor *Varanus varius*), whereas seven species were more abundant in this habitat (Cunningham *et al.* 2007). Two species in particular, the Olive Legless Lizard (*Delma inornata*) and the fossorial Three-toed Skink (*Hemiergis decresiensis*), were more abundant in tree plantings than in surrounding remnant vegetation. We suggest the absence of livestock grazing is an important attribute of fenced tree plantings in which terrestrial and litter-dwelling species respond positively.

Of the reptiles found in tree plantings, 53% were terrestrial, 26% were fossorial and 20% were saxicolous (or arboreal), and they included species from the families Agamidae, Scincidae, Pygopodidae, Elapidae and Gekkonidae (Cunningham *et al.* 2007). Findings from Lesson #2 have implications for the design of future tree plantings. In rocky environments, we suggest low tree density plantings should be established but which incorporate a high proportion of small shrubs. This will allow enough sunlight to reach the ground-layer and minimise the likelihood of planted vegetation shading basking sites and foraging areas. Tree plantings will also have greater value for reptiles if they incorporate fallen timber, standing dead trees, and ribbon-bark tree species. If future plantings are to have maximum biodiversity benefits, different planting recipes will need to be developed for different parts of landscapes. For example, dense overstorey plantings could be established along watercourses and lowland areas, and low density, shrub plantings established around rocky outcrops.

4. Rocky outcrops and surface rocks play an important role in structuring reptile assemblages in patches of remnant native vegetation and represent reptile hotspots

Rocky outcrops and surface rocks provide important habitats in agricultural landscapes where vegetation loss has been extensive. In the South-western Slopes, granite outcrops devoid of overstorey vegetation can support >10 species of reptiles (Michael *et al.* 2008b). The presence of rocky outcrops and surface rocks in remnant vegetation plays an important role in structuring reptile assemblages and can significantly increase species richness (Michael *et al.* in review). Species such as the Crevice Skink (*Egernia striolata*) and Southern Marbled Gecko (*Christinus marmoratus*) are often associated with attributes of old-growth vegetation (e.g. crevices and cavities formed by exfoliating bark, dead trees and fallen timber) but can occur in patches of regrowth when rocks are present (Michael *et al.* in review). This finding suggests the presence of rocks can reduce the time required for regrowth vegetation to acquire complex reptile assemblages equivalent to old-growth vegetation. Stands of regrowth, especially if rocks are present, are valuable components of the landscape and are local 'hotspots' for reptile diversity. Regrowth is often cleared, however, during agricultural activities or urban development. We recommend regrowth vegetation is protected from clearing and its condition enhanced by controlling invasive plants and pest animals to avoid negative impacts on the integrity of rocky outcrops (and their associated patterns of vegetation cover) as habitat for reptiles (Michael *et al.* 2010c).

5. Grazing landscapes can provide habitat for reptiles and matrix conditions can influence patterns of aggregation in the rock-dwelling Crevice Skink (*Egernia striolata*)

In heavily cleared agricultural landscapes, reptile habitat is not necessarily patches of remnant vegetation. Some species can persist in grazing landscapes, especially if fallen timber, surface rocks, native tussock grasses and rocky outcrops are present. Species such as the Olive Legless Lizard, Boulenger's Skink (*Morethia boulengeri*) and Southern Rainbow Skink (*Carlia tetradactyla*) can be common in grazing landscapes with retained microhabitat (Michael *et al.* 2010a). Rocky outcrops as small as a few square metres in size can support populations of Cunningham's Skink (*Egernia cunninghami*) and Southern Marbled Gecko (Michael *et al.* 2008b, 2010b), and isolated dead trees and fallen timber can support populations of the Ragged Snake-eyed Skink (*Cryptoblepharus pannosus*). The importance of these often-overlooked habitats for reptile fauna emphasises the need to recognise the role of the matrix in conserving reptile diversity in agricultural landscapes.

Conditions in the agricultural matrix can significantly influence reptile diversity in granite outcrops (Michael *et al.* 2008b). Reptile diversity decreased as patterns of vegetation in the matrix changed from continuous native cover to highly modified landscapes dominated by crops and pasture. Landscape context can also significantly affect patterns of aggregation in social species such as the Crevice Skink (Michael *et al.* 2010b). The probability of this species forming aggregations (family groups) was greatest on granite outcrops in landscapes with high amounts of native vegetation, tree plantings and native pasture in the surrounding environment. In contrast, the probability of the Crevice Skink forming aggregations was significantly lower in heavily cleared, modified landscapes (Michael *et al.* 2010b). Improved matrix management therefore plays a fundamental role in the conservation of insular reptile populations.

Conclusions

Our empirical work in the South-western Slopes of New South Wales provides new insights into the ecology and conservation of reptiles in agricultural landscapes. Remnant vegetation on private property is important; however, some species are rare or absent from agricultural landscapes and are currently restricted to a few nature reserves and travelling stock reserves. Our work highlights the value of rocky outcrops in preserving reptile diversity (Michael *et al.* 2008b), although these areas need urgent attention if their full potential in conserving reptiles in agricultural landscape is to be realised (Michael *et al.* 2010c). We have identified the habitat value of regrowth but maintaining these values is a major challenge for landholders and natural resource managers (Michael *et al.* in review).

Our work has covered many facets of reptile ecology although we recognise many knowledge gaps remain. For example, what will be the long-term benefits of tree plantings in the recovery of reptile fauna? What are the relationships between environmental conditions and phylogenetic variability or morphological attributes in wide ranging species? How quickly will reptiles respond to changes in vegetation condition initiated by government incentive schemes aimed at enhancing biodiversity? These and other questions form the basis of our ongoing, long-term monitoring and investigations on the ecology of reptile fauna in southern New South Wales.

Acknowledgements

Our research over the past decade has been possible only because of outstanding statistical collaboration from Ross Cunningham, Emma Knight, Jeff Wood, Alan Welsh and Christine Donnelly. Mason Crane, Rebecca Montague-Drake, Christopher MacGregor and Lachlan McBurney played a pivotal role in assisting with aspects of field surveys associated with this work. Many organisations and funding schemes have provided key financial assistance that has facilitated our ongoing work. These include: Land and Water Australia, the Australian Research Council, Rural Industries Research and Development Corporation, the Natural Heritage Trust, Caring for Our Country, the Kendall Foundation, the Thomas Foundation, the Herman Slade Foundation and the Murray Catchment Management Authority.

Bios

Damian Michael is a Senior Research Officer at the Fenner School of Environment & Society. He has been part of David Lindenmayer's research team since 2000, and has special expertise in botany, ecology and herpetology. Damian completed a PhD on the ecology and conservation of reptiles in agricultural landscapes with a focus on managing rocky outcrops.

David Lindenmayer is a Research Professor in the Fenner School of Environment and Society at the Australian National University. He has worked in the temperate woodlands of southern New South Wales for more than a decade and published many books and scientific articles on their conservation and management.

References

Beebee, T.J.C. and Griffiths, R.A. (2005). The amphibian decline crisis: a watershed for conservation biology? *Biological Conservation* **125,** 271–285.

Brown, G.W., Bennett, A.F. and Potts, J.M. (2008). Regional-fauna decline – reptile occurrence in fragmented rural landscape of south-eastern Australia. *Wildlife Research* **35**, 8–18.

Caughley, J. and Gall, B. (1985). Relevance of zoogeographical transition to conservation of fauna: amphibians and reptiles in the South-western Slopes of New South Wales. *Australian Zoologist* **21**, 513–529.

Cunningham, R.B., Lindenmayer, D.B., Crane, M., Michael, D. and MacGregor, C. (2007). Reptile and arboreal marsupial response to replanted vegetation in agricultural landscapes. *Ecological Applications* **17**, 609–619.

Lindenmayer, D.B., Cunningham, R., Crane, M., Michael, D.R. and Montague-Drake, R. (2007). Farmland bird responses to intersecting replanted areas. *Landscape Ecology* **22**, 1555–1562.

Michael, D.R. (2004). Distribution, habitat preferences and conservation status of reptiles in the Albury-Wodonga region. *The Victorian Naturalist* **121**, 180–193.

Michael, D.R., Crane, M., Holliday, S., Lindenmayer, D.B., McBurney, L., Montague-Drake, R. and MacGregor, C. (2009). A range extension for the Prong-snouted Blindsnake *Ramphotyphlops biturberculatus* (Typhlopidae) in the South-western Slopes of New South Wales. *Herpetofauna* **39**, 115–116.

Michael, D.R., Crane, M., Montague-Drake, R. and Lindenmayer, D.B. (2008a). A survey for herpetofauna on VISY pulp and paper mill, Tumut. *Herpetofauna* **38**, 116–122.

Michael, D.R., Cunningham, R.B. and Lindenmayer, D.B. (2008b). A forgotten habitat? Granite inselbergs conserve reptile diversity in fragmented agricultural landscapes. *Journal of Applied Ecology* **45**, 1742–1752.

Michael, D.R., Cunningham, R.B. and Lindenmayer, D.B. (2010a). Microhabitat relationships among five lizard species associated with granite outcrops in fragmented agricultural landscapes of South-eastern Australia. *Austral Ecology* **35**, 214–225.

Michael, D.R., Cunningham, R.B. and Lindenmayer, D.B. (2010b). The social elite? Habitat heterogeneity, complexity and quality in granite inselbergs influences social organisation in *Egernia striolata* (Scincidae: Lygosominae). *Austral Ecology* doi:10.111/j.1442-9993.2009.02029.x

Michael, D.R., Lindenmayer, D.B. and Cunningham, R.B. (2010c). Managing rock outcrops to improve biodiversity conservation in Australian agricultural landscapes. *Ecological Management and Restoration* **11**, 43–50.

Michael, D.R., Cunningham, R.B. and Lindenmayer, D.B. (in review). Regrowth and revegetation in temperate Australia presents a conservation challenge for reptile fauna in agricultural landscapes. *Biological Conservation.*

Michael, D.R. and Herring, M.W. (2005). Habitat of the Pink-tailed Worm-lizard *Aprasia parapulchella* in Albury, NSW. *Herpetofauna* **35**, 103–111.
Michael, D.R. and Lindenmayer, D.B. (2008). Records of the Inland Carpet Python, *Morelia spilota metcalfei* (Serpentes: Pythonidae), from the South-western Slopes of New South Wales. *Proceedings of the Linnean Society of New South Wales* **129**, 253–261.
Michael, D.R. and Lindenmayer, D.B. (2010). *Reptiles of the NSW Murray Catchment: A Guide to Their Identification, Ecology and Conservation.* CSIRO Publishing: Melbourne.
Montague-Drake, R.M., Lindenmayer, D.B. and Cunningham, R.B. (2009). Factors affecting site occupancy by woodland bird species of conservation concern. *Biological Conservation* doi:10.1016/j.biocon.2009.07.009.
Sass, S. (2003). A survey of the reptiles of Wagga Wagga, NSW. *Herpetofauna* **33**, 18–23.
Whitfield Gibbons, J., Scott, D.E., Ryan, T.J., Buhlmann, K.A., Tuberville, T.D., Metts, B.S., Greene, J.L., Mills, T., Leiden, Y., Poppy, S. and Winne, C.T. (2000). The global decline of reptiles, déjà vu amphibians. *Bioscience* **50**, 653–666.

27

TEMPERATE WOODLANDS: LEARNING FROM THE NORTH

Sue McIntyre

1. It is important for us to regularly question the assumptions that underpin our incomplete knowledge of woodlands.
2. We all need to spend time in woodlands, reflecting, enjoying and learning.
3. Knowledge breakthroughs can be achieved by asking the simple questions.
4. The disturbance history of a site dictates future management options.
5. Low nutrient levels are vital for eucalypt woodlands as they evolved on relatively infertile soils.
6. New thinking on the management of woodland structure is needed to identify how tree density, size and distribution can be managed optimally for plants and animals in contemporary landscapes.

Introduction

I have spent most of my working life investigating the ground-layer vegetation of eucalypt box-gum grassy woodlands in eastern Australia (see map on next page). Beginning in the mid-1970s in the urban grassy woodlands of Studley Park, near the centre of Melbourne, our interest as students was that of weed invasion and native plant species decline (McIntyre and Yugovic 1984). I later studied the effects of disturbance on exotic and native plant diversity on the northern tablelands of New South Wales. Here, the observation of continuous variation in human impacts developed into the concept of habitat variegation (McIntyre and Barrett 1992) and around the same time, a 20-year collaboration on plant functional types and disturbance was initiated with Sandra Lavorel (McIntyre *et al.* 1995; Lavorel *et al.* 1997).

In the 1990s I worked in subtropical grassy woodlands grazed by cattle, and managed most commonly as native pasture. Here, I was able to see the co-existence of native plant diversity with livestock grazing (McIntyre *et al.* 2003), and the clear differentiation of intensive sown pastures and crops from unfertilised pastures in terms of their effects on the flora (McIntyre and Martin 2002). I worked with a range of scientists who contributed to the translation of these concepts into landscape design principles, using an interpretation of habitat response of organisms, hybridised with an interpretation of minimum areas of habitat required for

connectivity. My current focus is on the grassy woodlands of the tablelands and slopes of southern New South Wales. Working with Josh Dorrough has further developed and reinforced my understanding of the vital role of fertilisation in the broadscale clearing of grassy woodlands in the temperate regions of Australia (Dorrough *et al.* 2006; McIntyre 2008). These experiences have all contributed to the observations below, which have been chosen as the key lessons emerging from my research.

Lessons

1. It is important for us to regularly question the assumptions that underpin our incomplete knowledge of woodlands

The research domain around grassy woodland management has operated within conceptual frameworks that are generally considered to represent the current state of knowledge. Researchers have worked to flesh out details within the paradigms, a perfectly sound and time-honoured way to advance knowledge (Kuhn 1962). It is useful, nevertheless, to question our assumptions periodically, lest we risk a lack of progress in our knowledge. Between the 1960s and 1990s, the theory of island biogeography reigned, and the paradigm was that temperate woodlands were isolated fragments (small reserves, roadsides, stock routes) in a sea of hostile agricultural land – non-habitat (McIntyre and Barrett 1992). Part of this thinking would have been derived from work in the basalt grasslands west of Melbourne, where intensive farming had led to very dramatic losses of the native flora (Stuwe and Parsons 1977). The paradigm of habitat *versus* non-habitat failed us in a number of ways. First, many landscapes had temperate woodlands in a range of condition states, but the model led to experiments which artificially divided the landscape in a binary way, regardless of its real condition. Second, the presence of woodland trees became the marker of 'habitat', meaning that many derived grasslands with a native grassy understory but no, or few trees, were discounted as habitat (McIntyre and Hobbs 1999). We now cast the net more widely across the landscape and have moved on to the question of managing entire landscapes to conserve the range of habitat states within.

2. We all need to spend time in woodlands, reflecting, enjoying and learning

This lesson I learned from the late David Ashton who taught plant ecology at the University of Melbourne. David Ashton was a brilliant field ecologist and when he was out in the bush, he thought deeply about what he was seeing, turning a snapshot view of the bush into an interpretation of vegetation dynamics. He used many cues – symptoms of health or disease, population size structure as an indication of regeneration patterns, soil and topographic patterning, etc. – all integrated with his past observations. This type of interpretation should be forming the basis of our current adaptive management and the development of new ideas in woodland conservation. The nature of our current working life discourages focussed meditative observation of the bush. How many field days involve more than a hundred metres of walking? How many people can identify the plants and birds in woodlands? We need to spend time in the field more than ever, reflecting, discussing, and learning from our observations. Much of our drive to conserve woodlands is derived from the pleasure that we gain from spending time in them.

3. Knowledge breakthroughs can be achieved by asking the simple questions

Understanding how woodlands work is an immensely complex task. It is the researchers' goal to reduce this complexity into simple but accurate statements and ask the obvious, simple questions. When we started working with cattle producers in south-east Queensland, the question was raised – how much woodland needs to be maintained on properties to conserve biodiversity adequately? While the process of answering that question was complicated, we were able to come up with a robust generalisation that still stands. We concluded that at least two-thirds of a woodland landscape should consist of land managed for conservation, or a mixture of conservation management and low input use (e.g. native pastures, timber harvesting). A maximum of one-third of the landscape can be managed intensively (e.g. crops, fertilised pastures, human settlement) (McIntyre *et al.* 2000, 2002). Such generalisations make my colleagues uncomfortable and most are squeamish about recommending numbers to managers (e.g. Lindenmayer *et al.* 2005; Radford *et al.* 2005). Nonetheless, these rather conservative recommendations have been widely used in planning and policy and have given managers a yardstick to operate against, which invariably leads to conservation actions on the ground, rather than further development.

4. The disturbance history of a site dictates future management options

Almost all grassy woodlands have been affected by livestock grazing and most have had an interesting array of other activities imposed upon them: cultivation, fertilisation, erosion, pasture sowing, rubbish dumping, and other things that are probably unmentionable. Sometimes the management history of a site can be determined from historical or physical evidence, but more often, the effects of these activities are evident from the presence of exotic species relative to the native species that remain. Recent fertiliser history can be determined from the measurement of available soil phosphorus. Removing grazing from a site that is nutrient-enriched is likely to result in the dominance of weeds at the expense of native species (Wilkins *et al.* 2003). It may be better to graze or burn a grassy woodland over the years that it takes phosphorus levels to decline. If there are abundant exotic annuals, it indicates that soil nitrate levels will be high at certain times of year. The use of sugar is a method of reducing nitrates and favouring perennial plants (Prober *et al.* 2005). Most native ground-layer plants do not produce much seed and are therefore poor dispersers and colonisers. The choice is therefore to introduce species that may have been previously lost, or to manage for the species that are already at the site, and which may be few in number.

5. Low nutrient levels are vital for eucalypt woodlands as they evolved on relatively infertile soils

The research community has been slow in recognising just how crucial nutrients are to the integrity of woodlands. Yet it is obvious in hindsight that an ecosystem that evolves on infertile soils will be adapted to these, rather than to elevated nutrient levels. We now know that even low levels of fertiliser use dramatically reduces the diversity of native plants (Dorrough *et al.* 2006). Eucalypts are also affected, being more prone to dieback in fertilised landscapes (Jurskis 2005) and having their regeneration disrupted by the presence of exotic plants in the ground-layer (Dorrough and Moxham 2005). The use of fertilisers also jeopardises the condition of soil, as it leads to loss of perennial groundcover; the more intensive grazing that is associated with fertilised paddocks further contributes to the loss of perennials (McIntyre 2008).

For a long time, the combination of initial tree clearing and livestock grazing was seen to be responsible for the demise of grassy woodlands, and the removal of grazing (e.g. Prober and Thiele 1995) and the planting for trees have been the primary management actions used to restore them. It is now clear that the use of introduced legumes and superphosphate in the late 20th century finalised the demise of both the grassy layer and trees. The fast-growing exotic species adapted to fertile soils out-compete tree seedlings, and when regeneration stops, stands of eucalypts eventually die out. These interacting factors cannot be seen readily in temperate grassy woodlands because grazing and fertilisation occur mostly together. A different story is apparent in the less intensively developed northern regions, where livestock grazing without fertilisation is common, and is compatible with a diverse ground-layer (McIntyre *et al.* 2003) and abundant tree regeneration. If we are to manage woodland sustainably, we need to limit the amount of fertilised pasture and cropping that occurs on the landscape (McIntyre *et al.* 2002).

6. New thinking on the management of woodland structure is needed to identify how tree density, size and distribution can be managed optimally for plants and animals in contemporary landscapes

In some grassy woodlands, getting trees back into the system is a limitation, either because the tree species on site are poor regenerators, or more often because the understorey has been fertilised and weed species disrupt seedling establishment. Planting trees in dense tree lines is a common response, but it produces a structure that is very unlike a woodland, which is defined as having trees that are open-grown, with wide, well spaced crowns. Where soils are low in nutrients and functional populations of trees occur, removing livestock can lead to dense regeneration, leading to loss of the woodland structure as well. We do not really understand what maintained the tree spacing in pre-European grassy woodlands, although fire is thought to have been a factor controlling dense regeneration. As yet, we do not know what the ideal structure for our woodland reserves is. We do know that mature trees with hollows are disproportionately important, irreplaceable and must be conserved at all costs (Gibbons and Lindenmayer 2002). Birds can respond positively to dense tree regeneration, but a forest is not a woodland. The question remains as to what configurations of trees should we be aiming for? Is there an ideal level of patchiness that includes open and densely treed areas? What is the best way of achieving tree thinning? Can we accelerate the development of tree hollows?

Conclusions

These lessons reflect experiences and learning gained over the course of my professional life, and each is linked to the other: questioning assumed knowledge, looking at the problem objectively, simplifying the complexity for good communication. As a result of these, I have realised how important the issue of intensive land use is, and can see the management of tree structure in woodlands as an emerging issue. Australia's tradition of conservation management has generally relied on a low intervention philosophy, with the control of weeds being the most important activity. There is much to be said for this approach, as bold management actions are often made in the absence of much technical knowledge. There is also scope, however, to be more proactive in our management, bearing in mind that mistakes are made, and no single management approach should be universally applied. In an ideal world, these bold moves should be done experimentally with proper assessments of their effects, giving us the opportunity to refine our approach.

Acknowledgements

I have always worked with supportive and positive colleagues and while all cannot be listed, I would particularly like to thank David Ashton, Pauline Ladiges, Jeff Yugovic, David Mitchell, John Lumb, Sandra Lavorel, Wal Whalley, John Taylor, Phil Price, Richard Hobbs, Neil MacLeod, Katina Heard, Tara Martin, Sandra Díaz, Jill Landsberg, Josh Dorrough, Adrian Manning, Ian Lunt, David Lindenmayer, and Suzanne Prober. I have received particularly strong institutional support from CSIRO, Land and Water Australia and Meat and Livestock Australia.

Bio

Sue McIntyre is a Senior Principal Research Scientist with CSIRO, where she has worked for 17 years. Her training was in botany (ecology) at the University of Melbourne, where she completed Honours and her PhD. She was awarded an ARC Postdoctoral Fellowship at the University of New England, and moved to Brisbane in 1993 to take up a research scientist position with CSIRO Tropical Crops and Pastures. She is currently in Canberra with CSIRO Sustainable Ecosystems.

References

Dorrough, J. and Moxham, C. (2005). Eucalypt establishment in agricultural landscapes and implications for landscape-scale restoration. *Biological Conservation* **123**, 55–66.

Dorrough, J., Moxham, C., Turner, V. and Sutter, G. (2006). Soil phosphorus and tree cover modify the effects of livestock grazing on plant species richness in Australian grassy woodland. *Biological Conservation* **130**, 394–405.

Gibbons, P. and Lindenmayer, D. (2002). *Tree Hollows and Wildlife Conservation in Australia.* CSIRO Publishing: Melbourne.

Jurskis, V. (2005). Eucalypt decline in Australia, and a general concept of tree decline and dieback. *Forest Ecology and Management* **215**, 1–20.

Kuhn, T.S. (1962). *The Structure of Scientific Revolutions.* University of Chicago Press: Chicago.

Lavorel, S., McIntyre, S., Landsberg, J. and Forbes, T.D.A. (1997). Plant functional classifications: from general functional groups to specific disturbance response groups. *Trends in Ecology and Evolution* **12**, 474–478.
Lindenmayer, D.B., Fischer, J. and Cunningham, R.B. (2005). Native vegetation cover thresholds associated with species responses. *Biological Conservation* **124**, 311-316.
McIntyre, S. (2008). The role of plant attributes in linking land use to ecosystem function and values in temperate grassy vegetation. *Agriculture, Ecosystems and Environment* **128**, 251–258.
McIntyre, S. and Barrett, G.W. (1992). Habitat variegation, an alternative to fragmentation *Conservation Biology* **6**, 146–147.
McIntyre, S., Heard, K.M. and Martin, T.G. (2003). The relative importance of cattle grazing in sub-tropical grasslands – does it reduce or enhance plant biodiversity? *Journal of Applied Ecology* **40**, 445–457.
McIntyre, S. and Hobbs, R.J. (1999). A framework for conceptualizing human impacts on landscapes and its relevance to management and research models. *Conservation Biology* **13**, 1282–1292.
McIntyre, S., Lavorel, S. and Trémont, R.M. (1995). Plant life-history attributes: their relationship to disturbance response in herbaceous vegetation. *Journal of Ecology* **83**, 31–44.
McIntyre, S. and Martin, T.G. (2002). Managing intensive and extensive land uses to conserve grassland plants in eucalypt woodlands. *Biological Conservation* **107**, 241–252.
McIntyre, S., McIvor, J.G. and Heard, K.M., eds. (2002). *Managing and Conserving Grassy Woodlands*. CSIRO Publishing: Melbourne.
McIntyre, S., McIvor, J.G. and MacLeod, N.D. (2000). Principles for sustainable grazing in eucalypt woodlands: landscape-scale indicators and the search for thresholds. In *Management for Sustainable Ecosystems*. (Eds P.A. Hale, A. Petrie, D. Moloney and P. Sattler) pp. 92–100. Centre for Conservation Biology, University of Queensland: Brisbane.
McIntyre, S. and Yugovic, J. (1984). A checklist for Studley Park and Yarra Bend Reserves. *Victorian Naturalist* **99**, 147–152.
Prober, S.M. and Thiele, K.R. (1995). Conservation of the grassy white box woodlands: relative contributions of size and disturbance to floristic composition and diversity of remnants. *Australian Journal of Botany* **43**, 349–366.
Prober, S.M., Thiele, K.R., Lunt, I.D. and Koen, T.B. (2005). Restoring ecological function in temperate grassy woodlands: manipulating soil nutrients, exotic annuals and native perennial grasses through carbon supplements and spring burns. *Journal of Applied Ecology* **42**, 1073–1085.
Radford, J.Q., Bennett, A.F. and Cheers, G.J. (2005). Landscape-level thresholds of habitat cover for woodland-dependent birds. *Biological Conservation* **124**, 317–337.
Stuwe, J. and Parsons, R.F. (1977). *Themeda australis* grasslands on the Basalt Plains, Victoria: floristics and management effects. *Australian Journal of Ecology* **2**, 467–476.
Wilkins, S., Keith, D.A. and Adam, P. (2003). Measuring success: evaluating the restoration of a grassy eucalypt woodland in the Cumberland Plain, Sydney, Australia. *Restoration Ecology* **11**, 489–503.

28

LANDSCAPE AND REGIONAL PERSPECTIVES FROM EASTERN AUSTRALIA

Clive A. McAlpine, Michiala E. Bowen and Jonathan R. Rhodes

1. There is a real danger of applying uniform conservation actions across regions; rather, priority management actions and targets for woodlands should be identified individually for different species and for different regions.
2. Stopping habitat loss needs to be a critical priority, but the spatial arrangement and functional connectivity of that habitat is also important.
3. Subtropical regrowth woodlands are important for the conservation and recovery of bird communities, and should be allowed to regenerate.
4. Mistletoes are an important resource for woodland birds in Brigalow; however, their persistence requires management of land-use disturbances combined with restoration programs.
5. Linear strips of Brigalow are an important resource and should not be allowed to degrade
6. Intact native woodlands can play an important biophysical role in mitigating climate change at a regional scale. This role needs to be explicitly considered in climate change mitigation policies, vegetation management and regional planning processes.

Introduction

Our research spans 15 years and a wide range of environments in eastern Australia (see map on next page). We have conducted studies in savanna woodlands of tropical north Queensland; eucalypt and acacia woodlands of southern Queensland and northern New South Wales; and temperate eucalypt woodlands of central Victoria. This work has addressed a broad range of problems including: 1) the effect of habitat loss and fragmentation on native biodiversity; 2) the ecological values of regrowth vegetation; 3) cross-regional variation in habitat relationships and habitat retention thresholds; and 4) the effect of historical land clearing on climate. We have studied a range of biota including kangaroos, koalas, woodland birds, reptiles, and arboreal and ground-dwelling mammals, and have worked at multiple spatial scales from the site (<1 ha), patch (1–1000s ha), landscape (100s–10 000s ha), region (10 000s km^2) and even the continental

scale. We have developed partnerships with the Queensland and New South Wales State Governments, local governments, regional natural resource management bodies and catchment management authorities, and with non-government organisations and industry bodies.

Lessons

1. There is a real danger of applying uniform conservation actions across regions; rather, priority management actions and targets for woodlands should be identified individually for different species and for different regions.

Understandably, there is considerable interest in developing general rules of thumb that can be applied to managing natural resources and biodiversity across different regions. In Australia, some general rules of thumb have been proposed, such as retaining at least 30–50% of the original extent of native vegetation in the landscape to conserve biodiversity (McAlpine *et al.* 2002). These simple rules are appealing, but evidence from some of our work on the Koala (*Phascolarctos cinereus*) in woodlands across eastern Australia suggests that it may be misleading to apply general management rules uniformly across different regions, even for the same species (McAlpine *et al.* 2008; Rhodes *et al.* 2008). McAlpine *et al.* (2008) show that some factors, such as the effect of tree species, are consistently important determinants of Koala presence. The importance of most factors, however, including those commonly used to guide management policies, such as the effect of road density and stand composition, vary among regions. Further, Rhodes *et al.* (2008) show that threshold responses of Koalas to forest loss vary considerably among regions. This has particularly important consequences for the setting of woodland and forest retention targets in Australia where uniform rules have regularly been applied; for example, the 15% targets in the Regional Forest Agreements (ANZECC/MCFFA 1997). In contrast to the uniform application of these rules across regions, priority management actions and targets for woodlands should be identified individually for different species and for different regions.

2. Stopping habitat loss needs to be a critical priority, but the spatial arrangement and functional connectivity of that habitat is also important

Habitat loss is recognised as the major driver of biodiversity decline worldwide (Fahrig 2003). A key lesson emerging from our research is that habitat amount is the critical factor limiting the occurrence of forest and woodland dependent fauna, especially mammals (McAlpine *et al.* 2006). This lesson has emerged primarily from our extensive Koala research, where the loss of high quality breeding habitat (eucalypt forests and woodlands growing on fertile soils) has triggered major contractions in this species' distribution and ultimately its persistence in the landscape (ANZECC 1998; McAlpine *et al.* 2006; Rhodes *et al.* 2006; Januchowski *et al.* 2008). The loss of low quality habitat also has a negative effect on species' occurrence; movement of individuals among small patches of high quality breeding habitat can be enhanced by low quality habitat in the surrounding landscape (Wiegand *et al.* 2005). Landscape configuration and the heterogeneity of the matrix also matter (Ricketts 2001). The isolation of patches is an important determinant of Koala occurrence, with Koalas more likely to occur in large well-connected patches than in isolated patches.

3. Subtropical regrowth woodlands are important for the conservation and recovery of bird communities, and should be allowed to regenerate

Naturally regenerating regrowth vegetation on formerly cleared land is a common feature of the sheep–wheat belt in north-western New South Wales and southern Queensland (Doherty 1998; Fensham and Guymer 2009). These regenerating woodlands are widely considered to be 'woody weeds' by landholders and regrowth vegetation management has traditionally focused on methods for controlling the development and spread of regrowth (e.g. Johnson 1964), with short-term (5–10 year) cycles of re-clearing. A growing body of evidence, however, suggests that a wide range of regrowth forests, if allowed to regenerate for 20–100 years, can support diverse fauna communities and have a vital role in the restoration of highly modified agricultural landscapes (Bowen *et al.* 2007).

In fragmented agricultural landscapes of southern Queensland, Brigalow (*Acacia harpophylla*) regrowth forests are capable of supporting a diverse suite of the original bird community of mature woodland within a period of at least 30–60 years (Bowen 2009). Similarly, the amount of regrowth vegetation in the surrounding landscape also has positive benefits for the diversity and abundance of woodland birds if it is permitted to regenerate for periods longer than 30–60 years (Bowen *et al.* 2009a). Younger Brigalow regrowth stands, however, particularly <15 years, support a depauperate bird community (Bowen *et al.* 2009a). Furthermore, some woodland bird species, (e.g. White-throated Treecreeper *Corombates leucophaeus*) are absent from Brigalow regrowth and will most likely remain dependent on mature woodlands. These absences are due to the much longer time periods (>100 years) required to develop critical habitat resources characteristic of mature woodlands, such as tree hollows and fallen woody debris (Wormington and Lamb 1999; Gibbons *et al.* 2000; Vesk *et al.* 2008). Thus the conservation and management of mature woodlands in highly modified agricultural landscapes remains an important priority for vegetation management and biodiversity conservation.

The key lesson arising from this research is that the ecological values of Brigalow regrowth should be viewed from a long-term perspective of their potential to recover woodland bird communities, rather than assessing this vegetation on current day similarity, or dissimilarity, to mature woodlands. There is an urgent need to increase landholder and land manager awareness of the ecological values of naturally regenerating subtropical woodlands and to start planning now for the long-term restoration of landscapes, because the decision to retain regrowth woodlands for conservation will take many decades to produce substantial biodiversity benefits.

4. Mistletoes are an important resource for woodland birds in Brigalow; however, their persistence requires management of land-use disturbances combined with restoration programs

A key lesson arising from our research in fragmented Brigalow woodlands is that mistletoes are an important resource for woodland birds. In addition, their increased abundance in narrow linear strips may facilitate the movement and persistence of some woodland birds in these landscapes. Mistletoes are hemi-parasitic plants that demonstrate complex responses to woodland fragmentation in agricultural landscapes of temperate Australia (Norton and Reid 1997). Mistletoes provide food (fruit, nectar and foliage), shelter and nesting substrate for many woodland animals, potentially representing a keystone resource (Watson 2001). Our research in Brigalow woodlands suggests that mistletoes, particularly Grey Mistletoe (*Amyema quandang*), are important resources for woodland birds (Bowen 2009; Bowen *et al.* 2009b). Brigalow woodlands with abundant mistletoe populations had higher species richness of woodland birds and greater numbers of nectarivorous/frugivorous birds, such as the Mistletoebird (*Dicaeum hirundinaceum*), Spiny-cheeked Honeyeater (*Acanthagenys rufogularis*) and the Painted Honeyeater (*Grantiella picta*).

Brigalow woodland fragmentation had a positive effect on mistletoe abundance, with the highest numbers of mistletoe plants occurring in highly modified landscapes and in narrow linear strips (Bowen *et al.* 2009b). Similar responses to fragmentation are documented for temperate woodlands, where increased mistletoe abundance is often attributed to reduced health and increased mortality of host trees (Calder and Bernhardt 1983). This link has not been proven for other woodland systems (e.g. Reid and Lange 1988; Reid *et al.* 1992). In subtropical Brigalow woodlands, mistletoes were more abundant in healthier stands with lower basal area of standing dead trees (Bowen *et al.* 2009b). Thus in Brigalow woodlands, mistletoes clearly represent an important resource for bird communities, and as yet, there is no clear detrimental effect of increased mistletoe abundance on local host tree health.

5. Linear strips of Brigalow are an important resource and should not be allowed to degrade

When it comes to conserving woodland patches, landscape ecology theory suggests that bigger is better. In the highly modified agricultural landscapes (e.g. >90% cleared) of the Brigalow Belt, however, woodland remnants are often restricted to narrow linear strips along roadsides, fences and as shadelines in paddocks, or as isolated paddock trees. A key lesson arising from our research is that linear strips of mature Brigalow woodland are a critical habitat resource and greater effort is needed to protect these patches from further degradation. These linear strips support a similar species richness of woodland-dependent birds to the larger mature forest blocks in National Parks (>900 ha) (Bowen 2009). Although the species composition of woodland birds in linear patches comprises more disturbance-favouring species (e.g. Noisy Miner *Manorina melanocephala*, Crested Pigeon *Ocyphaps lophotes*, Galah *Cacatua roseicapilla*) compared to large areas of woodland, many species that are in decline in New South Wales and Victoria can also be found in these linear strips, particularly those wider than 50 m (e.g. Red-capped Robin *Petroica goodenovii*, Eastern Yellow Robin *Eopsaltria australis*, Jacky Winter *Microeca fascinans*, Grey-crowned Babbler *Pomatostomus temporalis*, White-browed Babbler *P. superciliosus*, Varied Sittella *Daphoenositta chrysoptera*).

Many linear strips of Brigalow, however, are subject to frequent and intense grazing disturbance by sheep and cattle, and chemical spray drift from adjacent croplands. In general, these types of disturbance contribute to increased rates of tree mortality, reduced structural complexity and lowered regenerative capacity due to removal of shrubs and young trees (Hobbs 2001; Briggs *et al.* 2008). In addition to these land-use disturbances, recent severe droughts are also causing low to moderate rates of mortality in Brigalow trees. This phenomenon is likely to increase under projected changes in the region's climate (CSIRO 2007). To remain viable in the long-term, many of these linear strips will require management of land-use disturbances. This will also need to be combined with some form of restoration, via passive (e.g. regrowth) or active (e.g. planting) methods, to widen these strips and buffer them from the synergistic effects of fragmentation, adjacent land-use, grazing disturbance and drought.

6. Intact native woodlands can play an important biophysical role in mitigating climate change at a regional scale. This role needs to be explicitly considered in climate change mitigation policies, vegetation management and regional planning processes

Australia is facing unavoidable warming due to increased concentrations of anthropogenic greenhouse gas emissions (Pittock 2009). An important consequence of this is an increase in the frequency of extremes such as droughts, severe storms and heatwaves. Climate change is a complex process, however, with multiple forcings and feedbacks affecting the climate system (IPCC 2007; McAlpine *et al.* 2009). Our recent research has shown that historical clearing of native vegetation in the intensive land-use zones of eastern Australia and south-west Western Australia is an important factor influencing both mean climate and climate extremes (McAlpine *et al.* 2007; McAlpine *et al.* 2009). Historical land clearing has contributed to hotter and longer droughts during El Niño years, such as the severe drought of 2002/2003. We also found an increase in the number of dry and hot days, a decrease in daily rainfall intensity and wet day rainfall, and an increase in drought duration for modified land cover conditions in eastern Australia (Deo *et al.* 2009). These changes were statistically significant for all years, and especially pronounced during strong El Niño events. They provide a timely warning of the risk of dangerous climate change stemming from the interaction of an extensively cleared land surface and a hotter and more drought-prone climate driven by increased anthropogenic greenhouse gases. They also highlight the need for stronger policy action to expand native vegetation cover as a mitigation option for climate change. The large-scale restoration of native forests and woodlands has the potential to ameliorate regional climate change while providing other ecological services such as carbon bio-sequestration, biodiversity recovery, and catchment hydrology. We currently do not know to what extent such actions will modify temperature and rainfall patterns directly under global warming. This question requires urgent consideration to inform climate change policies, if they are to lead toward more integrated climate mitigation and sustainable land-use outcomes.

Conclusions

What are the knowledge gaps for research and management?

Our work has significantly advanced understanding of a range of species' responses to forest and woodland change across a range of spatial scales. We have also been able to identify important feedbacks between forest loss and climate change. These aspects are critical components for better management of woodland and forest ecosystems at the local, landscape and regional scales. Several knowledge gaps still remain.

Our finding that you cannot apply uniform conservation actions or targets across different regions poses a significant challenge for broadscale research and management of forest and woodland ecosystems. This is because it suggests that we need to conduct region-specific research to identify appropriate management actions individually in each region. This is a daunting prospect for the management of woodlands across large areas. As a consequence, it is critical that we aim to understand under what circumstances we can apply general rules and under what circumstances we have to focus our efforts on understanding the idiosyncrasies of individual regions.

While our research has highlighted the importance of retaining regrowth vegetation for woodland birds, this research is largely based on species' occurrence data and needs to be expanded to address a broader range of animal taxa. It also needs to focus on population processes of recovering fauna (e.g. dispersal, movement, breeding, mortality), which are poorly understood. For example, at what stage does regrowth woodland provide the necessary resources to support breeding populations, and does regrowth provide all the resources necessary to support fauna populations, or is it merely providing a supplementary resource for species that remain dependent on nearby mature woodlands? This requires an understanding of the dynamics of the system. For land managers, regrowth vegetation needs to be considered from a long-term perspective of potential values rather than remain based on the current condition. Regeneration trajectories may vary considerably among regrowth patches, and long-term monitoring of fauna and flora in regrowth and mature forests will therefore be essential for informing landscape restoration efforts.

Our discovery that historical land clearing has resulted in a hotter and drier climate in eastern Australia raises the question of the potential for large-scale revegetation programs to help mitigate the regional impacts of climate change, and if they can, how much vegetation is required and where should it be located? Restoring native woodlands and forests has the potential to at least partially reverse the climate impacts of historical clearing and help buffer increased extremes resulting from global warming. There is an increasing expectation that native vegetation should be used to sequester atmospheric CO_2, as well as for biodiversity and other ecosystem services. The strategic revegetation of highly modified agricultural landscapes, either through active planting or passive recovery of regrowth, can provide enhanced opportunities for mitigation of climate change and biodiversity conservation (McAlpine *et al.* 2009). A major research challenge, however, is to develop integrated landscape designs that restore these multiple functions in production landscapes.

What are the challenges and frontiers for management and policy?

Coordinated Commonwealth, State and regional strategies are needed urgently to reduce land-use pressures and restore the condition and extent of native vegetation cover in order to increase the resilience of woodland landscapes to climate change (McAlpine *et al.* 2009). Firstly, this requires tighter legislative controls on the clearing of native vegetation in all woodland communities, including the clearing of regrowth native vegetation in previously cleared subtropical landscapes. Secondly, a comprehensive evaluation is required of woodland ecosystem dynamics under a projected hotter, drier and more variable climate, and their capacity to sustain biodiversity, sequester carbon and maintain beneficial feedbacks of the land surface on the climate system. Finally, based on this evaluation, strategic revegetation programs are required that integrate the conservation, production, climate, ecohydrological and carbon sequestration values of native woodland vegetation at the farm, landscape and catchment scales.

Implications of inaction and/or failure to learn

The long-term future of our native woodland ecosystems and their native biodiversity is uncertain. The effects of habitat loss and fragmentation on biodiversity and ecosystem function are now being compounded by climate change, especially the increased frequency of severe droughts. Although we now have some understanding of the impact of forest loss, fragmentation and modification on biodiversity and climate, there are still large uncertainties. These uncertainties are not an excuse for inaction; decisions should be made within the context of existing knowledge while accounting for the level of uncertainty (e.g. Possingham *et al.* 2001). A failure to do this runs the risk of failing to act appropriately to conserve our unique biodiversity and to reduce the risk of catastrophic climate change.

It is not widely appreciated that if global greenhouse gas emissions stopped tomorrow, the Australian climate would continue to undergo significant changes well into the future. We may already be committed to significant ecosystem changes due to climate change (*sensu* Jones *et al.* 2009). If current emission trends continue, we are heading towards a 5°C temperature increase and up to 30% decline in rainfall by 2070 in southern Australia (CSIRO 2007; Smith and Chandler 2009). Feedbacks from a hotter and drier land surface are likely to accentuate these temperature and rainfall changes, resulting in increased aridity (McAlpine *et al.* 2009). This represents a major risk for woodland ecosystems and their biodiversity. A hotter and drier landscape will also increase the risk of crossing abiotic and biotic thresholds that significantly alter regeneration pathways (*sensu* Hobbs and Harris 1998) and prevent restoration of woodland ecosystem processes on previously cleared lands. Policy actions are required now that aim to reduce land-use pressures and invest in strategic revegetation programs to reduce the risk of irreversible loss of biodiversity and dangerous climate change.

Bios

Clive McAlpine is an Associate Professor in the School of Geography, Planning and Environmental Management and The Ecology Centre at the University of Queensland. His research is at the interface of landscape ecology, biodiversity conservation, and climate change. He is particularly interested in the processes driving landscape change, the conservation of biodiversity in human-modified landscapes and the ecological and climatic consequences of landscape change.

Michiala Bowen is a postdoctoral research fellow in the School of Geography, Planning and Environmental Management at the University of Queensland. Her main research interest is in understanding how landscape modification influences the spatial and temporal dynamics of faunal communities. She is particularly interested in understanding the recovery processes of regrowth forest communities on agricultural lands and how these forests can contribute to broadscale landscape restoration of fauna habitat.

Jonathan Rhodes is a Lecturer in the School of Geography, Planning and Environmental Management and The Ecology Centre at the University of Queensland. His research focuses on understanding the spatial dynamics of biological populations and the implications of this for decision making in biodiversity conservation. He works across a range of terrestrial systems in Australia and overseas, but with a focus on highly human-modified urban and production landscapes.

References

ANZECC/MCFFA (1997). *Nationally agreed criteria for the establishment of a comprehensive, adequate and representative reserve system for forests in Australia*. Commonwealth of Australia: Canberra.

ANZECC (1998). *National Koala conservation strategy*. Environment Australia: Canberra.

Bowen, M.E. (2009). Quantifying the ecological values of brigalow regrowth for woodland birds: a hierarchical landscape approach. PhD thesis, The University of Queensland.

Bowen, M.E., McAlpine C.A., House A.P.N. and Smith G.C. (2007). Regrowth forests on abandoned agricultural land: a review of their habitat values for recovering forest fauna. *Biological Conservation* **140**, 273–296.

Bowen, M.E., McAlpine C.A., House A.P.N. and Smith G.C. (2009a). Agricultural landscape modification increases the abundance of an important food resource: mistletoes, birds and brigalow. *Biological Conservation* **142**, 122–133.

Bowen, M.E., McAlpine C.A., Seabrook L., House A.P.N. and Smith G.C. (2009b). The age and amount of regrowth forest in fragmented brigalow landscapes are both important for woodland dependent birds, Queensland, Australia. *Biological Conservation* **142**, 3051–3059.

Briggs, S.V., Taws, N.M., Seddon, J.A. and Vanzella, B. (2008). Condition of fenced and unfenced remnant vegetation in inland catchments in south-eastern Australia. *Australian Journal of Botany* **56**, 590–599.

Calder, D.M. and Bernhardt, P. (1983). *The Biology of Mistletoes*. Academic Press: Sydney.

CSIRO (2007). 'Climate change in Australia: technical report 2007'. Commonwealth Scientific and Industrial Research Organisation, Canberra.

Deo, R.C., Syktus, J., McAlpine, C.A., Lawrence, P.J., McGowan, H.A. and Phinn, S. (2009). Impact of historical land cover change on daily indices of climate extremes including droughts in eastern Australia. *Geophysical Research Letters* **36**, L08705, doi:10.1029/2009GL037666.

Doherty, M. (1998). 'The conservation value of regrowth native plant communities: a review'. Report by CSIRO Division of Wildlife and Ecology for the New South Wales Scientific Committee: Canberra.

Fahrig, L. (2003). Effects of habitat fragmentation on biodiversity. *Annual Review of Ecology, Evolution and Systematics* **34**, 487–515.

Fensham, R.J. and Guymer, G.P. (2009). Carbon accumulation through ecosystem recovery. *Environmental Science & Policy* **12**, 367–372.

Gibbons, P., Lindenmayer, D.B., Barry, S.C. and Tanton, M.T. (2000). Hollow formation in eucalypts from temperate forests in Southeastern Australia. *Pacific Conservation Biology* **6**, 218–228.

Hobbs, R.J. (2001). Synergisms among habitat fragmentation, livestock grazing, and biotic invasions in southwestern Australia. *Conservation Biology* **15**, 1522–1528.

Hobbs, R.J. and Harris, J.A. (1998). Restoration ecology: repairing the Earth's ecosystems in the new millennium. *Restoration Ecology* **9**, 239–246.

IPCC (2007). *Climate Change 2007: The Physical Science Basis. Contribution of Working Group I to the Fourth Assessment Report of the Intergovernmental Panel on Climate Change*. (Eds S. Solomon, D. Qin, M. Manning, Z. Chen, M. Marquis, K.B. Averyt, M. Tignor and H.L. Miller). Cambridge University Press: New York.

Januchowski, S.R., McAlpine, C.A., Callaghan, J.G., Griffin, C.B., Bowen, M.E., Mitchell, D. and Lunney, D. (2008). Identifying multiscale habitat factors influencing koala

(*Phascolarctos cinereus*) occurrence and management in Ballarat, Victoria, Australia. *Ecological Management and Restoration*, **9**, 134–142.

Johnson, R.W. (1964). *Ecology and control of brigalow in Queensland.* Queensland Department of Primary Industries: Brisbane.

Jones, C., Lowe, J., Liddicoat, S. and Betts, R. (2009). Committed terrestrial ecosystem changes due to climate change. *Nature Geoscience Letters* **2**, 484–487.

McAlpine, C.A., Fensham, R.J. and Temple-Smith, D.E. (2002). Biodiversity conservation and vegetation clearing in Queensland: principles and thresholds. *Rangelands Journal* **24**, 36–55.

McAlpine, C.A., Rhodes, J.R., Callaghan, J.G., Bowen, M.E., Lunney, D., Mitchell, D.L., Pullar, D.V. and Possingham, H.P. (2006). The importance of forest area and configuration relative to local habitat factors for conserving forest mammals: a case study of koalas in Queensland, Australia. *Biological Conservation* **132**, 153–165.

McAlpine, C.A., Rhodes, J.R., Bowen, M.E., Lunney, D., Callaghan, J.G., Mitchell, D.L. and Possingham, H.P. (2008). Can multi-scale models of species' distribution be generalised from region to region? A case study of the koala. *Journal of Applied Ecology* **45**, 558–567.

McAlpine, C.A., Syktus, J., Deo, R.C., Lawrence, P.J., McGowan, H.A., Watterson, I.G. and Phinn, S.R. (2007). Modeling the impact of historical land cover change on Australia's regional climate. *Geophysical Research Letters* **34**, L22711, doi:10.1029/2007GL031524

McAlpine, C.A., Syktus, J., Ryan, J.G., Deo, R.C., McKeon, G.M., McGowan, H.A. and Phinn, S.R. (2009). A continent under stress: interactions, feedbacks and risks associated with impact of modified land cover on Australia's climate. *Global Change Biology* **15**, 2206–2223.

Norton, D.A. and Reid, N. (1997). Lessons in ecosystem management from management of threatened and pest loranthaceous mistletoes in New Zealand and Australia. *Conservation Biology* **11**, 759–769.

Pittock, A.B. (2009). *Climate Change: The Science, Impacts and Solutions.* 2nd edn. CSIRO Publishing: Melbourne.

Possingham, H.P., Andelman, S.J., Noon, B.R., Trombulak, S. and Pulliam, H.R. (2001). Making smart conservation decisions. In *Conservation Biology: Research Priorities for the Next Decade.* (Eds M.E. Soulé and G.H. Orians). Island Press: Washington.

Reid, N. and Lange, R.T. (1988). Host specificity, dispersion and persistence through drought of two arid zone mistletoes. *Australian Journal of Botany* **36**, 299–313.

Reid, N., Smith, D.M.S. and Venables, W.N. (1992). Effect of mistletoes (*Amyema preissii*) on host (*Acacia victoriae*) survival. *Australian Journal of Ecology* **17**, 219–222.

Ricketts, T.H. (2001). The matrix matters: effective isolation in fragmented landscapes. *The American Naturalist* **158**, 87–99.

Rhodes, J.R., Wiegand, T., McAlpine, C.A., Callaghan, J., Lunney, D., Bowen, M. and Possingham, H.P. (2006). Modelling species distributions for improving conservation in semi-urban landscapes: a koala case study. *Conservation Biology* **20**, 449–459.

Rhodes, J.R., Callaghan, J.G., McAlpine, C.A., de Jong, C., Bowen, M.E., Mitchell, D.L., Lunney, D. and Possingham, H.P. (2008). Regional variation in habitat-occupancy thresholds: a warning for conservation planning. *Journal of Applied Ecology* **45**, 549–557.

Smith, I. and Chandler, E. (2009). Refining rainfall projections for the Murray Darling Basin of south-east Australia: the effect of sampling model results based on performance. *Climate Change*, doi:10.1007/s10584-009-9757-1.

Vesk, P.A., Nolan, R., Thomson, J.R., Dorrough, J.W. and Mac Nally R. (2008). Time lags in provision of habitat resources through revegetation. *Biological conservation* **141**, 174–186.

Watson, D.M. (2001). Mistletoe – a keystone resource in forests and woodlands worldwide. *Annual Review of Ecology and Systematics* **32**, 219–249.
Wiegand, T., Revilla, E. and Moloney, K.A. (2005). Effects of habitat loss and fragmentation on population dynamics. *Conservation Biology*, **19**, 108–121.
Wormington, K. and Lamb, D. (1999). Tree hollow development in wet and dry sclerophyll eucalypt forest in south-east Queensland, Australia. *Australian Forestry* **62**, 336–345.

29

RETAINING AND RESTORING WILDLIFE VALUES OF WOODLANDS

Martine Maron

1. Trying to achieve all management objectives everywhere can be inefficient.
2. Details can be deal breakers.
3. Noisy Miners are the most useful indicator for woodland bird communities.
4. Paddock trees are important, and they are disappearing.
5. 'Gains' on paper can be losses in reality.
6. Achieving 'naturalness' is not always a useful restoration goal.
7. To design better landscapes, we need a better understanding of the mechanisms behind species decline.

Introduction

My focus is biodiversity conservation in human-dominated landscapes, particularly understanding the processes driving species declines and the implications of this for natural resource management and conservation policy. I am interested in developing strategies to restore and retain biodiversity (particularly fauna) in landscapes ranging from extensive forest areas with a history of disturbance, to relictual landscapes (*sensu* McIntyre and Hobbs 1999) which are, and will continue to be, dominated by intensive agriculture. Most of the research in which I have been involved has been in partnership with regional natural resource management organisations, government departments and conservation non-government organisations. The systems in which I have worked include two endangered non-eucalypt woodlands which are at the extreme end of the fragmentation spectrum: Buloke (*Allocasuarina luehmannii)* in Victoria and Brigalow (*Acacia harpophylla)* in Queensland and New South Wales (see map on next page). Both of these woodland types grow on fertile grey clays and were therefore preferentially cleared for agriculture (~3% and ~10% vegetation cover left, respectively).

Both Buloke and Brigalow woodland have unusual features. Unlike most fragmented eucalypt woodland in eastern Australia, pure Buloke woodland has no aggressive Noisy Miners (*Manorina melanocephala*), and therefore even tiny remnants support many of the small passerine birds which are declining in other landscapes (Maron 2007). These fragmented remnant woodlands demonstrate emphatically that we cannot afford to overlook exceptions to our

generalisations about the effects of habitat modification—for several woodland bird species, a small patch of Buloke woodland is more valuable than a large patch of eucalypt woodland. The Noisy Miner does occur in the Brigalow woodlands, but these more recently cleared landscapes still regenerate in the absence of cultivation, thus opening up significant opportunities for large-scale restoration which would be almost impossible to achieve in more southerly regions (Bowen *et al.* 2009). Our research group is currently examining the influence of habitat and landscape factors, including the role of regrowth retention, on a range of vertebrate taxa in Brigalow woodlands, and developing an approach for the delivery of this information to land managers in the form of web-based decision-support tools.

My experience working in more intact landscapes is from the subtropical woodlands of the Brigalow Belt that stretch over several million hectares from southern Queensland up to the Carnarvon range. These woodlands are dominated by Spotted Gum (*Corymbia citriodora),* ironbarks (*Eucalyptus* spp.) and White Cypress Pine (*Callitris glaucophylla).* The sheer area of almost contiguous remnant woodlands suggests the promise of a significant refuge for fragmentation-sensitive woodland species, but the reality is different, with grazing by feral and domestic herbivores, logging, and burning affecting their conservation value (Maron and Kennedy 2007; Eyre *et al.* 2009; Howes and Maron 2009). For example, even in these extensive areas of woodland, the Noisy Miner (previously considered an edge specialist) dominates, with considerable effects on the rest of the bird community (Eyre *et al.* 2009; see Lesson #2). There is clearly no part of Australia's temperate and subtropical woodland zones where wildlife has avoided the significant impacts of human activity.

Lessons

1. Trying to achieve all management objectives everywhere can be inefficient

Considering the multitude of biodiversity management and restoration objectives that land managers must try to address, it is inevitable that some will be in conflict. Thus, in trying to achieve a suite of outcomes in a given region, it is as critical to identify where there is potential for conflict as it is to look for potential synergies. The existence of ecological thresholds means that a compromise on one objective to achieve another can result in suboptimal outcomes for both objectives (Maron and Cockfield 2008). In the Buloke woodlands of the Victorian Wimmera, the Noisy Miner invades woodlands with five or more eucalypts per hectare, leading to the almost complete exclusion of small woodland birds. Yet revegetation of Buloke woodland in the region has typically included a high density of eucalypts amongst the plantings, partly because of the faster growth rates and perceptions that eucalypts are more pleasing aesthetically than Bulokes (Maron 2007). Such negative tradeoffs highlight the critical importance of clearly identifying specific objectives at the project design and prioritisation stage of restoration planning. Having vague objectives increases the risk that restoration will attempt to restore a suite of conflicting habitat requirements, leading to inefficient compromises. For example, the use of a single metric which attempts to encapsulate multiple aspects of 'habitat value' to assess competing projects across a region can result in all projects in the portfolio representing a similar compromise, but none providing optimal habitat for any given species. Diverse metrics for diverse goals should be used to achieve a balanced portfolio of projects, each with a dominant objective (Maron and Cockfield 2008).

2. Details can be deal breakers

The rapid increase in the ability of remotely sensed data to characterise landscapes has led to a strong reliance on vegetation mapping products to inform restoration planning and habitat assessment. For many species, however, the most important things may not be represented in vegetation mapping data (Kath *et al.* 2009). While consideration of the role of spatial patterns in habitat suitability is essential, there tends to be an emphasis on factors such as structural connectivity that is stronger than can be justified by the literature (Fahrig 1997). Often the details of habitat structure and composition can be the most important determinants of habitat suitability of a site. For example, the extensive eucalypt forests of Queensland's Brigalow Belt South would appear to be ideal refugia for species which have declined in more highly-fragmented landscapes further south. Reduced shrub and groundcover due to the presence of feral herbivores such as horses (*Equus caballus)* in conservation areas has, however, created suitable habitat conditions for the Noisy Miner. This has resulted in a depauperate bird community similar to that expected in small, degraded woodland remnants in agricultural landscapes (Howes and Maron 2009). In the large State Forests of the region, frequent prescribed burning plays a similar role in facilitating high Noisy Miner abundance (Maron and Kennedy 2007). Paying adequate attention to such factors is particularly important considering the potential for ecological thresholds in some systems, which mean that small changes in habitat structure or condition can have big impacts on wildlife, such as the threshold eucalypt density described in Lesson #1 (Huggett 2005; Maron 2007).

3. The Noisy Miner is the most useful indicator for woodland bird communities

In the landscapes with which I am familiar, the Noisy Miner has been the strongest and most consistent predictor of the composition of woodland bird assemblages. Even after accounting for the effect of habitat structure, they remain strongly influential (Maron 2007; Maron and Kennedy 2007; Eyre *et al.* 2009; Kath *et al.* 2009; Howes and Maron 2009). The difference in the abundance of small-bodied passerines between sites with and without Noisy Miner colonies, in landscapes ranging from the Wimmera of western Victoria to the Carnarvon Range of central Queensland, is between 200% and 1000%. Few indicators of biodiversity fulfil so comprehensively the criteria of being quick, easy and cheap to monitor, and having a reliable and strong causal relationship with a large number of other species. They also occur widely throughout eastern Australia's woodlands, and so are potentially useful in many woodland systems. If the Noisy Miner is abundant, much can be accurately inferred about the bird fauna of a site. The importance of the Noisy Miner in structuring woodland bird assemblages also allows some generalisations: 1) woodlands comprising plant species with few food resources suitable for Noisy Miners, such as *Allocasuarina* and *Callitris* spp., tend to be disproportionately important for small passerines; and 2) reducing grazing pressure and encouraging dense understorey can reduce Noisy Miner impacts in eucalypt/*Corymbia* and Brigalow woodlands.

4. Paddock trees are important, and they are disappearing

So much native vegetation has already been cleared from agricultural landscapes that the disappearance of remaining paddock trees will represent a relatively small proportion of the total loss. Yet such trees are disproportionately valuable for wildlife as they typically are large, highly productive, and likely to have hollows important for roosting and nesting (Maron 2005; Manning *et al.* 2006). One paddock Desert Stringybark (*Eucalyptus arenacea)* can represent up to 27 times the food value of a tree in continuous forest for a south-eastern Red-tailed Black Cockatoo (*Calyptorhynchus banksii graptogyne*) (P. Koch unpublished data). A single paddock tree could also transform the space for some metres around it to habitat for pouncing birds. Once habitat value multipliers such as these are applied to paddock trees, it becomes clear that the ongoing and inevitable loss of these trees over the next 50 years is potentially catastrophic (Maron and Fitzsimons 2007). The maintenance of landscapes with habitat components scattered through production-dominated areas is recognised as an essential part of biodiversity management in Europe, but in Australia, this is still an emerging concept (Attwood *et al.* 2009). Recruiting paddock trees in agricultural landscapes is challenging as it can require at least temporary exclusion or reduction of grazing and cultivation from many small, disconnected areas. This may be fiddly, it may be expensive and it definitely requires a new mindset, but if we don't get scattered trees recruiting in our landscapes then we will lose much more biodiversity than we might expect (see Figure 29.1).

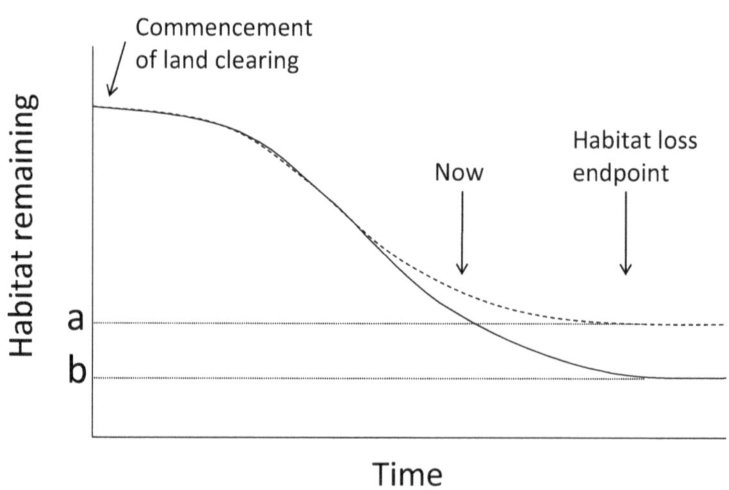

Figure 29.1 Theoretical impact of future loss of all paddock trees on habitat availability in an agricultural landscape. The dashed line shows expected future habitat availability (a) assuming equivalent value of paddock trees and trees in woodland; solid line indicates expected future habitat availability (b) assuming disproportionately high habitat value associated with paddock trees. The projected future loss of wildlife habitat assuming the latter scenario is considerably greater than would otherwise be anticipated.

5. Gains on paper can be losses in reality

The level of threat to remnant habitat is not adequately taken into account before we decide on the most efficient way to make conservation gains (or avoid losses) via incentive payments, reserve purchase or offsetting (Pressey and Taffs 2001). Investment, either through incentive payments to landholders or purchase of land for the reserve system, tends to be skewed towards larger areas of higher quality habitat. This approach, however, neglects the level of threat to the ecological values of the site. More gains (in quality gain and avoided loss) may be able to be made in smaller, poorer quality sites than in large, high quality ones, as smaller patches of habitat are more likely to be cleared. This concept is particularly important as applied to offset calculations. A 2:1 ratio of habitat protected to habitat cleared potentially places a 33% cap on clearing of already depleted woodlands, but only achieves an offset effect if half the values of the protected habitat would otherwise certainly be lost. The reality of offsetting approaches as currently applied involves a tendency to trade the sorts of resources which are at high risk of loss (scattered trees, small patches of vegetation) for resources which are (at least theoretically) already well protected (larger patches of endangered, intact woodland) (Gibbons and Lindenmayer 2007; Maron *et al.* 2010). In cases where we trade development which destroys some habitat for the protection of other already existing habitat, we have to consider more carefully whether the avoided loss is genuinely equivalent to the realised loss.

6. Achieving 'naturalness' is not necessarily a useful restoration goal

The non-random pattern of clearing and habitat modification, and the novel disturbance regimes which now characterise both remnant woodland and the surrounding matrix, mean that people attempting to restore woodland landscapes do not have the luxury of restoring them to a 'natural' or 'pristine' state. In this immensely changed context, it makes sense to do whatever we can to retain and reinforce the ecological values which are most at risk, and these may not necessarily be those which were represented within a given site prior to European influence. Indeed, whether an ecological restoration endpoint is 'natural' or not may be irrelevant. Some weeds can be important habitat and we may sometimes be better off investing in other actions rather than removing them, given limited restoration budgets. For example, Lantana (*Lantana camara)* can provide important habitat for small birds, and areas from which it is been removed can become dominated by open-country bird species at the expense of woodland-dependent species if not replaced with shrubs of similar structure (Kath *et al.* 2009).

We hear often that the wooded landscapes of parts of the continent were open enough to 'gallop a horse through', and are now unnaturally dense with woody growth. Whether or not this was the case for any given region, many species which today benefit from more open habitat are less likely to be of conservation concern, while areas of denser woodland with a shrubby understorey are in short supply and can provide useful refuge for woodland-dependent species (Howes and Maron 2009). Therefore, in some landscapes, we may be better off adopting habitat management regimes which simply aim to create an ecologically valuable landscape than attempting to achieve a 'natural' landscape which provides habitat for few species of concern (Maron and Kennedy 2007).

7. To design better landscapes, we need a better understanding of the mechanisms behind species decline

We do understand one important mechanism of woodland bird decline in Australia – interspecific competition from hyperagressive honeyeaters (Lessons #1–2). But in many cases, our understanding of how species respond to landscapes is limited to statements about patterns such as 'requires large remnant patches'. Although important information, these do not tell us what the species are in fact responding to—an individual or a local population responds not to the area of a patch but to the resources (food, nesting sites) and threats (predators, competitors) it contains. Identifying which of these more proximate mechanisms is at work is important for two reasons. First, if we don't know what the mechanism is, we can't generalise to other landscapes, even for the same species. For example, if food availability is the critical factor, a large patch of woodland on less fertile soil might be required whereas a smaller patch area requirement might apply in woodland on more fertile soil. Second, often we can do little to alter substantially factors such as patch area. We may, however, be able to manipulate the proximate factor – providing a higher density of fallen timber within the patch may increase foraging opportunities to the level that a larger patch might represent, for example. By relying solely on relating patterns of species occurrence at one particular time to landscape pattern, we may be missing some important opportunities for efficient restoration investment.

Conclusions

In Australia, we face an urgent need to restore many woodland landscapes which do not have enough wildlife habitat left to support sustainable populations, yet even what is left is being eroded rapidly in both quality and quantity. We still have not stopped clearing woodland habitat, and although clearing is occurring at much lower rates than those of last century, the loss of a few hectares is now far more important, simply because so little is left (see Lesson #3). We are at the accelerated end of the spectrum of clearing impact. With pressures on wildlife likely to be compounded by climate change, continuing to allow vegetation removal is irresponsible. When we talk about balance in allowing some ongoing vegetation removal for development, we need to remember that when dealing with the last few per cent of native vegetation in a landscape, the balance was tipped long ago. We must have some vision of a desirable future for our landscapes, and work towards it, rather than dealing with potential developments on an individual case by case basis.

Despite a large amount of landscape ecological research relating habitat amount and configuration to patterns of species occurrence, there still is little understanding of the processes behind these relationships. An effective strategy for managing and restoring the biodiversity value of landscapes will need a more detailed knowledge of how species perceive landscapes. Identifying the most effective levers to pull is critical, given the scale of the challenges we face in comparison with the resourcses we have available to deal with them.

Bio

Martine Maron is an ecologist and Lecturer in Environmental Management at the University of Queensland. Her focus is providing research which informs the development of sound policy for conservation and natural resource management. She completed her PhD at Monash University on bird assemblages of Buloke (*Allocasuarina luehmannii)* woodlands before lecturing in landscape ecology at the University of Southern Queensland in Toowoomba. She serves on committees for conservation non-government organisations and community groups and has current research projects in woodlands in western Victoria, northern NSW, and southern and central Queensland.

References

Attwood, S. J., Park, S. E., Maron, M., Collard, S. J., Robinson, D. Reardon-Smith, K. M. and Cockfield, G. (2009). Declining birds in Australian agricultural landscapes may benefit from aspects of the European agri-environment model. *Biological Conservation* **142**, 1981–1991.

Bowen, M.E., McAlpine, C.A., Seabrook, L., House, A.P.N. and Smith, G.C. (2009). The age and amount of regrowth forest in fragmented brigalow landscapes are both important for woodland dependent birds, Queensland, Australia. *Biological Conservation* **142**, 3051–3059.

Dwyer, J. M., Fensham, R. J., Butler, D. W. and Buckley, Y. M. (in press). Carbon for conservation: assessing the potential for win-win investment in an extensive Australian regrowth ecosystem. *Agriculture, Ecosystems and Environment.*

Eyre, T. J., Maron, M., Mathieson, M. T. and Haseler, M. (2009). Impacts of grazing, selective logging and hyper-aggressors on diurnal bird fauna in intact forest landscapes of the Brigalow Belt, Queensland. *Austral Ecology* **34**, 705–716.

Fahrig, L. (1997). Relative effects of habitat loss and fragmentation on population extinction. *The Journal of Wildlife Management* **61**, 603–610.

Gibbons, P. and Lindenmayer, D.B. (2007). Offsets for land clearing: no net loss of the tail wagging the dog? *Ecological Management and Restoration* **8**, 26–31.

Howes, A. L. and Maron, M. (2009). Interspecific competition and conservation management of continuous subtropical woodlands. *Wildlife Research* **36**, 617–626.

Kath, J., Maron, M. and Dunn, P. K. (2009). Interspecific competition and small bird diversity in an urbanizing landscape. *Landscape and Urban Planning* **92**, 72–79.

Manning, A. D., Fischer, J. and Lindenmayer, D. B. (2006). Scattered trees are keystone structures – implications for conservation. *Biological Conservation* **132**, 311–321.

Maron, M. (2005). Agricultural change and paddock tree loss: implications for an endangered subspecies of red-tailed black-cockatoo. *Ecological Management and Restoration* **6**, 206–211.

Maron, M. (2007). Threshold effect of eucalypt density on an aggressive avian competitor. *Biological Conservation* **136**, 100–107.

Maron, M. and Cockfield, G. (2008). Managing tradeoffs in landscape restoration and revegetation projects. *Ecological Applications* **18**, 2041–2049.

Maron, M., Dunn, P. K., McAlpine, C. A. and Apan, A. (2010). Can offsets really compensate for habitat removal? The case of the endangered red-tailed black-cockatoo. *Journal of Applied Ecology* **47**, 348–355.

Maron, M. and Fitzsimons, J.A. (2007). Agricultural intensification and loss of matrix habitat over 23 years in the West Wimmera, south-eastern Australia. *Biological Conservation* **135**, 587–593.

Maron, M. and Kennedy, S. (2007). Roads, fire and aggressive competitors: determinants of bird distribution in subtropical production forests. *Forest Ecology and Management* **240**, 24–31.

McIntyre, S. and Hobbs, R. (1999). A framework for conceptualizing human effects on landscapes and its relevance to management and research models. *Conservation Biology* **13**, 1282–1292.

Pressey, R. L. and Taffs, K. H. (2001). Scheduling conservation action in production landscapes: priority areas in western New South Wales defined by irreplaceability and vulnerability to vegetation loss. *Biological Conservation* **100**, 355–376.

30

GRAZING AWAY OUR WOODLAND BIRD DIVERSITY

Tara G. Martin

1. Make predictions *a-priori* using historic and/or expert knowledge and test these predictions with field data.
2. Livestock grazing and tree clearing often occur together but the effects of each on bird species are different.
3. To conserve bird species highly sensitive to any level of livestock grazing, exclude livestock grazing from a portion of the landscape to encourage the recovery of a diverse vegetation structure including native shrubs and tussock grasses. This conservation area should include riparian habitats.
4. Consider landscape context when restoring grazed riparian habitats. Minimise the influence of surrounding land uses on riparian habitats by situating intensive land uses away from riparian habitats.
5. Minimise intensive land uses such as cropping, fertilisation and irrigation.
6. The woodland bird extinction debt from past agricultural and pastoral development is likely to persist for a long time.

Introduction

The grassy eucalypt woodlands of south-east Queensland and northern New South Wales have been my laboratory for over 15 years (see map on next page). Through observation and experimentation, I have gained an understanding of the complex interactions between herbivore grazing, mineral fertilisation and disturbance on both grassy woodland plant species richness and composition (McIntyre and Martin 2001; McIntyre *et al.* 2002; McIntyre and Martin 2002; McIntyre *et al.* 2003; see also Figure 30.1) and the bird fauna that live in these grassy woodlands (Martin *et al.* 2005; Martin and McIntyre 2007).

Over a century ago, Australian scientists were writing about the ecological impacts of extensive alteration of grassy eucalypt woodlands to accommodate pastoralism and agriculture (Tepper 1896; Wood 1924). Studies also warned of the impact of pastoralism on Australian woodland bird species (Barnard 1925; Barnard 1934). Grazier and ornithologist Charles Barnard (1925, p. 4) discussed the decline in birds from 1870 to 1925 in the grassy woodlands

Figure 30.1 Sue McIntyre and Katina Heard surveying grassland plants in low grazed grassy woodland, south-east Queensland. (Photo: T. Martin)

of central Queensland, proposing pastoralism as the primary cause, the effects of which were further exacerbated by severe drought.

> *'In those days the bush was a babel of bird notes. Sit beside a small waterhole anywhere and observe the numbers of birds coming to drink. One was never lonely in the bush, there was so much of interest. Now this has all changed.'*

Today over 70% of Australia's land mass has been affected by commercial levels of grazing (Freudenberger and Landsberg 2000) with grassy eucalypt woodland ecosystems incurring a disproportionate impact. Reviewing the literature on the impact of livestock grazing on birds, I realised that most studies have been confounded by other factors. Grazing is invariably linked with other processes which also affect bird species abundance such as tree clearing and thinning (e.g. Arnold and Weeldenburg 1998; Jansen and Robertson 2001) and in arid landscapes, the provision of artificial watering points (James *et al.* 1999). A further challenge for research has been finding large areas of vegetation on fertile land types without a history of commercial grazing to be used as controls for assessing the impacts of pastoralism.

There are three main limitations of previous grazing studies. First, studies had been correlational and exploratory and lacked a theoretical framework. Second, the direct short-term impact of grazing was invariably confounded with reduction in tree density. Third, studies had so far failed to consider the influence of the broader landscape context when assessing the impact of grazing on birds. The following lessons draw largely from my work in grassy eucalypt woodlands of south-eastern Queensland, where I investigated the impact of livestock grazing on woodland birds with the aim of building on and overcoming the limitations of earlier research.

Lessons

1. Make predictions a-priori using historic and/or expert knowledge and test these predictions with field data

I took a more predictive approach than previous studies to investigating the impact of grazing on bird species. Most research on the effects of grazing on woodland bird fauna uses a classical null hypothesis testing approach, assuming there is no impact, and attempting to reject that hypothesis (e.g. Woinarski and Ash 2002). These studies do not make predictions *a priori* but rather make *post hoc* inferences regarding the impact of grazing on birds. Ecology has come under criticism for not being a predictive science (Peters 1991). If we have to learn everything about every location, then progress in conserving our grassy woodlands will be slow. Only through the formulation of hypotheses with testable predictions will we be able to forecast the likely impact of different land use scenarios on a wide variety of species across different grassy eucalypt woodland ecosystems, facilitating their management and conservation (e.g. Mac Nally and Bennett 1997).

With over 100 years of accessible, published, exploratory and descriptive studies at our fingertips, and a wealth of expert knowledge covering a vast range of woodland ecosystems, taxa and land use scenarios, there is now ample prior information on which to make testable predictions in ecological and conservation research. We used two approaches to make such predictions regarding the impact of livestock grazing on Australian woodland birds. The first generated predictions using the expert knowledge of 20 ornithologists and researchers across Australia (Martin *et al.* 2005), and the second used published literature on bird foraging behaviour (Martin and Possingham 2005).

Using expert knowledge as informative priors in conjunction with field data in a Bayesian model significantly increased our ability to infer the impact of grazing on all but one woodland bird species examined (Martin *et al.* 2005). When experts were in agreement with each other and with our field data, confidence intervals surrounding predictions of grazing effects were substantially reduced. When expert opinions did not agree with each other and also disagreed with the field data, predictions remained uncertain with wide confidence intervals. In each case, the degree of uncertainty was propagated through the model. In situations where expert data is available, yet time and monetary constraints limit the amount of data that can be collected, the use of expert information in ecological models is a cost-effective way of making more informed predictions about the impact of management on woodland biodiversity (Martin *et al.* 2005).

Another way to make predications *a priori* is to base them on a mechanism. Habitat structure has long been associated with bird diversity (MacArthur 1964). Most bird species, especially insectivores, have specific heights at which they forage. It follows then that impacts that preferentially remove certain strata will preferentially affect those species that use those strata. One mechanism by which habitat structure is altered is livestock grazing. Using previously published information on where in the vegetation strata bird species forage and the vegetation strata most affected by livestock grazing, we made predictions about the effect of specific grazing regimes on individual bird species (Martin and Possingham 2005). These predictions were then tested against field data. We demonstrated that foraging height was a good predictor of a species' susceptibility to livestock grazing and avoided *post hoc* interpretation from impact data. Species which forage predominantly between >0–5 m generally declined with increasing grazing pressure, whereas those that foraged on the ground or in the canopy increased or showed little response to increasing grazing pressure, respectively. The good performance of our foraging model was surprising given the simplicity of the predictive model and all the unaccounted factors that also influence bird distribution and abundance. It provides real hope that we can predict the effects of habitat alteration, whether it be grazing or habitat restoration, on bird species and provide cost-effective management solutions when limited resources prevent large-scale data collection.

2. Livestock grazing and tree clearing often occur together, but the effects of each on bird species are different

Disentangling the effect of grazing with tree clearing, we found that the maintenance of tree cover was the single largest determinant of bird species richness, abundance and composition. Second to tree clearing, the alteration and loss of understory vegetation as a result of livestock grazing was the most important determinant of the woodland bird species assemblage (Martin and McIntyre 2007). Our research suggests that any level of commercial livestock grazing is detrimental to some woodland birds, but extensively managed, moderately grazed grassy eucalypt woodland and riparian habitats provide for a rich and diverse woodland bird assemblage, including bird species that are declining in more intensively managed parts of Australia. This is an important finding as authors have tended to attribute woodland bird decline to any level of livestock grazing.

Highly grazed habitats supported few management-sensitive woodland bird species. With the exception of the Noisy Miner (*Manorina melanocephala*), habitats with high levels of grazing were dominated by ground-foraging birds which are increasingly abundant nationally (e.g. Masked Lapwing *Vanellus miles*, Crested Pigeon *Ocyphaps lophotes*, Galah *Cacatua roseicapilla*, Australian Wood-duck *Chenonetta jubata* (Barrett *et al.* 2003). Lightly grazed woodland and riparian habitats supported the richest and most abundant bird assemblages compared with moderate and high grazed sites, suggesting that the increased structural complexity of the understorey and lack of disturbance provided through partial stock exclusion are of benefit to many woodland birds (e.g. Brown Thornbill *Acanthiza pusilla*, Variegated Fairy-wren *Malurus lamberti*, White-browed Scrub-wren *Sericornis frontalis*, Brown Gerygone *Gerygone mouki*, Scarlet Honeyeater *Myzomela sanguinolenta*). A further group of species were found at highest relative abundance under moderate levels of grazing and included the Jacky Winter (*Microeca fascinans*), Sacred Kingfisher (*Todiramphus sanctus*) and Brown Treecreeper (*Climacteris picumnus*) (Martin and McIntyre 2007).

Removing bird species that forage predominantly in or around water, the riparian bird assemblage was very similar to that found in adjacent woodland; however, these same birds were far more sensitive to livestock grazing than in woodland habitat. This may be due to riparian habitats being more structurally diverse than non-riparian woodland habitats to begin with, and therefore the influence of grazing through the simplification of vegetation structure is likely to be more pronounced in riparian than woodland habitats. From this work, we recommended avoidance of non-selective heavy grazing in favour of selective grazing at moderate levels to maintain native grassland diversity and structure including large native perennial tussock grasses.

3. To conserve bird species highly sensitive to any level of livestock grazing, exclude livestock grazing from a portion of the landscape to encourage the recovery of a diverse vegetation structure including native shrubs and tussock grasses. This conservation area should include riparian habitats

Less than 10% of the grassy eucalypt woodlands within our study region (Crows Nest Shire, south-east Queensland) was excluded from livestock grazing, yet these areas were providing habitats for several bird species (e.g. Brown Thornbill, White-throated Treecreeper *Cormobates leucophaeus*) found nowhere else. While our research demonstrates that a rich bird assemblage can coexist with moderate levels of grazing, for the entire assemblage of woodland birds to be conserved in the long term, more woodland and riparian habitat needs to be excluded from livestock grazing (Martin and McIntyre 2007). Ideally a minimum of 30% should be protected from commercial levels of grazing.

4. Consider landscape context when restoring grazed riparian habitats. Minimise the influence of surrounding land uses on riparian habitats by situating intensive land uses away from riparian habitats

The management of riparian zones in grazed grassy woodland landscapes is a highly emotive and controversial issue. On the one hand, riparian habitats contain a disproportionately large component of the biodiversity, including birds, and perform critical ecosystem functions such as processing of organic matter and nitrogen uptake. On the other hand, they form an integral part of the grazing enterprise, providing water for stock, shade, flat ground and fertile productive pastures (Martin and Green 2002). The relative impact of local habitat characteristics and context on riparian habitats has received little attention. Yet the high edge-to-area ratio of riparian habitats is likely to make them susceptible to changes in the surrounding land use. We examined the influence of three landscape contexts representing low (grazed native pasture) to high land-use intensity (cropping) and three riparian habitat types representing low (lightly grazed uncleared) to high (highly grazed and cleared of trees) levels of disturbance (Martin *et al.* 2006).

Local riparian habitat characteristics significantly affected the relative abundance of over 80% of bird species, while landscape context significantly affected relative abundance of species for over half of the bird assemblage (Martin *et al.* 2006). Results from this research can help target where conservation management and restoration resources could be best spent for the greatest gain in bird conservation across grazing enterprises.

Regardless of landscape context, local riparian habitat condition as a result of grazing and tree clearing was the primary determinant of bird species composition and abundance. Restoring trees along cleared riparian habitat will result in a dramatic increase in bird species richness, relative abundance and composition. To allow such restoration, stock would have to be excluded from the riparian zone, at least in the short to medium term, to allow trees to establish and grow to a stage where they would no longer be vulnerable to damage by stock. Given the choice of whether to restore a riparian zone cleared of trees surrounded by native pasture or by crops, our study suggests that restoring a riparian zone surrounded by native pasture is likely to result in a bird assemblage more closely resembling that found in intact riparian habitat.

5. Minimise intensive land uses such as cropping, fertilisation and irrigation

While our work and that of others demonstrates that livestock grazing can have deleterious impacts on birds, particularly at high levels, overall a rich and abundant bird fauna has been able to persist in extensively managed, grazed grassy eucalypt woodland despite over 200 years of livestock grazing. This is a similar finding to that reported for grassland plants in the same study region (McIntyre and Martin 2001). Extensive land uses, which constitute little or no fertiliser use, no cropping, little tree clearing, together with low to moderate stocking rates and specific protection from grazing in some parts of the landscape, will foster bird conservation. Even extensive land use without these extras would be better for bird conservation than intensively using the land (Chamberlain 2002). Conservation land management with complete livestock exclusion, however, will lead to the greatest benefits for woodland bird species (Martin and Green 2002; Martin and McIntyre 2007).

In my view, two threatening processes could result in drastic changes to extensive grazed landscapes by increasing intensification of land use with detrimental effects on the bird fauna. First, rising global demand for livestock products is increasing the pressure on extensive grazing enterprises to become more intensive. This equates to more tree clearing, replacement of native pastures with sown pastures, broadscale use of fertilisers and irrigation, and increased pressure to graze properties at high levels. A second threat, peri-urban encroachment, is consuming grazing land close to major urban centres and coastlines, transforming the land into small 'lifestyle' blocks. This not only has the potential to intensify the land use of these areas but forces agriculture and grazing into more marginal landscapes (Martin and McIntyre 2007).

6. The woodland bird extinction debt from past agricultural and pastoral development is likely to persist for a long time

While our research has taken care to compare grazing regimes in woodland with similar tree densities, current tree size and density is likely to be different to that prior to European settlement in the early 1800s (Fitzgerald 1982). Even lightly grazed sites are likely to have sustained some level of tree removal and livestock grazing over the past 200 years. Fire regimes have been altered. Other alterations include removal of large hollow-bearing trees, and overall changes in tree health and demography are likely, regardless of physical tree clearing or thinning. Bird species most sensitive to these changes are likely to have already been lost from these landscapes: what we survey today is the bird assemblage that remains after two centuries of commercial grazing (Woinarski and Ash 2002; Woinarski and Catterall 2004). Species such as the Paradise Parrot (*Psephotus pulcherrimus*), Crested Bellbird (*Oreoica gutturalis*), Diamond

Firetail (*Stagonopleura guttata)* and Black-throated Finch (*Poephila cincta cincta)* are likely to have disappeared from the study region long ago (Chisholm 1922; Barnard 1925; Barnard 1934; Blakers *et al.* 1984). For other species, the impact of pastoral management may take decades to come to light (Recher 1999). For example, Ford *et al.* (2009) report on the progressive loss of the Hooded Robin (*Melanodryas cucullate)* and the Brown Treecreeper from woodland habitats despite most habitat loss occurring over 100 years ago. What our research as well as that of others demonstrates is that tree clearing and grazing are major contributors to an extinction process which will continue to unfold over the following century.

Documenting the changes brought about by livestock grazing on his central Queensland property 'Coomooboolaroo' and the impacts on birds and riparian habitats, Charles Barnard (1925, p. 5) wrote:

> *'With the advent of white man the native [Aborigine] rapidly decreased, and so we started to upset the balance. We introduced flocks and herds, we ringbarked millions of acres of forest trees, and cut down thousands of acres of rain forests, thus destroying not only the shelter for birds, but also the food for billions of insects in which birds fed. Small wonder our birds found it hard to exist.*
>
> *On top of this, we introduced rabbits, and then scattered poison broadcast. Still not satisfied, we let loose the fox and domestic cat, animals of destruction of which Australia had no parallel.*
>
> *The herds of cattle and sheep fed down the grasses till they completely killed out many of the best kinds, thus destroying the food of the flocks of seed-eating parrots and finches, also quail and pigeons. By making pads (miles in length) going to the water-holes to drink, the earth washed into the swamps and water-holes till they gradually silted up, till now, in all those fine creeks and swamps there is not a drop of water left.'*

Reading Barnard's account, one wonders how woodland birds have survived at all. But many have. To ensure a rich assemblage of grassy woodland birds remains into the future, we need to transform our research lessons into conservation land management action and policy.

Acknowledgements

Three eminent scientists have been pivotal in challenging and fostering my understanding of Australian grassy woodlands, their fauna and conservation. Early on, Carla Catterall nurtured my passion for bird conservation and encouraged me to explore the dynamics of habitat loss and fragmentation. Second, working alongside Sue McIntyre developed my ability to read landscapes as continuum rather than binary habitat and non-habitat. Finally, the annoyingly brilliant Hugh Possingham instilled in me the power and necessity of good models to inform conservation decision-making.

Bio

Tara Martin is a research scientist with CSIRO Sustainable Ecosystems and adjunct Professor with University of British Columbia, Canada and University of Queensland, Australia. She is a pioneer in the field of optimal conservation resource allocation, using decision theory to develop frameworks for making more efficient and effective decisions with respect to conservation land management, design of conservation areas, conserving migratory species, managing invasive species, recovering endangered species, and designating critical habitat for endangered species. For the past 15 years, Dr Martin has been researching the complex interactions between livestock grazing systems and the persistence of native bird and plant species assemblages throughout Australia.

References

Arnold, G.W. and Weeldenburg, J.R. (1998). The effects of isolation, habitat fragmentation and degradation by livestock grazing on the use by birds of patches of Gimlet *Eucalyptus salubris* woodland in the wheatbelt of Western Australia. *Pacific Conservation Biology* **4**, 155–163.

Barnard, C.A. (1925). A review of the bird life on Coomooboolaroo Station, Duaringa district, Queensland, during the past fifty years. *Emu* **24**, 252–265.

Barnard, H.G. (1934). Observations of the disappearance and probable cause of many of our native birds in central Queensland. *Queensland Naturalist* **9**, 3–7.

Barrett, G., Silcocks, A., Barry, S., Cunningham, R. and Poulter, R. (2003). *The New Atlas of Australian Birds*. Royal Australasian Ornithologists Union: Victoria.

Blakers, M., Davies, S.J.J.F. and Reilly, P.N. (1984). *The Atlas of Australian Birds*. Royal Australian Ornithologists Union, Melbourne University Press: Melbourne.

Chamberlain, D.E. (2002). Effects of agricultural intensification on birds. In *Birds and Agriculture*. (Eds N.D. Boatman, N. Carter, A.D. Evans, P.V. Grice, C. Stoate and J.D. Wilson). pp. 1–10. Association of Applied Biologists: Warwick.

Chisholm, A.H. (1922). The 'lost' Paradise Parrot. *Emu* **22**, 4–17.

Fitzgerald, R. (1982). *From the Dreaming to 1915 : A History of Queensland*. University of Queensland Press: St. Lucia.

Ford, H.A., Walters, J.R., Cooper, C.B., Debus, S.J.S. and Doerr, V.A.J. (2009). Extinction debt or habitat change? Ongoing losses of woodland birds in north-eastern New South Wales, Australia. *Biological Conservation* **142**, 3182–3190.

Freudenberger, D. and Landsberg, J. (2000). Management of stock watering points and grazing to maintain landscape function and biological diversity in rangelands. In *Management for Sustainable Ecosystems*. (Eds P. Hale, A. Petrie, D. Molony and P. Sattler). pp. 71–77. Centre for Conservation Biology, The University of Queensland: Brisbane.

James, C.D., Landsberg, J. and Morton, S.R. (1999). Provision of watering points in the Australian arid zone: a review of effects on biota. *Journal of Arid Environments* **41**, 87–121.

Jansen, A. and Robertson, A.I. (2001). Relationships between livestock management and the ecological condition of riparian habitats along an Australian floodplain river. *Journal of Applied Ecology* **38**, 63–75.

Mac Nally, R. and Bennett, A.F. (1997). Species-specific predictions of the impact of habitat fragmentation: local extinction of birds in the Box-Ironbox forests of central Victoria, Australia. *Biological Conservation* **82**, 147–155.

MacArthur, R.H. (1964). Environmental factors affecting bird species diversity. *American Naturalist* **98**, 387–397.

Martin, T.G. and Green, J.G. (2002). Wildlife and Core Conservation Areas. In *Managing and Conserving Grassy Woodlands*. (Eds S. McIntyre, J.G. McIvor and K.M. Heard). pp. 111–142. CSIRO Publishing: Melbourne.

Martin, T.G., Kuhnert, P.M., Mengersen, K. and Possingham, H.P. (2005). The power of expert opinion in ecological models using Bayesian methods: impact of grazing on birds. *Ecological Applications* **15**, 266–280.

Martin, T.G. and McIntyre, S. (2007). Livestock grazing and tree clearing: impacts on birds of woodland, riparian and native pasture habitats. *Conservation Biology* **21**, 504–514.

Martin, T.G., McIntyre, S., Catterall, C.P. and Possingham, H.P. (2006). Is landscape context important for riparian conservation? Birds in grassy woodland. *Biological Conservation* **127**, 201–214.

Martin, T.G. and Possingham, H.P. (2005). Predicting the impact of livestock grazing on birds using foraging height data. *Journal of Applied Ecology* **42**, 400–408.

McIntyre, S., Heard, K.M. and Martin, T.G. (2002). How grassland plants are distributed over five human-created habitats typical of eucalypt woodlands in a variegated landscape. *Pacific Conservation Biology* **7**, 274–289.

McIntyre, S., Heard, K.M. and Martin, T.G. (2003). The relative importance of cattle grazing in sub-tropical grasslands – does it reduce or enhance plant biodiversity? *Journal of Applied Ecology* **40**, 445–457.

McIntyre, S. and Martin, T.G. (2001). Biophysical and human influences on plant species richness in grasslands: comparing variegated landscapes in subtropical and temperate regions. *Austral Ecology* **26**, 233–245.

McIntyre, S. and Martin, T.G. (2002). Managing intensive and extensive land uses to conserve grassland plants in sub-tropical eucalypt woodlands. *Biological Conservation* **107**, 241–252.

Peters, R.H. (1991). *A Critique for Ecology*. Cambridge University Press, Cambridge.

Recher, H.F. (1999). The state of Australia's avifauna: a personal opinion and prediction for the new millennium. *Australian Zoologist* **31**, 11–27.

Tepper, J.G.O. (1896). Trees and their role in nature. *Agricultural Gazette of NSW* **7**, 29–37.

Woinarski, J.C.Z. and Ash, A.J. (2002). Responses of vertebrates to pastoralism, military land use and landscape position in an Australian tropical savanna. *Austral Ecology* **27**, 311–323.

Woinarski, J.C.Z. and Catterall, C.P. (2004). Historical changes in the bird fauna at Coomooboolaroo, northern Australia, from the early years of pastoral settlement (1873) to 1999. *Biological Conservation* **116**, 379–401.

Wood, W.E. (1924). Increase of salt in soil and streams following the destruction of the native vegetation. *Journal of the Royal Society of Western Australia* **X**, 35–47.

PART C

Policy: national

31

TOWARDS A NATIONAL FRAMEWORK FOR DESCRIBING AND MAPPING VEGETATION CONDITION: OBSERVATIONS FROM TEMPERATE WOODLANDS

Richard Thackway

1. Develop nationally consistent terms and practical tools to compile spatial vegetation condition datasets at various scales.
2. Promote collection of new site-based vegetation condition data relative to a benchmark.
3. Use nationally consistent vegetation and land use information systems to provide context.
4. Recognise how land management practices can be used to change vegetation condition.
5. Develop a hierarchy of scales for mapping and monitoring – patch, surrounds and landscape.

Introduction

The development of a consistent national framework for describing and mapping the condition of Australia's vegetation has its roots in the development of several national information systems including the Interim Biogeographic Regionalisation for Australia, National Vegetation Information System and the Australian Collaborative Land Use and Management Program. These national information systems have involved extensive collaboration and research across Australia's bioregions extending well beyond the area defined by Australia's temperate woodlands. The compilation of nationally consistent data and information on vegetation and land cover management is used to support the development of policy and programs in natural resources management and sustainable agriculture.

Australia's temperate woodlands are the areas where much of our agricultural production and urban areas occur (see map on next page). It is estimated that 80–90% of temperate woodlands have been removed from major agricultural regions since the European settlement of Australia due to those activities and other factors, such as changes in grazing and fire regimes

(McIntyre *et al.* 2002). In some regions, the remaining temperate woodland vegetation is a mosaic of small patches that are vulnerable to disturbance from surrounding land uses.

Across these ecosystems, considerable work has been done to describe and map the condition of temperate woodlands; however, this information has not been compiled because of a lack of consistent framework for describing vegetation extent and condition. This chapter outlines five lessons which describe the processes used to develop, test and use the Vegetation Assets States and Transitions (VAST) framework, with particular reference to temperate woodlands.

Lessons

1. Develop nationally consistent terms and practical tools and compile spatial vegetation condition datasets at various scales

The development of consistent terms and practical tools in consultation with vegetation management agencies has enabled considerable spatial data on vegetation condition to be translated and compiled. Previously these data have been poorly coordinated and could not readily be compared. Access to these datasets in a consistent form is assisting decision makers to investigate conservation and sustainable production options, e.g. investigating least-cost options to improve the connectivity of native vegetation between remnant patches.

Before 2005 there was no consistent national framework to guide the development and compilation of vegetation condition datasets, or to evaluate whether existing datasets could contribute to our collective understanding of the extent and condition of temperate woodlands at a regional or national level. Similarly, there was no apparent general understanding of what was meant by vegetation condition in a particular context, how condition can be repeatedly measured and at what spatial and temporal scales data should be collected, modelled and mapped for different applications (Thackway *et al.* 2006a). Benchmarks were not widely understood nor was the need to define diagnostic attributes associated with defined condition classes. Despite the significance of temperate woodlands for conservation and sustainable production, terms such as 'cleared', 'modified' and 'unaltered' were poorly defined and used inconsistently (Thackway and Lesslie 2006, 2008).

In 2005, work commenced on the development of a national framework to describe and map the extent and condition of native vegetation types as indicators of sustainable, productive ecosystems. The aim was to describe and map the proportion of remaining native vegetation in specified condition classes (NLWRA 2008). Several explanatory models and tools were investigated for their ability to describe and map increasing levels of habitat fragmentation and modification of vegetation extent and condition at a landscape level. A preferred approach for reporting on the extent and condition of sustainable and productive ecosystems was a state and transition model (e.g. McIntyre and Lavorel 2007), which involves identifying the vegetation states, and determining which of the states are linked and describing the transitions between the states (Thackway and Lesslie 2008).

The VAST framework was initially developed in 2005 and has been revised since then. Figure 31.1 describes the framework, which consists of seven classes, where the residual class I is the benchmark with which classes II–VI are compared, (i.e. largely unchanged through to totally removed). Naturally bare cover types are classified as Class 0. Intensively managed, native temperate woodlands are described by VAST class III and areas where these woodlands have been replaced with cropping and horticulture are described by VAST class V. VAST can be applied to the full spectrum of vegetation, native and non-native, and encapsulates concepts of changes in condition over time. It provides a useful tool to assist decision-makers assess and communicate the actions and effort needed to plan realistic goals for vegetation management, and to monitor and report progress toward desired management outcomes.

The numerous datasets related to temperate woodland that have been translated and compiled into the VAST framework include DIPNR (2002), ACT Government (2004), Newell *et al.* (2006), Westaway *et al.* (2004), DECCW (2009) and Zerger *et al.* (2009). These datasets

Increasing vegetation modification from left to right

		Native vegetation extent Dominant structuring plant species indigenous to the locality and spontaneous in occurrence – i.e. a vegetation community described using definitive vegetation types relative to estimated pre1750 state i.e. Class I			Non-native vegetation extent Dominant structuring plant species indigenous to the locality but cultivated; alien to the locality and cultivated; or alien to the locality and spontaneous		
	Class 0: NATURALLY BARE	Class I: RESIDUAL	Class II: MODIFIED	Class III: TRANSFORMED	Class IV: REPLACED - *ADVENTIVE*	Class V: REPLACED - *MANAGED*	Class VI: REMOVED
Vegetation condition class described using diagnostic attributes	areas where native vegetation does not naturally persist and recently naturally disturbed areas where native vegetation has been entirely removed. (i.e. open to primary succession)	native vegetation community structure, composition, and regenerative capacity intact – no significant perturbation from land use/land management practice	native vegetation community structure, composition and regenerative capacity intact - perturbed by land use/land management practice	native vegetation community structure, composition and regenerative capacity significantly altered by land use/land management practice	native vegetation replacement – species alien to the locality and spontaneous in occurrence	native vegetation replacement with cultivated vegetation	vegetation removed - alienation to non-vegetated land cover

Figure 31.1 The VAST framework.

represent a convergence of similar approaches, which when compiled using the VAST Framework provide valuable insights into the condition of Australia's temperate woodlands at a range of scale. Successfully compiling these datasets depended on each dataset having documented information on: 1) an implicit or explicit benchmark for vegetation types; 2) the relationship between effects of land management practices on structure, composition and regenerative capacity; and 3) patterns and processes operating at the site and landscape level.

Additional sub-classes were added to improve the level of discrimination of condition for some case studies. For example, a 250 ha farm-level application, utilised land management information for a grazing property in the southern tablelands of Australia to describe and map changes and trends in vegetation condition between 1960 and 2004 (Thackway *et al.* 2006b). That study showed that VAST can be used to provide a simple metric for classifying and reporting the condition and trends in vegetation across the landscape (Figure 31.1).

2. Promote collection of new site-based vegetation condition data relative to a benchmark

Development of an agreed national site-based approach for collecting data and information on the condition of vegetation relative to a fully mature benchmark has been fundamental to broader scale mapping, modelling and monitoring. Generally there is now widespread support for assessing the condition of, and threats to, temperate woodlands using this national approach.

The vegetation component of the CSIRO field handbook (Hnatiuk *et al.* 2009) provides agreed national guidelines for field survey at sites. These guidelines include how to collect data on structure (height, cover, growth forms), species composition and regenerative capacity and link directly to the VAST framework. Attributes described in that handbook can be benchmarked using the Executive Steering Committee for Australian Vegetation Information (ESCAVI) site-based vegetation approach (ESCAVI 2007), and modelled and mapped to create maps of native vegetation condition at the landscape level.

ESCAVI coordinated a series of State and Territory site-based pilots, which are generally based on Parkes *et al.* (2003), to assess biodiversity relative to a reference condition for a vegetation type. While the ESCAVI pilots only surveyed native vegetation, VAST applies to all vegetation, native and non-native, thus enabling comparison with mapped pre-1750 vegetation maps. The VAST framework defines native and non-native vegetation separately and establishes a strong basis for building an agreed nationally consistent definition of native vegetation from a site to the landscape level. An interim national native vegetation baseline (for 2004), representing VAST classes I-III (Figure 31.1), has been developed (Thackway *et al.* 2009); this should lead to more consistent mapping/modelling of site-based scores of native vegetation condition at the landscape level.

VAST requires information on floristic composition, vegetation structure and regenerative capacity, which are all significantly affected by the intensity of land use and land management practices. There are several ways to collect that information, including those described by ESCAVI (2007) and McIntyre and Lavorel (2007).

Change in floristic composition and vegetation structure is defined relative to the benchmark for each vegetation association (i.e. NVIS Level V), which is categorised on the basis of the dominant species in three strata (ESCAVI 2003). Experience shows that there is no substitute for detailed site-based surveys for collecting those data. Repeated site visits are required to ascertain whether changes in land management practices have changed vegetation structure and species composition. Remote sensing (e.g. Quickbird, SPOT and Landsat imagery) is cost-effective and reliable for mapping and modelling of some of the site-based attribute condition scores, particularly the overstorey (Zerger *et al.* 2009). Remote sensing is also of value in detecting loss of groundcover where the tree canopy is open or has been removed.

Experience gained in modelling vegetation condition scores shows the benefits of using a combination of site-based, remote sensing data and other ancillary datasets. Knowledge of the location and intensity of past and present land management practices in different parts of the landscape has been used to model the likely regenerative capacity present at the landscape level (Zerger *et al.* 2009).

3. Use nationally consistent vegetation and land use information systems to provide context

Using the VAST framework to show the relationship between the intensity of land use and land management practices and vegetation types has helped decision makers to improve their understanding of how to practically report on the extent and condition of native vegetation. Establishing these relationships has led to improved levels of accuracy in reporting the condition of temperate woodlands at regional and national levels.

National level data for vegetation cover types is available from the National Vegetation Information System database and for land use from the Australian Collaborative Land Use and Mapping Program database. Prior to the development of the VAST framework in 2005, these datasets could not be integrated.

The national VAST dataset was compiled by comparing vegetation extent and condition with an assumed pre-1750 vegetation condition benchmark (i.e. VAST class I) (Lesslie *et al.* 2009) This dataset integrated the national land use and vegetation cover datasets to provide a status of the condition of Australian vegetation cover types.

The value of the VAST framework at the regional level is illustrated in Table 31.1, which shows a comparison between vegetation types and condition classes for the Nandewar bioregion in NSW. The table highlights the extent to which temperate woodlands in this bioregion have been modified for agricultural production (i.e. VAST classes III and V). The vegetation types shown in the table are derived from the NVIS database.

4. Recognise how land management practices can be used to change vegetation condition

Recognising the role that land management practices have on changing vegetation condition has led to an improved understanding by land managers and researchers that changes in land management practices can be used to alter a class of vegetation condition over time, relative to a benchmark. In turn this knowledge is leading to more consistent methods for mapping, modelling and monitoring.

Modelling and mapping the extent and condition of native vegetation at a landscape level requires a sound understanding of the links between site-based measures of condition and how to extrapolate and interpolate these observations at broader scales.

The process used to create the national VAST dataset also required a thorough working knowledge of the relationships between land management practices and vegetation condition (Lesslie *et al.* 2009). For example, where native vegetation is primarily intact, (i.e. the effects of land management practices are minor), VAST class I was assigned. In another example, classes 4 and 5 of the biophysical naturalness (BN) indicator were reclassified as VAST class I, the latter describing areas of native vegetation cover where there has been no grazing or logging land uses (BN 5), low-intensity timber harvesting and livestock grazing more than 60 years ago (BN 4 – 'non-arid'), or intermittent exposure to arid/semi-arid pastoralism (BN 4 – 'arid')

Table 31.1 Relative condition of vegetation types in the Nandewar bioregion of NSW (regional scale 100 m). Temperate woodlands are shaded grey.

	Area by VAST class (ha)						
Pre 1750 Major Vegetation Group types	**Residual Class I**	**Modified Class II**	**Transformed Class IIIa**	**Transformed Class IIIb**	**Replaced – managed Class V**	**Replaced – removed Class VI**	**Total (ha)**
Rainforests and vine thickets	715	1044	450	191	3	0	2403
Eucalypt tall open forests	242	2307	2354	672	46	17	5638
Eucalypt open forests	101 601	282 646	238 779	268 544	15 017	4914	911 501
Eucalypt low open forests	15 447	25 779	28 613	37 215	1739	11	108 804
Eucalypt woodlands	32 276	110 293	262 137	922 631	297 999	11 175	1 636 511
Eucalypt open woodlands	1060	5868	19 066	41 269	4931	46	72 240
Mallee woodlands and shrublands	640	4131	251	179	155	0	5356
Acacia forests and woodlands	0	0	18	58	53	0	129
Acacia open woodlands	6	18	15	51	1	0	91
Callitris forests and woodlands	6313	4192	5523	1228	20	710	17 986
Casuarina forests and woodlands	1149	5935	2400	2840	578	2	12 904
Other forests and woodlands	25	18	57	0	0	0	100
Other shrublands	5	17	6	4	1	0	33
Heathlands	2784	274	367	95	10	0	3530
Other grasslands, herblands, sedgelands and rushlands	0	22	4	10	0	0	36
Tussock grasslands	527	2524	5256	85 904	9633	147	103 991
Naturally bare – sand, rock, claypan, mudflat	7	95	21	2	0	0	125
Unknown/no data	7	29	80	135	22	11	284
Total	162 804	445 192	565 397	1 361 028	330 208	17 033	2 881 662

Source of condition classes (VAST): Westaway *et al.* (2004).

Source of vegetation classes (Pre-1750 Major Vegetation Groups): Department of Environment, Water Heritage and the Arts.

(Lesslie *et al.* 2009). Similar rules were established to derive each of the VAST classes from each of the input datasets and their respective attribute classes.

At the local and regional level, on-ground information on land management practices and actions can be compiled either from discussions with landholders or where available from published information. As an example, Thackway *et al.* (2006) obtained detailed records of on-ground land management practices and corresponding patterns of vegetation structure observed in a range of places and in large format aerial photography to classify and map changes and trends in vegetation from 1980 to 2004 for a 250 ha property in southern New South Wales. Observed states and transitions in the vegetation condition over the 24 years were compared with undisturbed vegetation (i.e. VAST class I).

5. Develop a hierarchy of scales for mapping and monitoring – patch, surrounds and landscape

Development of a hierarchy of vegetation condition datasets within a designated region (i.e. the patch – within which sites may be sampled – and the immediate surrounds and the broader landscape) has enabled natural resource managers to develop least cost options to improve the connectivity of native vegetation between remnant patches. This information is also enabling the development of decision support systems to assess the potential significance of parcels of land with the broader landscape.

Work is underway to assess how the VAST framework can be used to help assess the status and condition of selected vegetation types that are currently used for agricultural production and to investigate conservation and sustainable production options. A hierarchy of mapping scales has been developed encompassing the patch (within which sites may be sampled), the immediate surrounds and the broader landscape (Thackway and Lesslie 2008).

The 'patch' and 'immediate landscape' context are described and mapped using a VAST dataset that is Intact (I), Modified (II), Transformed (III), Replaced-Adventive (IV), Replaced-Managed (VI) and Removed (VI). The broader landscape was described and mapped using a four landscape alteration levels dataset McIntyre and Hobbs (1999), (i.e. Intact, Fragmented, Variegated, and Relictual). This broader landscape level is derived using VAST dataset (e.g. 100 m × 100 m raster) and GIS spatial analysis. Landscape alteration levels are defined as the proportions (i.e. % area) of each condition state measured in a 500 m moving window radius using FRAGSTATS software (Thackway and Lesslie 2008).

Data for the three levels enables decision makers to develop priorities and options for vegetation management and investment options. For example, least-cost options can be developed by targeting land parcels with multiple patches of vegetation (i.e. VAST II and/or III) which are large in size, relatively well connected to other large patches, and least modified. Within the broader context (i.e. landscape alteration levels), the same parcels can be assessed to determine which are located in least fragmented and modified landscapes.

Conclusions

Since 2005 major progress has been achieved in describing and mapping the extent and condition of native vegetation in specified condition classes at a range of scales. Prior to this, researchers and land managers did not have access to a simple metric which provided consistent national descriptions of condition classes that related to vegetation condition attributes measured at sites, and mapped and/or modelled at national and regional scales. These products are now being used to help assess the status and condition of vegetation types that are currently used for agricultural production, and to investigate conservation and sustainable production options. For example, investigating least-cost options to improve the connectivity of native vegetation between remnant patches. This information also is supporting the development of land use planning options.

Acknowledgements

The national framework has been developed and evaluated with input from many people including Australian and State and Territory Government agency researchers, non-government agencies, academics, and regional natural resource bodies as well as land managers.

Bio

Richard Thackway is a senior research scientist with extensive experience across several natural management agencies in the Australian Government as a landscape ecologist. In the Bureau of Rural Sciences he coordinates the compilation of nationally consistent vegetation and land cover information to support the development of policy and programs in natural resources management and sustainable agriculture.

References

ACT Government (2004). *Woodlands for Wildlife: ACT Lowland Woodland Conservation Strategy.* Action Plan No. 27. Environment ACT: Canberra.

DECCW (Department of Environment, Climate Change and Water) (2009). *Planning for catchment biodiversity targets at a local landscape scale: a proposal for the South-west Slopes Bioregion of NSW.* Department of Environment, Climate Change and Water: Sydney.

DIPNR (Department of Infrastructure Planning and Natural Resources) (2002). *North-Lachlan Bogan Vegetation Mapping Survey 1: 100 000 Map Sheets.* Department Infrastructure Planning Natural Resources: Newcastle.

ESCAVI (Executive Steering Committee for Australian Vegetation Information) (2003). Australian Vegetation Attribute Manual, Version 6.0. Department of Environment and Heritage: Canberra.

ESCAVI (Executive Steering Committee for Australian Vegetation Information) (2007). Native vegetation condition in Australia: a synthesis of outcomes from pilot projects on national monitoring and evaluation indicators. Department of the Environment and Water Resources: Canberra.

Hnatiuk, R.J., Thackway, R. and Walker, J. (2009). Vegetation. In *Australian Soil and Land Survey Field Handbook.* 3rd edn. (Ed. National Committee on Soil and Terrain). CSIRO Publishing: Melbourne.

Lesslie, R., Thackway, R. and Smith, J. (2009). 'A national-level Vegetation Assets, States and Transitions (VAST) dataset for Australia (version 2), BRS Technical Report'. Bureau of Rural Sciences: Canberra.

McIntyre, S. and Hobbs, R.J. (1999.) A framework for conceptualising human impacts on landscapes and its relevance to management and research models. *Conservation Biology* **13**, 1282–1292.

McIntyre, S. and Lavorel, S. (2007). A conceptual model of land use effects on the structure and function of herbaceous vegetation. *Agriculture, Ecosystems and Environment* **199**, 11–21.

McIntyre, S., McIvor, J.G. and Heard, K.M., eds. (2002). *Managing and Conserving Grassy Woodlands.* CSIRO Publishing: Melbourne.

Newell, G., White, M., Griffioen, P. and Conroy, M. (2006). Vegetation condition mapping at a landscape-scale across Victoria. *Ecological Management and Restoration* 7, S65–S68.

NLWRA (2008). The National Land and Water Resources Audit 2002–2008. Achievements and Challenges. NLWRA: Canberra.

Parkes, D., Newell, G. and Cheal, D. (2003). Assessing the quality of native vegetation: the 'habitat hectares' approach. *Ecological Management and Restoration* **4**(Supp. l), 29–38.

Thackway, R., Frakes, I. and Lesslie, R. (2006). Reporting trends in vegetation assets, states and transitions at the farm level – a southern tablelands case study. Veg Futures Albury Greening Australia.

Thackway, R. and Lesslie, R. (2006) Reporting vegetation condition using the vegetation assets, states, and transitions (VAST) framework. *Ecological Management and Restoration* 7(Suppl 1), 53–62.

Thackway, R. and Lesslie, R. (2008). Describing and mapping human-induced vegetation change in the Australian landscape. *Environmental Management* **42**, 572–590.

Thackway, R., Sonntag, S. and Keenan, R. (2006a). Measuring Ecosystem Health and Natural Resource Productivity – Vegetation condition as an indicator of sustainable, productive ecosystems. Science for Decision Makers Series. Bureau of Rural Sciences: Canberra.

Thackway, R., Wilson, P., Hnatiuk, R., Bordas, V. and Dawson, S. (2009). Establishing an interim national baseline 2004 to assess change in native vegetation extent. Bureau of Rural Sciences: Canberra.

Westaway, J., Drielsma, D. and Ferrier, S. (2004). 'Nandewar WRA Landscape Conservation. Report for the Resource and Conservation Assessment Council, NSW Western Regional Assessment, Report no. NAND01'. Department of Environment and Conservation, Coffs Harbour.

Zerger, A., Gibbons, P., Seddon, J., Briggs, S. and Freudenberger, D. (2009) A method for predicting native vegetation condition at regional scales. *Landscape and Urban Planning* **91**, 65–77.

32

WOODLANDS CONSERVATION: SOME LESSONS FROM MANAGING RESEARCH AND OBSERVING POLICY

Stuart Pearson and Jim Donaldson

1. Conservationists, scientists and policy makers must demonstrate a more compelling case for woodland conservation if we are to reverse the long-term decline in their extent and condition in Australia. Demonstrating a compelling case will require more than a plea to preserve species and ecological communities for their intrinsic values alone – it must address the range of ecosystem services provided by woodlands and tackle the underlying economic and social drivers of how we use and manage natural resources.
2. Better woodlands management is possible using the knowledge we already have.
3. The drivers of woodland degradation are many and complex, so we need more sophisticated, multifaceted and workable responses from policy and management. There will be no 'mission accomplished': the long-term view is critical to woodland management and that long-term view is under threat in science, policy and community. Optimism drives innovations and improvement and the quick steps between problem identification and other actions are critical.

Introduction

We have worked at the national level in research management and policy, investing public money in vegetation and biodiversity research and communication to inform policy, regional planning and on-ground management in agricultural landscapes. Much of this work has been done through projects funded by Land & Water Australia, and some of these are reported in this publication.

Australia's temperate woodlands are the most stressed landscapes in the country, with the highest levels of decline in the condition of ecosystems and species (National Land and Water Resources Audit 2001; 2002). They have been extensively cleared and fragmented over the past 200 years, but predominantly in the last 60 years, to support the development of new agricultural systems to produce food (wheat, beef, lamb, dairy) and fibre (wool, cotton). This change in land cover and use has provided the springboard for developing the modern Australian economy, but in recent decades the contribution of agriculture to Gross Domestic Product has declined from 14% in 1960s (Productivity Commisison 2005) to less than 2% in 2009 (ABARE 2009). In contrast, the apparent value placed by the wider community on environmental outcomes, such as conserving biodiversity, halting land degradation, maintaining water quality and sequestering carbon continues to grow (Australian Bureau of Statistics 2006; Department of Environment and Climate Change 2007).

Hamblin (2009) provides an excellent overview of the complexities of agriculture and the environment in Australia. The temperate woodlands are particularly problematic to manage for conservation outcomes for several reasons. They are extensive and diverse, subject to multiple threatening processes such as ongoing fragmentation, lack of regeneration, grazing, weed invasion and changed fire regimes. Most significantly, they are valued by different stakeholders for multiple, often conflicting, reasons (Lindenmayer *et al.* 2005; McIntyre *et al.* 2002) with many of the benefits derived from woodland conservation accruing 'offsite' to the broader public whilst most of the costs are borne by landholders. Woodlands are frequently managed by landholders with an attitude that their proprietary rights entitle them to maintain a 'sole and despotic dominion' over their landholdings and the ecosystems situated on them (Blackstone 1809; Raff 2005); these landholders are often vulnerable to external pressures yet lack the capacity to change or integrate conservation into agricultural production systems (National Land and Water Resources Audit 2002; Nelson *et al.* 2007).

This chapter describes the need for a compelling case for action that includes governments investing in research to protect and rehabilitate Australia's temperate woodlands. It rests on a firm foundation of research and evidence that is adequate for informing action. The complexity of woodlands management and other impediments to action need to be understood, and the longer-term, integrated view will win out in the end because it provides actionable knowledge for sustainable improvements. The perspective of people working to use research to inform policy is presented in this chapter.

Lessons

1. Conservationists, scientists and policy makers must demonstrate a more compelling case for woodland conservation if we are to reverse the long-term decline in their extent and condition in Australia

Government intervention and external investment in woodland conservation is essential but to-date has been insufficient. While the benefits of agriculture to farmers are clear, most environmental benefits derived from woodland ecosystems are public goods and are not priced or traded in conventional economic markets so they tend to be undervalued and neglected (Schofield *et al.* 2007). If a reversal of the decline of woodlands is to occur, it will require significant government intervention and external investment. Justification of the case for the necessary intervention remains an overwhelming priority.

The need to reduce the rate of biodiversity loss has been well recognised at the highest levels of national and global governance since the late 1980s, most notably at the Rio Earth Summit in 1992. Since then, the idea that protecting biodiversity is essential to maintaining the ecological processes upon which our life-support systems depend has formed a cornerstone of Australian environmental policy. This has been expressed, for example, in the National Strategy for Ecologically Sustainable Development (1992), the Convention on Biological Diversity (1993), the National Strategy for the Conservation of Australia's Biodiversity (1996) and the Environment Protection and Biodiversity Conservation Act (1999). These policy commitments have been translated into a range of programs such as the Natural Heritage Trust (1997–2008) and most recently Caring for our Country (2009–2011) to achieve policy goals. These programs have supported some innovative initiatives designed to provide practical incentives to influence the way landholders manage their properties, such as the covenanting of titles organised by private NGOs, Bush Tenders and the Box Gum Grassy Woodland Stewardship Project (Australian Government Natural Resource Management Team 2008).

While society's concern for biodiversity loss has manifested itself in these various ways, and despite the multiplicity of national and state strategies and programs, the raft of State of the Environment reports, Biodiversity Assessments and audits of national programs reveal, we continue to steadily lose ground (Australian National Audit Office 2008; Department of the Environment 2009; National Land and Water Resources Audit 2002; State of the Environment 1996; State of the Environment 2006).

Demonstrating a compelling case for woodland conservation will require more than a plea to preserve species and ecological communities for their intrinsic values alone. It must address the range of ecosystem services woodlands provide and tackle the underlying economic drivers of how we use and manage natural resources. It is trite (but true) to say human well being fundamentally depends on the flow of ecosystem services provided by nature (Rapport *et al.* 1998). It is a daunting challenge, however, to demonstrate precisely which of these services is important in temperate woodlands and demonstrate how important. As McIntyre *et al.* (2002) noted, when the process of biodiversity loss is the result of a multitude of small, unconnected land management decisions, it is difficult to identify any particular action as the one to stop. Attribution of specific decisions and actions to specific impacts is intractable – and invites the tyranny of small decisions (Kahn 1966).

To-date, the policies, institutions and practices for managing temperate woodlands have proved ineffective in conserving their biodiversity. Given the nature and scope of the problem, it would be reasonable to conclude that woodland conservation is a 'wicked' problem not amenable to simple technical solution (Australian Public Service Commission 2007; Ison *et al.* 2007). As the 1996 State of the Environment report noted, achieving progress towards sustainability depends on widespread recognition that it requires the integration of ecological thinking into all our social and economic planning processes. More than this, it requires addressing a range of complex, underlying, systemic social, economic and institutional factors that drive the pattern of use and management of natural resources. Making substantive progress will rely on broadscale behavioural change and coordinated action by multiple organisations both within and outside government – not an easy task by any means!

To achieve significant change in this context, the challenge for scientists is to go beyond demonstrating incremental biodiversity loss and ecosystem degradation to provide a compelling case that the benefits of woodland conservation, though difficult to measure in conventional economic terms, exceed the opportunity costs, particularly to agriculture. The summary in Table 32.1 is a first attempt to evaluate the social, environmental and economic benefits of five research investments that sought to achieve outcomes in woodlands and agricultural environments. Space precludes a fuller description but the key message is that a conservative counting of the net present value (benefits less the costs) from investing in research to inform woodland management exceeded all the costs many (four to 19) times (Schofield *et al.* 2007). Often the quantified benefits represented a 'win-win' for productivity and conservation.

As Garnaut (2008) observes, in relation to making the case for mitigation of climate change, under conditions of potentially significant risk and great uncertainty, dealing with how to measure the costs of not taking preventative action is a highly complex and difficult issue. Demonstrating a convincing case for woodland conservation is challenging when the consequences of delaying action are not certain, immediate or tangible. The various appetites for risk, inter-

Table 32.1 Summary of results for case studies involving woodlands (in 2008–09 dollar terms and discounted to June 2009; analyses over 40 years). (Source: Pearson *et al.* in prep)

Investment	Period of Land & Water Australia (LWA) research	Net Present Value (all investors) ($m)	Proportion (%) of total benefits attributable to LWA by their dollar contribution	Benefit:cost ratio (LWA only)
ECOGRAZE ecosystem-based grazing	1997–2001	117.32	4.3	19.1
Joint Venture Agroforestry Program (25 projects)	1994–2003	82.60	9.4	4.9
Land Water and Wool	2002–2010	139.03	no $ contribution – no attribution	
Native vegetation (30 projects)	1995–2007	135.64	31.5	4.1
Tax incentives for native vegetation	1997–1999	34.63	32.5	18.2

ventions and even attitudes towards science that are playing out in climate change policy disagreements (Hulme 2009) also provide valuable insights into woodlands conservation.

For Australia's temperate woodlands, it is unlikely the benefits of biodiversity conservation per se will exceed those of agriculture when measured in terms of direct economic impacts, and even measuring indirect economic benefits, such as from recreation and tourism, are often problematic for a range of reasons. Even where intensification–conservation mixes are possible (Dorrough *et al.* 2007) or return-on-investment analyses have suggested they represent a 'win-win' (Schofield *et al.* 2007), few woodlands farmers can or do adopt conservation as a means to make a living for reasons described by Pannell *et al.* (2006).

Much of the case for conserving temperate woodlands will need to rely on a combination of several factors, including the community's growing demand (and investment) for sustaining the long-term productive capacity of the land, controlling waterlogging and salinity, improving water quality, sequestering carbon and providing a range of spiritual, ethical, aesthetic and intrinsic outcomes. As with climate change, a significant issue relates to how much insurance value we place on conserving biodiversity in woodlands now in order to deal with uncertainty and avoid the risk of potentially highly damaging future events brought about by ecosystem collapse. Demonstrating the potential consequences of such collapses and their implications for human well being is vital. The reality is that the explicit case for investment needs to be made against next-best options for government investment.

Without developing a compelling case, the level of investment in woodlands is likely to remain modest and piecemeal at best. If the case can be made, however, the imperative for economists and policy makers to find mechanisms to translate the public conservation benefits into effective on-ground action will gather momentum.

2. Better woodlands management is possible using the knowledge we already have

Australia has many of the world's leading researchers in environment and ecology (Esty *et al.* 2005) and their work should be used with confidence. These include:

- promoting the taking of an informed, integrated and long-term view to natural resource management and conservation programs
- raising the expectations of farmers about their duty of care
- using stewardship payments from the public and private sector to secure additional outcomes
- raising the quality of technical extension support that woodland owners receive to manage these valuable assets
- using appropriate mixes of volunteer and persuasive approaches along with market-like instruments to nurture change
- paying attention to knowledge management
- relaxing central control while strengthening local and regional groupings to develop new landscape approaches.

While more information may be needed to generate a compelling case for greater levels of policy action and investment, the current state of temperate woodlands management is not due to a lack of scientific information. The contents of this book and the careers of many leading ecologists are testament to the provision of information that could or should have changed behaviour or management or policy. A brief reflection on the research investments made by Land & Water Australia over 20 years shows that neither the weight of good quality evidence, nor new research published in prestigious journals, nor the finest of accessible glossy publications have resulted in rapid and comprehensive behaviour change (Schofield *et al.* 2007). Nevertheless, given the relatively low cost of research (compared to ignorance), the huge woodlands areas over which desirable changes can be made, the innovative people that manage the woodlands and the availability of some strong levers, we are optimistic that improvements can be made.

Some of the problems in temperate woodlands are related to adoption of behaviours supported by scientists and policy makers for different ends. It is commonplace to hear of those required to clear, graze or burn woodlands to be expected, a decade later, to plant, protect and manage values that were not entirely unimaginable. It seems prudent to expect the future will include further surprising changes and so people in communities, government and research should be building relationships that are sustainable, sensible and supportive. As uncertainty increases, the need for powerfully interconnected responses such as adaptive management, resilience thinking and subsidiarity (Marshall 2008) that empowers and engages citizens will need to be better understood (Campbell 2009; Marshall 2009). Some of what we learnt managing research investment and informing policy for this uncertain future makes the next lesson.

3. The drivers of woodland degradation are many and complex, so we need more sophisticated, multifaceted and workable responses from policy and management

While the range of threats to woodlands is known (e.g. fragmentation, grazing, fire and invasive species; see, for example, Burgman *et al.* 2007, and other chapters in this book) it is less clear what can be done about addressing these threats in a substantive way. The contest for resources between production and conservation is real and requires management. For example, with respect to the management of grazing and scattered trees, scientists can readily identify practices which are more beneficial for conserving woodlands, and even quantify the relative benefits of alternative strategies, but changing the practices requires more than science alone (Fischer *et al.* 2009; House *et al.* 2008). These are contested landscapes over which multiple values and many interested parties operate at a range of scales to fulfill their objectives.

It is not generally a lack of scientific information that is holding back change; it is more a question of the requisite economic drivers being absent (Pannell 2008). At the local scale, the taking of appropriate, efficient and effective conservation action is not likely to be without regret; in fact the paradox of choice (Schwartz 2004) means that regret is highly likely and so the use of market-like mechanisms to reveal options to achieve preset goals (Stoneham 2008) is very attractive to policy makers. Yet goal-setting is often the most deeply problematic and fraught issue – something often unappreciated by people who are not working in public policy – and so simple or concrete goals are rarely available.

Similarly, the diversity of land managers needs to be recognised and cultivated as sophisticated and informed consumers of policy and science. For example, the Master Tree Grower's model has enabled over 1600 people to come together in local communities and share their aspirations and define their questions relating to agroforestry (Reid 2008). Despite making many scientists and bureaucrats uncomfortable, this decentralised approach has demonstrated the benefit of empowering groups of farmers to be 'experts' and to pose their management questions to the research community.

These multiple and emerging threats to woodlands, the complexity of issues and diversity of views demands a variety of political, scientific, local and other knowledge. We think supporting those ready to change now is important. We recognise that democracy, inclusion, human rights, property rights, inter-generational, intra-generational and interspecific rights may be non-negotiable and constantly at odds with the needs and knowledge of others. Because there will never be a 'mission accomplished' in managing Australian woodlands, it needs ongoing policy and management effort to solve the competing values placed on natural resources. This effort should include ways of resolving continually those values and making decisions about the net community welfare while not doing irreparable harm to either the ecosystem or the social system.

This approach needs a landscape-scale focus that recognises these are already highly fragmented landscapes used predominantly for grazing and cropping. The mosaic of conservation as a land use with other productive land uses has been well researched (Lindenmayer and Hobbs 2007). The challenge is to embrace that diversity to improve biodiversity outcomes in production forestry and agriculture. The different values of remnant, native, restoration, native plantings and exotic non-native ecosystems are playing out across landscapes undergoing change without people being clear about the objectives we would like to attain from them. Further work needs to be done to relate conservation and biodiversity to the various trade-offs involved in managing landscapes.

There will be no 'mission accomplished' in woodland conservation and management, so the long-term view is critical to woodland management and that long-term view is under threat in science, policy and community. Surprisingly (and annoyingly for people rewarded for quick results) it has become very clear that while commercial concerns are important, decisions made on farms include opportunities for expressing altruistic and collaborative behavior that spins off into conservation and research support. Short-term programs frustrate effective delivery of on-ground outcomes. Research, capacity building and relationship-based approaches should be sustained through a sensitive mixture of indirect and direct support for voluntary and paid activities. The stakes are too high and problems too pressing to rely on rafts of short-term projects with fast-churning contractors – this is simply dissipating the research investment, trust and experience of all involved. Projects run by Greening Australia, Birds Australia and others like Master Tree Growers (Reid 2008) cope with short-term funding cycles through individual resilience, although this seems an inappropriately risky strategy for the country as a whole. The loss of community and individual resilience following the frequent battles for government funding holds a huge (undiscussed) political risk and threatens long-term damage to Australian community capacity to work collaboratively and to manage the deteriorating set of natural resources more sustainably.

The lack of long-term leadership in policy and research arenas is alarming and results in programs of projects that hold the future so uncertainly and do not reward learning, engagement, collaboration or improvement but instead support private knowledge, consultants, competition and game playing. The natural resource managers of Australia expect better and that new call for professionalism begins with articulate leadership and an acute duty of care. A functionally consistent support network that includes research and knowledge management is urgently needed to support that leadership (Campbell 2009; Campbell and Schofield 2006) – it needs a stronger grassroots perspective and closer connection between science and community to drive adequate public and private investments in whole-of-landscape solutions to issues such as woodlands conservation.

Integrated responses are needed to achieve the various and formidable challenges of woodland management. Woodland managers have learnt from grand experiments on the social, political and biophysical landscape and it is reasonable to support and expect that to continue. Market-based instruments, developed by resource economists to address missing markets for environmental goods, have been used for many years in Victoria (Stoneham 2008), and more recently form the backbone approach for White Box Woodlands conservation (Australian Government Natural Resource Management Team 2008). They direct public funds with great efficiency to achieve articulated goals and targets and in doing so have revealed much otherwise private information about the woodlands (Stoneham 2008).

Challenges, improvements and new approaches will continue to emerge so work is urgently needed to understand how individuals and communities can manage landscapes to satisfy many values and to strengthen communities of land managers (Morrison *et al.* 2009). Recent theoretical work (Hertzler 2009) suggests that a citizen jury, where a small representative group of non-experts try the evidence for a proposed action (Kenyon and Nevin 2001; Niemeyer and Blamey 2004) is a particularly useful approach because non-market valuation approaches had understated the importance of people seeing themselves as owners rather than consumers of the environment. There is also strong evidence, summarised in Table 32.1 from Schofield *et al.* (2007), that legal and tax measures (Binning and Young 1998; Martin 2007) are very efficient drivers of change in woodlands management (Schofield *et al.* 2007).

The context of environmental policy is changing and the level of expectation about the science is becoming at once richer and more contestable. This has implications for integrated responses to woodlands management that should be preempted. First, Garnaut's (2008) analytical framework for climate shows that the types of evidence needed to justify and direct intervention in Australia's woodlands will include the trade-off of current costs against immediate profits; insurance against breaching catastrophic tipping points; protection of non-market environmental qualities; and avoiding large costs in the next century. Second, the social landscape of environmental value is more complex than most people realise – for example, the importance of personality types and spirituality/religion and falling environmental values (as identified in the United States in SRI Consulting and ecoAmerica 2006) – and they have been poorly integrated into woodlands management.

Managers are urgently needed who can manage the complexity of agricultural productivity and sustainability through periods of stability and rapid change in the temperate woodlands. Policy and science needs to likewise engage in building their own skills and patience to see that as opportunities come and go they can contribute and benefit from new policy opportunities or new spending and then endure the inevitable drought. The woodlands are subject to a range of fads and innovations (and threats) such as: use as biomass generators and plantations; pre-emptive clearing prior to protective legislation; threats such as charcoal-production plants, coal mining, carbon emission trading schemes, fuel price variations, irrigation using groundwater and/or centre pivots. New technology, including communication and gene technology, is likely to transform many of the options farmers have in woodland areas in the next few years, and it is likely to raise productivity against a background of increased fuel and agrochemical costs.

The sustained attention of policy and public interest is something that people in both the woodlands and the rangelands crave unrealistically. It is possible to use return-on-investment measures on woodland research to see that programs have a maximum impact value nearly 20 years later (Schofield *et al.* 2007). So woodlands require sustained investment with strong returns that can be expected in the future – although we have seen recent evidence that this long-term view is not appreciated in the policy arena and future scenarios of declining public investment seem highly credible. Therefore, it seems entirely appropriate for government policy to encourage and facilitate private sector investment in this emerging market and well-developed models to do this already exist (Martin 2007).

Actionable knowledge or information that guides 'why' government should intervene as well as 'how' and 'what' it should intervene to conserve, is a useful axiom for those considering 'what next?' We believe the woodland problem to be an action rather than an information problem. We have already suggested action is not occurring because there are many options and many possible regrets (and critics at every corner). We think it is unlikely that the subsidiary problems of managing incentives, duty of care or community management issues will be resolved. Perhaps a crisis such as the further intensification of drought, capital, oil or food shortages or the nascent carbon market may trigger concerted action but we feel deeply uncertain about the direction that action may take. In the interim, the knowledge of what to do and capacity to do it needs to be shared and developed.

Setting explicit, preferably evidence-based and scientifically-based, objectives is vital to measure whether we are making progress. Getting broader agreement on those objectives that converts to commitment and action is usually much harder than is portrayed in the naïve recommendations of individuals or single issue advocates. Better engagement of the research, social and institutional processes and sharing of tools for determining conservation objectives and understanding the share of costs and trade-offs may help integrate ideas, evidence and action (for example, Hajkowicz 2008, using multicriteria approaches). The demonstrated inability to set correct and acceptable goals and objectives is particularly frustrating (for example, views expressed at the 2009 Fenner Conference, Applied Environmental Decision Analysis Research Hub 2009). It is a complicated business that can be informed by biophysical scientists and needs further support from those who understand values and social processes. Researchers and scientists are sometimes unhelpful because they engage only through their specialties, and are perhaps overly critical of the policy and community response, without making a sensible contribution to the situation.

There is a need, in natural resource management, to foster partnerships between scientists and practitioners as part of an ongoing adaptive management approach. Zammit (2009) has observed that two key places that scientific input is needed are in efficient policy implementation and in framing the context of new policy development. Particularly in woodlands framing the biodiversity conservation challenge and having input into setting objectives and monitoring to measure progress seems critical. There are useful research integration approaches that could be more widely used (Bammer *et al.* 2005; McDonald *et al.* 2009).

We believe that appropriate scientific knowledge is available but more investment in sharing this knowledge is needed, particularly between researchers and practitioners. At the program and project implementation end, there is a criticism that while much good science is out there, it is not used effectively in regional planning or on-ground projects (Lovett *et al.* 2009). The science that is 'out there' is generated by researchers who are generally well rewarded for contributions at the international scale and poorly rewarded for work at the local scale. On the other hand, project investments that are not underpinned by sound science can be inefficient or (worse) ineffective, and will often fail to measure the degree of their success, so there is little learning for the future. Greater collaboration between universities and research end-users needs to be more systematically encouraged. Scientists need to engage more directly with regional groups delivering biodiversity conservation programs; for example, to assist in identifying credible project objectives and designing scientifically robust monitoring protocols. Research funders, policy makers and program implementers need to create the institutional arrangements, cultures and incentives which encourage and reward such collaborative behaviours.

There are some excellent collations of what is currently known (e.g. Salt and Lindenmayer 2008) and we have evidence that these will permeate the natural resource management community. Evidence of strong returns on investing in vegetation management knowledge can occur over periods of 30 years and, as mentioned earlier, some of the strongest returns come from surprising investments in tax reform that fostered broadscale change in behaviour (Schofield *et al.* 2007).

On-ground action happens at the scale of the local site or patch of vegetation, yet planning to enable and invest in this action often occurs at the regional, state and national levels. Long time frames and nested scales are part of the woodlands management challenge. They demand longer time horizons for implementing projects and programs to make it worthwhile for regional and community groups to invest in capacity building and to plan, monitor and make improvements. That the term 'ecosystem services' had been given lip-service by some regional bodies in their competitive efforts to secure funding (Plant 2009) is evidence that omitting credible engagement of citizens and their local representatives working on government-funded programs places the trust between people and government at risk.

Finally, we believe that optimism drives improvement and innovations and that quick steps between problem identification and other actions are critically important. It is important to reaffirm the value of ongoing scientific research to improve the fundamental understanding of how our ecosystems work. It also helps target our conservation and restoration efforts more effectively and efficiently. The science of woodland ecology, management and restoration has brought great returns but we can, and must, do better in the way that knowledge is put to use. Improvements in woodland management can result in a multiplier-effect across an enormous part of Australia. A small improvement across most of the woodlands adds up to a big difference for all Australians. With sustained attention, a big and positive difference could be made in the conservation and management of the woodlands. The optimism needed for action, even in the face of uncertainty and incomplete knowledge, is not found widely in research or policy arenas where problem definition and problem solving are the dominant ways of doing business. We also recognise that Australian farmers and the community have demonstrated their capacity to innovate and that they are willing to accept there is a rising duty of care expected in their stewardship of woodland ecosystems. This is a source of great optimism to us, for there is much to do.

Acknowledgements

Jim Donaldson and Stuart Pearson recognise the role of Land & Water Australia in making this chapter possible.

Bios

Stuart Pearson's background in palaeoecology research and experience teaching environmental management and biogeography helped him take the long view and a scientific perspective into Land & Water Australia (one of the Australian Government's primary industry Research and Development Corporations). As a knowledge broker and program manager he was able to see the machinations of researchers, research investment and extension in woodlands management. He has recently returned to the academic life.

Jim Donaldson has spent most of his career working at the interface of policy, research and extension in vegetation management in various incarnations of the Australian Government departments responsible for agriculture and forestry, natural resource management and the environment. Jim managed the first national Farm Forestry Program, worked with the National Forest Inventory, tried to develop a Bush for Greenhouse program, developed national strategies for implementing Bushcare, and was closely associated with the Joint Venture Agroforestry Program for many years. At Land & Water Australia, he had the pleasure of managing the Native Vegetation and Biodiversity R&D Program from 2006 to 2009.

References

Applied Environmental Decision Analysis Research Hub (2009). 'The art and science of environmental decision making'. Fenner Conference: Canberra.

Australian Bureau of Agricultural and Resource Economics (2009). 'Australian commodities: June quarter'. Australian Bureau of Agricultural and Resource Economics: Canberra.

Australian Bureau of Statistics (2006). 'What do Australians think about protecting the environment?' Department of the Environment and Heritage: Canberra.

Australian Government Natural Resource Management Team (2008). 'Environmental Stewardship Box Gum Grassy Woodand Project Implementation Plan.' Australian Government: Canberra.

Australian National Audit Office (2008). 'Regional Delivery Model for the Natural Heritage Trust and the National Action Plan for Salinity and Water Quality.' Australian National Audit Office: Canberra.

Australian Public Service Commission (2007). 'Tackling Wicked Problems: A Public Policy Perspective.' Australian Public Service Commission, Canberra.

Bammer, G., O'Connell, D., Roughley, A. and Syme, G. (2005). Integration research for natural resource management in Australia: an introduction to new challenges for research practice. *Journal of Research Practice* **1**, E1.

Binning, C. and Young, M. (1998). 'Talking to the Taxman about Nature Conservation'. Land & Water Australia: Canberra.

Blackstone, W. (1809). *Commentaries on the Laws of England.* 15th edn. Cadell and Davies: London.

Burgman, M., Keith, D., Hopper, S., Widyatmoko, D. and Drill, C. (2007). Threat syndromes and conservation of the Australian flora. *Biological Conservation* **134**, 73–82.

Campbell, A. (2009). It's time to renew Landcare. *Agricultural Science* **2**.

Campbell, A. and Schofield, N. (2006). 'The Getting of Knowledge – a guide to funding and managing applied research.' Land & Water Australia: Canberra.

Department of Environment and Climate Change (2007). 'Who Cares about Water and Climate Change in 2007?' Department of Environment and Climate Change: Sydney.

Department of the Environment, Water, Heritage and the Arts (2009). 'Assessment of Australia's Terrestrial Biodiversity 2008'. Department of the Environment, Water, Heritage and the Arts: Canberra.

Dorrough, J., Moll, J. and Crosthwaite, J. (2007). Can intensification of temperate Australian livestock production systems save land for native biodiversity? *Agriculture, Ecosystems and Environment* **121**, 222–232.

Esty, D., Levy, M., Srebotnjak, T. and de Sherbinin, A. (2005). *Environmental Sustainability Index: Benchmarking National Environmental Stewardship.* Yale Center for Environmental Law and Policy: New Haven.

Fischer, J., Stott, J., Zerger, A., Warren, G., Sherren, K. and Forrester, R. (2009). Reversing a tree regeneration crisis in an endangered ecoregion. *Proceedings of the National Academy of Science* **106**, 10386–10391.

Garnaut, R. (2008). 'The Garnaut climate change review: final report'. Cambridge University Press Australia: Melbourne.

Hajkowicz, S. (2008). Supporting multi-stakeholder environmental decisions. *Journal of Environmental Management* **88**, 607–614.

Hamblin, A. (2009). Policy directions for agricultural land use in Australia and other post-industrial economies. *Land Use Policy* **26**, 1195–1204.

Hertzler, G. (2009). 'Dynamic non-market valuation of ecosystem services'. Land & Water Australia: Canberra.

House, A., MacLeod, N., Cullen, B., Whitbread, A., Brown, S. and McIvor, J. (2008). Integrating production and natural resource management on mixed farms in eastern Australia: the cost of conservation in agricultural landscapes. *Agriculture, Ecosystems and Environment* **127**, 153–165.

Hulme, M. (2009). *Why We Disagree About Climate Change.* Cambridge Books: Cambridge.

Ison, R., Röling, N. and Watson, D. (2007). Challenges to science and society in the sustainable management and use of water: investigating the role of social learning. *Environmental Science and Policy* **10**, 499–511.

Kahn, A. (1966). The tyranny of small decisions: market failures, imperfections, and the limits of economics. *Kyklos* **19**, 23–47.

Kenyon, W. and Nevin, C. (2001). The use of economic and participatory approaches to assess forest development: a case study in the Ettrick Valley. *Forest Policy and Economics* **3**, 69–80.

Lindenmayer, D., Crane, M. and Michael, D. (2005). *Woodlands: Disappearing Landscape.* CSIRO Publishing: Melbourne.

Lindenmayer, D.B. and Hobbs, R.J. (2007). *Managing and Designing Landscapes for Conservation: Moving from Perspectives to Principles.* Blackwell Publishing: Malden, Massachusetts.

Lovett, S., Lambert, J., Williams, J.E. and Price, P. (2008). 'Restoring Landscapes with Confidence – an Evaluation of the Science, the Methods and their On-ground Application'. Land & Water Australia: Canberra.

Marshall, G. (2008). Nesting, subsidiarity, and community-based environmental governance beyond the local level. *International Journal of the Commons* **2**, 75–97.

Marshall, G. (2009). Polycentricity, reciprocity, and farmer adoption of conservation practices under community-based governance. *Ecological Economics* **68**, 1507–1520.

Martin, P. (2007). 'Improving private sector conservation using tax-effective instruments'. Land & Water Australia: Canberra.

McDonald, D., Bammer, G. and Deane, P. (2009). *Research Integration Using Dialogue Methods.* ANU Press: Canberra.

McIntyre, S., McIvor, J.G. and Heard, K.M. (2002). *Managing and Conserving Grassy Woodlands.* CSIRO Publishing: Melbourne.

Morrison, M., Durante, J., Greig, J. and Ward, J. (2009). 'Encouraging Participation in Market-based Instruments and Incentive Programs'. Land & Water Australia: Canberra.
National Land and Water Resources Audit (2001). 'Landscape Health in Australia: A rapid assessment of the relative condition of Australia's bioregions and subregions.' National Land and Water Resources Audit: Canberra.
National Land and Water Resources Audit (2002). 'Australian Terrestrial Biodiversity Assessment'. National Land and Water Resources Audit: Canberra.
Nelson, R., Brown, P.R., Darbas, T., Kokic, P. and Cody, K. (2007). 'The Potential to Map the Adaptive Capacity of Australian Land Managers for NRM Policy Using ABS Data'. National Land and Water Resources Audit: Canberra.
Niemeyer, S. and Blamey, R. (2004). 'Deliberation in the Wilderness: the Far North Queensland Citizen's Jury'. Land & Water Australia: Canberra.
Pannell, D. (2008). 'Effective Environmental Policy'. University of Western Australia: Perth.
Pannell, D., Marshall, G., Barr, N., Curtis, A., Vanclay, F. and Wilkinson, R. (2006). Understanding and promoting adoption of conservation practices by rural landholders. *Australian Journal of Experimental Agriculture* **46**, 1407–1424.
Plant, R. (2009). 'Ecosystem Services and NRM Practice: where the rubber hits the road'. University of Technology: Sydney.
Productivity Commisison (2005). 'Trends in Australian Agriculture'. Productivity Commission: Canberra.
Raff, M. (2005). Toward an ecologically sustainable property concept. In *Modern Studies in Property Law. Vol. III.* (Ed. E. Cooke). Hart Publishing: Oxford.
Rapport, D., Costanza, R. and McMichael, A. (1998). Assessing ecosystem health. *Trends in Ecology and Evolution* **13**, 397–402.
Reid, R. (2008). 'Tree Change – The Australian Master TreeGrower Phenomenon'. Rural Industries Research and Development Corporation: Canberra.
Salt, D. and Lindenmayer, D. (2008). 'The Bowral Checklist: A framework for ecological management of landscapes'. Land & Water Australia: Canberra.
Schofield, N., Chudleigh, P., Simpson, S. and Pearson, S. (2007). 'Land & Water Australia's Portfolio Return on Investment and Evaluation Case Studies.' Land & Water Australia: Canberra.
Schwartz, B. (2004). *The Paradox of Choice*. Harper Collins: New York.
SRI Consulting and ecoAmerica (2006). The American Environmental Values Survey. New York University: New York.
State of the Environment (1996). 'Australia: State of the Environment 1996.' Department of the Environment, Sport and Territories: Canberra.
State of the Environment (2006). 'Australia State of the Environment 2006.' Department of the Environment, Water, Heritage and the Arts: Canberra.
Stoneham, G. (2008). 'Creating Markets for Environmental Goods and Services: A Mechanism Design Approach.' Land & Water Australia: Canberra.
Zammit, C. (2009). 'Evidence, values and trade-offs in environmental decision-making.' The Fenner Conference: Canberra.

33

FROM PATTERN AND PROCESS TO PEOPLE, POLICY AND PRAGMATISM: INSIGHTS ON TEMPERATE WOODLAND CONSERVATION AND MANAGEMENT

Jann Williams

1. We know what needs to be done, so should focus on implementing the solutions.
2. More effective integration is needed within and between research, policy and management sectors.
3. Better knowledge exchange and communication can lead to changed on-ground management.
4. A long-term perspective is essential – reap the benefits of hindsight and foresight.
5. Monitor, monitor, monitor – learn from the rangelands experience.
6. Establish an effective and practical ecosystem services framework and payment scheme for temperate woodland landscapes.

Introduction

Describing the patterns native species, communities and ecosystems display, and examining the fundamental processes that underpin their distribution and abundance, provides a robust basis for managing biodiversity across multiple scales (Williams and Eamus 1997; Mackey *et al.* 2005). Biophysical science can only take us so far, however. The human dimension of natural resource management is also of critical importance. To achieve more effective management outcomes, there is an urgent need to improve the links between science, policy and practice (Williams 2007; Lovett *et al.* 2008; Bridle and Price 2009). Improved integration is relevant to all ecosystem management across all land tenures, but particularly so for temperate woodlands that mostly occur on private land. As demonstrated by the depth of knowledge and understanding found throughout this book, we already know what needs to be done to improve the conservation and management of temperate woodlands. The challenge is how to make the necessary changes while we still have the opportunity to do so. This chapter draws on my long-term experience in biodiversity conservation at the national level to identify a number of lessons that should help us move forward. This experience spans a diversity of woodland types across southern and eastern Australia (as well as many other ecosystems across the nation), covers a range of biota and scales, and involves working with a range of land managers (e.g. Williams 2000; Mortlock *et al.* 2002; Williams and Goodacre 2008).

Lessons

1. We know what needs to be done, so should focus on implementing the solutions

There have been literally hundreds of papers and several books written on the conservation and management of temperate woodland landscapes (Hobbs and Yates 2000; Lindenmayer *et al.* 2005; Lindenmayer and Hobbs 2007; Williams 2008). One recent analysis, for example, of the steps needed to help ensure the future of temperate eucalypt woodlands was written by Prober and Hobbs (2008). That work identifies three steps that focus on how to avoid further damage to temperate woodlands, including putting an end to broadscale woodland clearing and securing the travelling stock route network. This is followed by six steps that focus on the repair and restoration of sites and landscapes which, while more expensive and less effective than maintaining existing woodland systems, is a critical component of what is needed.

We know what needs to be done at a broad level to improve the conservation and management of temperate woodlands from an ecological perspective, as noted in other chapters in this book and more widely. The barriers to implementing the required actions are related to the social, cultural, economic and institutional context in which the management of temperate woodlands is embedded. Several of these barriers and the solutions to them are identified in Lindenmayer *et al.* (2008), particularly in the chapters on agriculture (Williams and McKenzie 2008), biodiversity (Possingham 2008) and policy and institutional reforms (Dovers 2008). While these authors address issues broader than temperate woodland conservation and management, implementing their recommendations as a priority, alongside those of Prober and Hobbs (2008) will go a long way towards a brighter future for these unique and iconic ecosystems. Preventative measures, such as securing the Great Western Woodlands in southern Western Australia, should be relatively easy to achieve, whereas measures such as redesigning agriculture will take considerable resources to implement. All actions are achievable however, with adequate resources and political will.

2. More effective integration is needed within and between research, policy and management sectors

It is widely accepted that good environmental management needs to be underpinned by good science. Relative to other ecosystems in Australia, there has been a considerable amount of research undertaken on temperate woodlands. Much of this has focused on the impacts of clearing and fragmentation of woodland ecosystems in southern and eastern Australia, with a major focus on mammals and birds (reviewed by Williams 2008). This science is world-class, but has suffered from a 'silo' approach where scientists from different disciplines have often worked in isolation. This has led, for example, to animal ecologists developing sets of management recommendations (such as 'bigger is better' for patches of woodland) that may inadvertently downplay recommendations made by plant ecologists, or to certain organisms being studied at the expense of others. Fortunately, plant and animal ecologists are beginning to work more closely together. There is still more room, however, for projects that bring together these and other ecological/environmental disciplines such as hydrological and climate scientists, and that integrates their findings.

To improve the conservation and management of temperate woodlands, better integration is also required across research disciplines such as the ecological, social and economic sciences. The spatial and planning sciences can also add considerable value to temperate woodland management. Much stronger interaction between environmental and agricultural scientists and advisors is required as well, as these are the individuals who work most closely with the farmers who manage temperate woodlands across our agricultural landscapes. Tailoring existing social research on managing native vegetation in agricultural systems, so that it is relevant and accessible for people working with the farming community (both in agriculture and natural resource management), is also likely to improve management outcomes.

Cross-disciplinary research can bring many benefits, but can be challenging because of the different paradigms and perspectives of each discipline. It is worth the effort. The recommended management actions arising from environmental research will have a higher likelihood of adoption if they recognise the social, cultural, economic and institutional drivers that influence the management of temperate woodland systems. This means that not only is more effective integration required within and between research disciplines, it is also needed between research, policy and management sectors.

3. Better knowledge exchange and communication can lead to changed on-ground management

Related to Lesson #2 is the need for more effective knowledge exchange, particularly between researchers, practitioners and policy makers (Lindenmayer and Hobbs 2007; Williams 2008). 'Knowledge exchange' is a term that has evolved over the last few years, and encompasses collaborative problem-solving between researchers and decision makers. Decision makers include both indigenous and non-indigenous policy makers and land managers. It involves more than improving the level and type of communication between scientists and the end-users of their research (e.g. Lindenmayer *et al.* 2005), although this is also crucial. Knowledge exchange also requires greater engagement between the research and farming/management community through, for example, participatory research models that have been used widely in the agricultural sector (Williams 2007; Bridle and Price 2009).

Lovett *et al.* (2008) identified a range of actions that could be put in place to improve both communication and knowledge exchange in agricultural landscapes. These included developing an online 'information hub' that would include several elements such as tailoring scientific information to different bioregions and synthesising scientific and case study material into a series of fact sheets, guidelines and manuals that address topics identified by end-users. Collaboration between scientists, science communicators and education experts would be essential to develop relevant online information. While these recommendations were in the context of landscape-scale restoration across a range of ecosystems, they have particular relevance to managing temperate woodland landscapes where much of the restoration activity is underway. Undertaking more effective communication and knowledge exchange requires ongoing, long-term support. It will, however, help ensure that the knowledge gained from the investment in research and development on temperate woodlands is more widely adopted and can lead to changed on-ground management.

4. A long-term perspective is essential – reap the benefits of hindsight and foresight

The mismatch between the issues involved with ecological sustainability which operate over decades and centuries, and the shorter time frames of politics and economics, is the cause of many sustainability problems (Dovers 2008). In the context of temperate woodland management, this expresses itself in many ways, including the short-term nature of funding for most research and on-ground management programs. Long-term environmental monitoring has been identified as an area of chronic under-investment and under-resourcing, and incentive programs have a history of stopping and starting. Even exceptions such as the Land for Wildlife (running for 28 years in Victoria) and LandCare programs (currently celebrating 20 years), can suffer from changes in funding arrangements (Williams and Price 2008). Such short-term investments and funding changes can lead to the loss of continuity and momentum, which can often negatively affect on-ground outcomes (Lovett *et al.* 2008). As noted by Prober and Hobbs (2008), off-reserve conservation models used in temperate woodland landscapes, such as Conservation Management Networks and Natural Resource Management councils, require ongoing development and integration.

The call for long-term investments in areas such as monitoring, research and on-ground management programs is not new, and is not restricted to temperate woodland systems. The fact that progress in this area has been limited, demonstrates that even with hindsight past mistakes keep on being made. Taking a long-term perspective is not only important in the context of funding cycles and program continuity. It is also relevant to understanding and managing woodland ecosystems. For example, actions such as broadscale clearing of temperate woodlands, even when carried out many decades in the past, can have impacts long into the future. Such is the nature of the extinction debt, a concept that is receiving increasing attention across temperate woodland landscapes (Possingham 2008).

5. Monitor, monitor, monitor – learn from the rangelands experience

The importance of monitoring species and ecosystems to explore how they respond to factors such as management actions and climate variability has been raised numerous times, over a long period (see Woinarski *et al.* 2000; Williams *et al.* 2001; Lindenmayer *et al.* 2008 for references to the history of monitoring). Developing effective monitoring systems for temperate woodlands, and related ecosystems such as temperate grasslands, could learn from the experience in the rangelands (Stafford-Smith 2008). There have been major advances towards the national integration of monitoring systems for rangelands over the last decade, resulting in the Australian Rangelands Collaborative Information System (ACRIS) (Bastin and the ACRIS Management Committee 2008). The Terrestrial Ecosystem Research Network is currently working with ACRIS to develop an integrated system that incorporates biodiversity in rangelands monitoring. There is still a long way to go to develop the single capital accounting system that Stafford-Smith (2008) identifies as crucial for the rangelands. Even so, the cross-sectoral approach that he outlines, which would include biodiversity as a high priority for improved measurement, is something that would be equally appropriate for temperate woodland systems. Such a system would eventually monitor indicators of health for environmental, social, human, cultural, institutional and built capital, recognising that sustainable landscapes, lifestyles and livelihoods are dependent on good report cards.

6. Establish an effective and practical ecosystem services framework and payment scheme for temperate woodland landscapes

Since Costanza *et al.* (1997) drew the attention of the world to ecosystem services and the economic value they provide to humankind, there has been much written about the concept, particularly at the theoretical level. In 2005, the publication of the Millennium Ecosystem Assessment (Millennium Ecosystem Assessment 2005) put the spotlight on ecosystem services once again, and the term is now widely referred to in policies and programs in Australia and across the globe. The relevance of this approach to temperate woodland conservation and management lies in its potential to recognise, and where appropriate pay, the managers of these ecosystems (which mostly occur on private land) for the services they provide the wider society. These include carbon, water, landscape values and biodiversity. Stewardship payments for managing for conservation outcomes on private land utilise the ecosystem services concept. There is scope to expand this 'single service' approach by developing a framework that encompasses a range of ecosystem services delivered by temperate woodland landscapes. This would build on the pioneering work in Australia of CSIRO (Abel 2003), as well as more recent projects (Eigenraam *et al.* 2006; Karanja *et al.* 2008). For this technique to be more widely adopted as part of the overall approach to temperate woodland conservation, as well as more broadly, the effectiveness of ecosystem service payments for improved environmental outcomes in Australia needs to be demonstrated in practice.

Conclusions

The title of this chapter is 'From pattern and process to people, policy and pragmatism'. This reflects my journey as an ecologist over the last 25 years, where I have come to appreciate the importance of the human dimension of landscape management. Developing a process-based understanding of ecosystems is still essential to improved management, a point that I have continued to make (e.g. Mackey *et al.* 2005). As important, however, is the need for integrated research into, and management of, linked socio-ecological systems. In turn, support from innovative institutions and insightful policy development is required. Improved communication and knowledge exchange are part of the way forward, as are developing systems that recognise, and where appropriate pay, temperate woodland managers for the services they provide to the wider community. Given the urgency of the situation, the need for such responses to be both practical and timely is where the pragmatic element arises. We know what needs to be done, so the focus should be on implementing the solutions.

Acknowledgements

I would like to thank all of the friends, colleagues, family members, farmers and funders who have supported and influenced me through my journey over the last few decades. In particular a big thank-you goes to my Dad, who passed away during the period this chapter was written. When I was a child, he predicted that I would become a scientist, so he has a lot to answer for! Thanks also to the two reviewers of this chapter, whose feedback was much appreciated.

Bio

Jann Williams' professional training is in vegetation and landscape ecology, with a BSc from Melbourne University and a PhD from ANU. Over her career she has worked in the university, government and private sectors, with 10 years spent as a coordinator of national Research and Development programs at Land & Water Australia. She is currently the Managing Director of NRM Insights Pty. Ltd. (jannw@bigpond.net.au) and retains strong academic links as an Adjunct Professor at the University of Tasmania (jann.williams@utas.edu.au). For nearly 20 years she has served the ecological profession through various roles in the Ecological Society of Australia, most recently as the Chair of the Editorial Board of *Ecological Management & Restoration*. 'Linking science and practice' is both the goal of this journal as well as Jann's personal goal, which she helps achieve by communicating widely to a diversity of audiences.

References

Abel, N. (2003). 'Natural Values: Exploring Options for Enhancing Ecosystem Services in the Goulburn Broken Catchment'. CSIRO Sustainable Ecosystems, Canberra.

Bastin, G. and the ACRIS Management Committee (2008). 'Rangelands 2008 – Taking the Pulse.' National Land & Water Resources Audit, Canberra.

Bridle, K. and Price, R. (2009). Undertaking participatory research at a national scale: the Biodiversity in Grain & Graze approach. *Animal Production Science* **49**, 916–927.

Costanza, R., D'Arge, R., De Groot, R., Farber, S., Grasso, M., Hannon, B., Limburg, K., Naeem, S., O'Neill, R.V., Paruelo, J., Raskin, R.G., Sutton, P. and Van Der Belt, M. (1997). The value of the world's ecosystem services and natural capital. *Nature* **387**, 253–260.

Dovers, S. (2008). Policy and institutional reforms. In *Ten Commitments: Reshaping the Lucky Country's Environment.* (Eds D. Lindenmayer, S. Dovers, M. Harris Olson and S. Morton) pp. 215–223. CSIRO Publishing: Melbourne.

Eigenraam, M., Strappazzon, L., Lansdell, N., Ha, A., Beverly, C. and Todd, J. (2006). 'EcoTender: Auction for Multiple Environmental Outcomes'. Department of Primary Industries, Victoria.

Hobbs, R.J. and Yates, C.J., eds. (2000). *Temperate Eucalypt Woodlands in Australia: Biology, Conservation, Management and Restoration.* Surrey Beatty & Sons: Chipping Norton, NSW.

Karanja, F., Reid, N. and Cacho, O. (2008). Economic valuation of ecosystem services from environmental flow provision in the Gwydir catchment, north-western NSW, Australia. In *IAIA08 Conference Proceedings.* pp. 1–6. IAIA Australia.

Lindenmayer, D., Crane, M. and Michael, D. (2005). *Woodlands. A Disappearing Landscape.* CSIRO Publishing: Melbourne.

Lindenmayer, D., Dovers, S., Harris Olson, M. and Morton, S., eds. (2008). *Ten Commitments: Reshaping the Lucky Country's Environment.* CSIRO Publishing: Melbourne.

Lindenmayer, D. and Hobbs, R.J. (Eds) (2007). *Managing and Designing Landscapes for Conservation: Moving from Perspectives to Principles.* Blackwell Publishing: Oxford.

Lovett, S., Lambert, J., Williams, J.E. and Price, P. (2008). 'Restoring Landscapes with Confidence — an Evaluation of the Science, the Methods and their On-ground Application. Final Report'. Land & Water Australia: Canberra.

Mackey, B.G., Soule, M.E., Nix, H.A., Recher, H.F., Lesslie, R.L., Williams, J.E., Woinarski, J.C.Z., Hobbs, R. and Possingham, H.P. (2005). Applying landscape-ecological principles to regional conservation: the WildCountry Project in Australia. In *Key Topics and Perspectives in Landscape Ecology.* (Eds J. Hu and R. Hobbs) pp. 192–213. Cambridge University Press: Cambridge.

Millennium Ecosystem Assessment (2005). *Ecosystems and Human Well-being: Synthesis.* Island Press: Washington.

Mortlock, W., Williams, J., Whiting, G. and Conroy, G., eds. (2002). 'Thinking Bush – Issue 1. Vegetation Research and Management'. Land & Water Australia: Canberra.

Possingham, H. (2008). Biodiversity. In *Ten Commitments: Reshaping the Lucky Country's Environment.* (Eds D. Lindenmayer, S. Dovers, M. Harris Olson and S. Morton) pp. 155–162. CSIRO Publishing: Melbourne.

Prober, S.M. and Hobbs, R.J. (2008). Temperate eucalypt woodlands. In *Ten Commitments: Reshaping the Lucky Country's Environment.* (Eds D. Lindenmayer, S. Dovers, M. Harris Olson and S. Morton) pp. 19–26. CSIRO Publishing: Melbourne.

Stafford-Smith, M. (2008). Rangelands. In *Ten Commitments: Reshaping the Lucky Country's Environment.* (Eds D. Lindenmayer, S. Dovers, M. Harris Olson and S. Morton) pp. 11–18. CSIRO Publishing: Melbourne.
Williams, J. and McKenzie, F. (2008). Agriculture. In *Ten Commitments: Reshaping the Lucky Country's Environment.* (Eds D. Lindenmayer, S. Dovers, M. Harris Olson and S. Morton) pp. 105–112. CSIRO Publishing: Melbourne.
Williams, J.E. (2000). 'Managing the Bush: recent research findings from the EA/LWRRDC National Remnant Vegetation R&D Program. Research Report 4/00.' National Research and Development Program on Rehabilitation, Management and Conservation of Remnant Vegetation. LWRRDC: Canberra.
Williams, J.E. (2007). Linking science and practice: the pros and cons of the participatory research model. *Ecological Management & Restoration* **8**(3), 158–159.
Williams, J.E. (2008). 'Restoring Landscapes with Confidence: State of Knowledge Discussion Paper on Landscape Restoration Science in Australia'. Land & Water Australia: Canberra.
Williams, J.E. and Eamus, D. (1997). Plant ecophysiology: linking patterns and process – a review. *Australian Journal of Botany* **45**, 351–357.
Williams, J.E. and Goodacre, M. (2008). Profitable and biodiverse wool production: the Land, Water & Wool Program. In *Biodiversity: Integrating Conservation and Production. Case studies from Australian Farms, Forests and Fisheries.* (Eds T. Lefroy, K. Bailey, G. Unwin and T. Norton) pp. 177–185. CSIRO Publishing: Melbourne.
Williams, J. and Price, R. (2008). 'Evaluation of Responses to Threats to Australia's Biodiversity'. National Land & Water Resources Audit: Canberra.
Williams, J.E., Read, C., Norton, T., Dovers, S., Burgman, M., Proctor, W. and Anderson, H. (2001). 'Biodiversity, Australia State of the Environment Report 2001 (Theme report)'. CSIRO Publishing on behalf of the Department of Environment and Heritage: Canberra.
Woinarski, J., Franklin, D., Fisher, A., Fensham, R. and Whitehead, P. (2000) 'Developing an Analytical Framework for Monitoring Biodiversity in Australia's Rangelands. Background paper 3. A review of information gathered from existing biodiversity monitoring programs'. Tropical Savannas Cooperative Research Centre: Darwin.

34

A CARBON VISION FOR THE RESTORATION OF EUCALYPT WOODLANDS

David Freudenberger

1. Woodland restoration is possible.
2. The past scale of restoration is inadequate.
3. Commercial-scale investment in natural infrastructure is needed.
4. The emerging carbon market could be a major investor in woodland restoration.
5. Poorly planned and executed investments in carbon biosequestration are a risk.
6. Design and implementation standards are needed for biosequestration.
7. Applied research is needed to reduce constraints to commercial investment and on-ground restoration.

Introduction

For the past 28 years, Greening Australia has been involved in woodland restoration throughout its distribution in Australia. For example in 2007, Greening Australia direct seeded nearly 4000 km of locally native trees and shrubs and erected over 1000 km of fencing to assist in the conservation of over 100 000 ha of remnant vegetation. We have collectively learned some positive, as well as some sobering lessons from these efforts in partnership with landholders, government agencies, commercial organisations and thousands of volunteers.

Lessons

1. Woodland restoration is possible

Greening Australia's restoration works have involved the planting of a modest diversity of local trees and shrubs into long-cleared land dominated by a ground-layer of exotic grasses and legumes. We now know how to successfully restore a large diversity of native plants, including groundcover (Gibson-Roy 2007). We know that if we plant it, most plants will survive and some wildlife will come (Taws 2007). These plantings are providing ecosystem services at the patch scale, including habitat for native birds (Barrett *et al.* 2008; Lindenmayer *et al.* 2010;

Munro *et al.* 2007) and native fungi (Barrett *et al.* 2009). The scale of these plantings, however, is unlikely to reverse ongoing declines in native biodiversity in agricultural landscapes.

2. The past scale of restoration is inadequate

To date, plantings have been generally small (a few hectares), narrow (less than 50 m) and a small proportion of the landscape (<5%) (Freudenberger and Harvey 2004). These plantings have been established for a wide variety of reasons including livestock and crop windbreaks, dryland salinity reduction, aesthetics, and the provision of wildlife habitat. To date, this scale of plantings is far from restoring native vegetation cover to a threshold of at least 30% hypothesised to be needed to reverse declining trends in biodiversity (Radford and Bennett 2007).

The fundamental challenge is to scale up ten-fold. The task is to move from narrow strips and little patch plantings covering less than 2% of a farm, to whole paddocks covering at least 20% of all farms suffering from a 150 year history of over clearing, erosion, salinity, tree dieback and the loss of the dawn chorus of woodland birds. The task is to restore larger and less costly areas of revegetation.

3. Commercial-scale investment in natural infrastructure is needed

Landscape scale restoration of woodlands requires billions of dollars over the next two decades. Yet, investment in the environment has suffered from too many decades of volunteerism. To date, a $10 000 restoration project is considered significant, whereas a truly significant project requires at least $10 million. Caring for Our Country (Australian Government 2009a), the primary program for the Australian Government's investment in the environment (including marine) reaches nowhere near the level of investment required. This five-year, $2 billion program, equates to just $0.52/ha per year if equally invested in every hectare of the Australian continent (768.2 M ha). This program is 0.13% of the 2008–09 Australian Government budget ($299 billion, Australian Government 2009b). In contrast, the 2009–10 Australian Government Budget has committed to a $22 billion investment in built infrastructure (e.g. roads) (Australian Government 2009c). The scale of private investment in built infrastructure dwarfs government investment. The limited Government investment in Caring for Our Country equates to just 800 new houses at an average individual cost of $0.5 M (on average 350 new houses are started each day in Australia; HIA 2009).

But all built infrastructure is dependent on robust and resilient natural infrastructure. Low cost, high quality and reliable water supplies are only delivered by catchments well covered by diverse native vegetation – a key form of natural infrastructure. The risk of infrastructure failure of roads, bridges and buildings increases with the catchment-scale loss of native vegetation that often results in increased flood severity, rising saline water tables, rising sea level and acid sulfate soil collapse in low-lying regions. There is a pressing need to garner investment in natural infrastructure, particularly native vegetation, on the same scale that Australia routinely invests in built infrastructure.

4. The emerging carbon market could be a major investor in woodland restoration

The emerging carbon market for biosequestration has the potential to restore at least the tree elements of woodlands at the required scale. For instance, if Australia offsets just 10% of its annual carbon emissions (approx. 600 M t $CO_{2\text{-}e}$) through carbon sinks based on native woodlands and forests, this level of investment would generate $1.5 billion per year assuming at a price of $25/t $CO_{2\text{-}e}$. At a generous planting cost of $6000/ha (including land access), this

$1.5 billion in carbon biosequestration could revegetate 250 000 ha annually. At this rate of establishment, 5 million ha of restored woodlands and forests could be established over the next 20 years. Such a restoration rate would be greater than that for the timber plantation industry which aims to have 3 million ha established by 2020 (Thompson 2008). Five million hectares may seem like a lot, but it pales against the 50 million hectares of land across the Australian continent that has already been cleared for agricultural purposes. We can have our carbon restoration and still produce a surplus of food and fibre.

Even if just 1% of Australia's emissions were offset through diverse native vegetation sinks, this could result in 25 000 ha of restoration per year. Over 20 years, such a rate could result in 500 000 ha of restored woodland and forest. Even at this low rate, the emerging carbon market could be a significant investor in restoration at a scale well beyond decades of Australian Government programs such as One Billion Trees, Landcare, the Natural Heritage Trust, and Caring for Country.

The opportunity to 'turbo-charge' woodland restoration through the carbon market is not fanciful (Bekessy and Wintle 2008). The international carbon market was estimated to be worth US$64 billion in 2008 (Capoor and Ambrosi 2008). To date, this rapidly growing market has invested mostly in low carbon energy generation (e.g. solar), energy reduction strategies and technologies (e.g. energy use efficiency), but investment in carbon biosequestration is increasing. In Australia, the commercial business, CO_2 Australia, has one of the largest pools of woodland sinks. They have over 12 500 ha in south-eastern and Western Australia planted with a small number of mallee (eucalypt) species (CO_2 Australia 2009). To date, Greening Australia has established 380 ha of carbon sinks with over 50 locally native species of trees and shrubs. These sinks for voluntary emissions offsets are likely to expand rapidly under the mandatory reductions required under the Australian Government's proposed Carbon Pollution Reduction Scheme.

5. Poorly planned and executed investments in carbon biosequestration are a risk

Markets tend to develop low-cost solutions to maximise short-term profits. The carbon market could support yet another monoculture of fast-growing tree crops poorly integrated with other land uses and values. The carbon market, including biofuels, could result in the next wave of woody weeds adding yet another threat to native biodiversity and ecosystem function. Markets, however, quickly differentiate products across a spectrum of price and quality. There is a need to create incentives, including tax benefits and regulation, to encourage the development of 'premium' quality carbon offsets based on the restoration of a diversity of native vegetation rather than monocultures of exotic tree crops.

6. Design and implementation standards are needed for biosequestration

In view of the risks and opportunities of the emerging carbon market, Greening Australia is developing a national business to offer large corporate emitters the opportunity to reduce a modest proportion of their greenhouse gas emissions through biosequestration plantings (Greening Australia 2009). Our product, *GA Biodiverse Carbon*, consists of sequestration plantings that restore a self-replacing diversity of regionally native vegetation on land cleared prior to 1990. Our carbon plantings are:

- at least 100 ha in size and more than 100 m wide
- actively managed for at least 100 years and are not harvested
- sourced from seed native to the bioregion in which they are planted

- suited to local (site by site) soil, slope and climatic conditions
- rich in diverse plantings, including a range of species from all layers of the plant community – from large trees to groundcover
- well-connected habitats for many species of native wildlife
- designed to cope with climate change including hotter temperatures, lower and more variable rainfall, and more frequent fires
- accredited by the Australian Government through a program existing at the time of planting, e.g. the *Greenhouse Friendly Program* managed by the Australian Government (2009d)
- planned to provide net environmental benefits at the landscape (sub-catchment) scale including reduction in the risk of salinity and improvement in water quality.

7. Applied research is needed to reduce constraints to commercial investment and on-ground restoration

Targeted research has a critical role in realising a vision of native trees, shrubs and grasses being restored across the vast rolling hills and plains of southern Australia. Research is urgently needed to accurately predict the likely carbon yield, resilience and persistence (longevity) of restored native vegetative. There is an equally urgent need to reduce the cost of large-scale revegetation and improve its quality (e.g. compositional, structural and functional diversity). Restoration research has a key role in facilitating investments in large scale restoration of temperate woodlands.

Conclusions

We are now close to an unprecedented opportunity to restore Australia's iconic woodlands. Greening Australia has developed direct seeding technology and incentive packages that are proving highly attractive to commercial farmers to re-integrate native trees, shrubs and grasses into their production systems that provide essential food and fibre to the Australian economy. The potential commercial carbon market for biosequestration can help finance this model for restoration at a scale far greater than public investment through government natural resource management programs.

Acknowledgements

David Williams, Rick Humphries, Michael Rooney, Robert Lambeck, Dave Carr and Di Dibley from the national office of Greening Australia have contributed to the concepts and lessons described in this chapter.

Bio

David Freudenberger is Greening Australia's Director of Science and Major Projects. He has 25 years of ecological research experience, the most recent 15 years with CSIRO. David is an author of over 130 research publications including 51 papers in peer-reviewed scientific publications. He has had research leadership roles in CSIRO and is a graduate of the Australian Rural Leadership Program. David Freudenberger's role in Greening Australia is to foster transformative landscape scale initiatives to conserve biodiversity, improve water quality in our rivers, contribute to sustainable agriculture and enhance community well being.

References

Australian Government (2009a). 'Caring for Our Country Business Plan 2008–2013'. http://www.nrm.gov.au/publications/books/caring-outcomes.html

Australian Government (2009b). 'Final Budget Outcome 2008–09'. http://www.budget.gov.au/2008-09/content/fbo/html/index.htm)

Australian Government (2009c). 'Budget at a Glance 2009–2010'. http://www.budget.gov.au/2009-10/content/at_a_glance/html/index.htm.

Australian Government (2009d). 'Greenhouse Friendly'. http://www.climatechange.gov.au/greenhousefriendly/

Barrett, G.W., Freudenberger, D., Drew, A., Stol, J., Nichols, A.O. and Cawsey, E. M. (2008). Colonisation of native tree and shrub plantings by woodland birds in an agricultural landscape. *Wildlife Research* **35**, 19–32.

Barrett, G., Trappe, J.M., Drew, A., Stol, J. and Freudenberger, D. (2009). Fungus diversity in revegetated paddocks compared with remnant woodland in a south-eastern Australia agricultural landscape. *Ecological Management and Restoration* **10**, 200–209.

Bekessy, S.A. and Wintle, B.A. (2008). Using carbon investment to grow the biodiversity bank. *Conservation Biology* **22**, 510–513.

Capoor, K. and Ambrosi, P. (2008). 'State and trends of the carbon market 2008'. World Bank, Washington D.C. http://siteresources.worldbank.org/NEWS/Resources/State&Trendsformatted06May10pm.pdf

CO_2 Australia (2010). 'Planting trees'. http://www.co2australia.com.au/?sectionID=6695&pageID=6696

Freudenberger, D. and Harvey, J. (2004). Predicting biodiversity benefits of the Saltshaker Project, Boorowa, NSW. *Ecological Management and Restoration* **5**, 5–14.

Gibson-Roy, P. (2007). Restoring Western (Basalt) Plains grassland. 2. Field emergence, establishment and recruitment following direct seeding. *Ecological Management and Restoration.* **8**, 123–132.

Greening Australia (2009). 'Combating climate change'. http://www.greeningaustralia.org.au/our-solutions/climate-change.

HIA (2009). 'Painful end to 2008 for housing starts'. economics.hia.com.au/…/abs%20dwelling%20commencements%20Dec%20qtr%2008.pdf

Lindenmayer, D.B., Knight, E.J., Crane, M.J., Montague-Drake, R., Michael, D.R. and MacGregor, C.I. (2010). What makes an effective restoration planting for woodland birds? *Biological Conservation* **143**, 289–301.

Munro, N.T., Lindenmayer, D. and Fischer, J. (2007). A faunal response to revegetation in agricultural areas of Australia: a review. *Ecological Management and Restoration* **8**, 199–207.

Radford, J.Q. and Bennett, A.F. (2007).The relative importance of landscape properties for woodland birds in agricultural environments. *Journal of Applied Ecology* **44**, 737–747.

Taws, N. (2007). 'Bringing Birds Back: A Glovebox Guide to Bird Identification and Habitat Restoration'. Greening Australia Capital Region: Canberra.

Thompson, D. (2008). 'Plantations for Australia: The 2020 vision progress report. Plantations 2020'. http://www.plantations2020.com.au/assets/acrobat/2020%20Vision%20Progress%20Summary%20Report%202008.pdf.

35

USING MARKETS FOR WOODLAND CONSERVATION ON PRIVATE LAND: LESSONS FROM THE POLICY–RESEARCH INTERFACE

Charlie Zammit, Simon Attwood and Emma Burns

1. Legislation and reservation alone will not be enough – a combination of legislative, market-based and new management approaches are needed for effective conservation of Australian temperate woodlands across tenures.
2. Market-based incentive schemes can be rapidly and effectively delivered to meet specific conservation policy goals.
3. The use of a state-and-transition model has helped policy makers and private land managers understand the complex ecology of woodlands in ways that shape on-ground management and investment planning.
4. The use of a purpose-built, conservation value metric has effectively guided investments and has helped land managers appreciate evidence-based decision making.
5. Market-based approaches to conservation provide significant opportunities for innovation across the policy–research interface

Introduction

Historically, the temperate eucalypt woodlands of Australia covered approximately 527 000 km^2.[1] Today they occupy about 132 000 km^2 of which approximately 60% is on private land and about 18% is protected in the National Reserve System. The level of reservation in some regions, however, is less. For example, in the New England Tablelands Bioregion of New South Wales about 10% is protected.

Temperate woodlands encompass many ecological communities, habitats and species of national and global significance. Land-use intensification and ongoing transformation to support agricultural and other developments has resulted in the loss of about 75% of these woodlands across the continent, with impacts on many communities and their fauna and flora. Today, 10 of the 46 ecological communities listed as threatened under the Commonwealth Government's *Environment Protection and Biodiversity Conservation Act (EPBC Act) 1999* are found in temperate eucalypt woodlands. In addition, about 38% of listed threatened species have been recorded in association with these woodlands, most being plants (Table 35.1).

Table 35.1 Estimated numbers of threatened species recorded in temperate eucalypt woodland, under the Commonwealth Government's *Environment Protection and Biodiversity Conservation Act (EPBC Act) 1999*. (Note: Fish are excluded from this summary)

Threatened spp. status	Taxonomic group	Number of spp.	Total
Critically Endangered	Plants	19	25
	Birds	2	
	Invertebrates	3	
	Reptiles	1	
Endangered	Plants	209	247
	Birds	13	
	Amphibians	8	
	Invertebrates	1	
	Mammals	12	
	Reptiles	4	
Vulnerable	Plants	348	397
	Birds	11	
	Amphibians	11	
	Invertebrates	2	
	Mammals	13	
	Reptiles	12	
All Threatened categories	Plants	576	669
	Birds	26	
	Amphibians	19	
	Invertebrates	6	
	Mammals	25	
	Reptiles	17	

Under Australia's federal system of government, the management of temperate woodlands is primarily a State and Territory Government responsibility. The Commonwealth Government becomes involved through: (i) its legal responsibilities under the Commonwealth EPBC Act and other legislation; (ii) working collaboratively with other governments through the relevant ministerial councils on national policy and program issues; and (iii) its various funding and investment programs.

For the temperate woodlands, iconic landscapes of scattered gum trees in grazing paddocks are as synonymous with Australian farming and rural life as hedgerows are with European agricultural landscapes. From a policy and scientific perspective, conserving biodiversity on private land in these agricultural landscapes presents a significant ongoing challenge (Productivity Commission 2001; McIntyre *et al.* 2002). This challenge is also an opportunity for innovative thinking and planning that engages researchers, policy makers and private land managers in conservation management of remaining woodlands from paddock to landscape scales.

Here we draw on our recent experiences in using markets to protect biodiversity on private land, in particular, lessons learnt when designing and implementing the Environmental Stewardship Program under the Commonwealth Government's *Caring for our Country* initiative (http://www.nrm.gov.au/stewardship/index.html). We note that the Stewardship Program is a work in progress which can be usefully considered as 'a landscape experiment' (*sensu* Herzog 2005).

Lessons

1. Legislation and reservation alone will not be enough – a combination of legislative, market-based and new management approaches are needed for effective conservation of Australian temperate woodlands across tenures

Temperate woodlands are typically working landscapes. They require management for sustainable production, biodiversity conservation and other needs, that engages rural communities and land managers and uses suitable policy approaches for particular purposes and at different scales. These approaches include legislation, market-based approaches, and more effective use of information and capacity building, to influence individual and sectoral management. They must be supported by credible and well targeted science that can improve decision making and monitor and evaluate the effectiveness of conservation initiatives.

The use of markets for improving biodiversity conservation has grown recently in countries where biodiversity conservation overlaps, and often competes, with extractive industries. While individual approaches vary across countries and schemes, a consistent premise is that long-term conservation of biodiversity is required to support a range of ecosystem goods and services which are fundamentally important to economic and social well being (Bishop *et al.* 2008).

In Australia, the use of market incentive schemes has been discussed for some time (Industry Commission 1997) and successfully used across a range of environment policy issues (Whitten *et al.* 2004). Well-designed, targeted and effectively delivered schemes can benefit both biodiversity and rural communities (Stoneham *et al.* 2003; Donald and Evans 2006). Advantages of these schemes include:

- identifying high conservation value habitats on private lands;
- encouraging the integration of production, conservation and other goals into on-farm management;
- paying land managers a competitive price for undertaking environmental activities above duty of care;
- developing mechanisms for buyers (e.g. governments) to select the best value-for-money proposals;
- developing mechanisms for sellers (private land managers) to price conservation management activities; and
- providing opportunities for differentiation of agricultural produce based on 'clean and green' marketing.

Effective policy responses to the conservation of temperate woodlands also require smarter approaches to environmental management, and in particular, approaches that clearly define the problem and outline costs and benefits of different solutions. In a world of constrained budgets and ongoing pressure to invest more in conservation, governments are increasingly driven to specify their priorities clearly and put in place more rigorous and accountable processes for ensuring that public funds are targeted to the highest priority issues and delivered cost-effectively.

2. Market-based incentive schemes can be rapidly and effectively delivered to meet specific conservation policy goals

In keeping with the global momentum to develop new markets for conservation, a number of approaches have been used by different Australian governments, including Victoria (e.g. BushTender, Plains Tender), Queensland (e.g. NatureAssist), NSW (e.g. BioBanking), South Australia (e.g. Bush Bids) and the Commonwealth (e.g. Forest Conservation Fund, the Environmental

Stewardship Program). While these approaches have each been designed to meet specific jurisdictional needs and priorities, common features include:

- governments' wish to purchase environmental and conservation outcomes using competitive approaches;
- the predominant use of reverse tenders as the mechanism to attract voluntary bids;
- providing opportunity for bidders (private land managers) to nominate a price for specific management actions that address the outcomes;
- use of a model of environmental change and/or ecological principles and understanding to anticipate the impacts of threats to the environment and likely benefits from remedial actions or other interventions;
- opportunities for securing long-term protection through use of covenants;
- use of environmental metrics to evaluate and rank bids;
- a time-bound financial contract with successful bidders to deliver specified management actions for an agreed price; and
- mechanisms for managing contractual compliance, including performance reporting and penalties for defaulting.

The Commonwealth Government's Environmental Stewardship Program (ESP http://www.nrm.gov.au/stewardship/) was designed to complement existing conservation grants programs by:

- providing clear guidance on priority conservation assets and objectives. The Program only targets particular matters of national environmental significance under the EPBC Act (see ESP Strategic Framework http://www.nrm.gov.au/publications/frameworks/environmental-stewardship.html).
- providing funding for up to 15 years to allow sufficient time for observable conservation improvements to be made, and to progressively build a culture of enduring private land conservation.
- developing management and monitoring actions in consultation with scientific specialists and experienced managers, thereby ensuring an adaptive approach to management based on sound evidence.

The Environmental Stewardship Program identified the critically endangered White Box–Yellow Box–Blakely's Red Gum Grassy Woodland and Derived Native Grassland (hereafter referred to as box gum grassy woodland) of south-eastern Australia as its first target for investment. Several regional NRM organisations have been contracted to deliver tender rounds in the Lachlan-Murrumbidgee, Central West, Namoi and Border Rivers-Gwydir NRM regions of NSW and the south-east Queensland, Condamine and Border Rivers-Maranoa-Balonne NRM regions of Queensland. The box gum grassy woodland project commenced its first tender round in August 2008, and by the end of 2009 completed four rounds, with a fifth closing in early 2010.

In less than two years, this project has designed and implemented a fully operational tender system for conserving remnants of these nationally endangered woodlands on private land. The project has received over 580 requests for site assessment and has completed 508, with the balance either withdrawing or not being eligible. From these, a total of 210 tender bids have been received resulting in 161 successful proposals. From these proposals contracted land

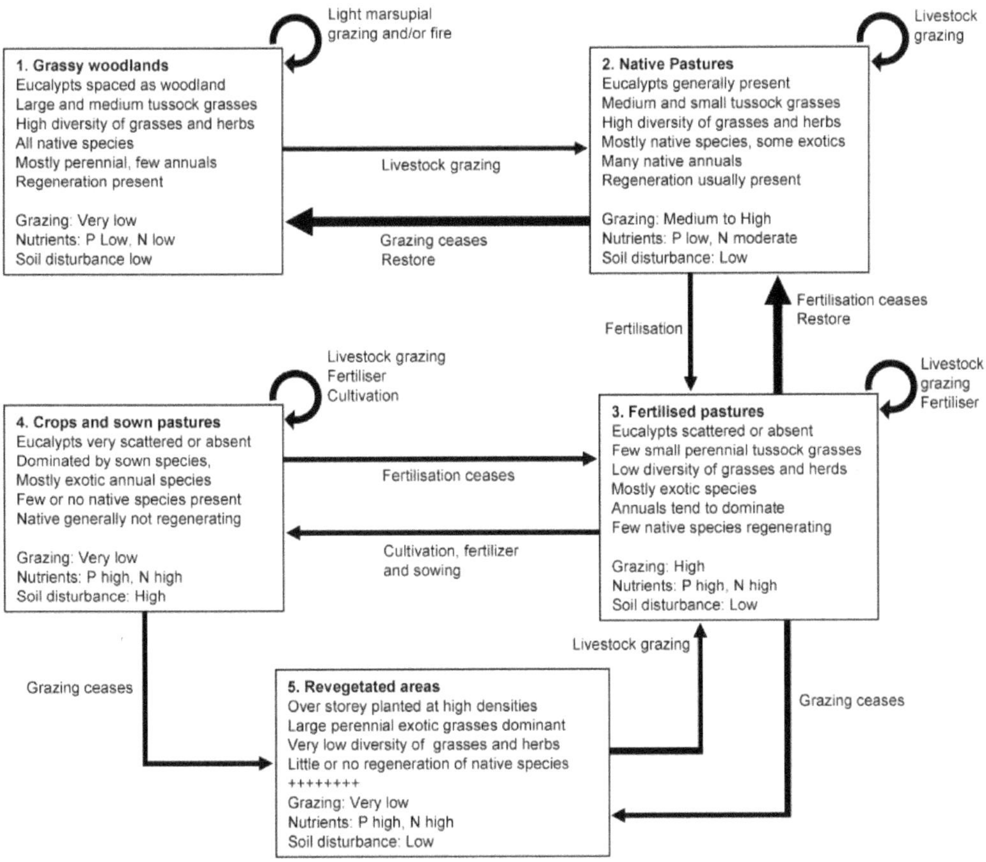

Figure 35.1 Environmental Stewardship's Box Gum Grassy Woodland State & Transition Model (developed by Gibbons and Ryan 2008 after McIntyre and Lavorel 2007). Arrows between boxes describe the key management inputs that drive transitions between states. Thin arrows represent relatively easy transitions, thick arrows more difficult transitions. Curved arrows indicate maintenance management inputs.

managers will conserve a total of approximately 16 000 ha of critically endangered woodlands for 15 years.

3. The use of a state-and-transition model has helped policy makers and private land managers understand the complex ecology of woodlands in ways that shape on-ground management and investment planning

State and transition models (S&TM) have a long pedigree in conceptualising non-equilibrium vegetation dynamics and as a heuristic tool to support land management (Westoby *et al.* 1989; Bestelmeyer *et al.* 2009). The need for a S&TM was driven by the requirement to understand the complexity of woodland ecology under different land-use regimes better and to codify the resulting variability in woodland attributes into a small set of easily recognisable 'states', with associated transitions to other states in the system.

The box gum grassy woodland S&TM depicts the different states of these woodlands under particular land uses and identifies the principal drivers of change from one state to another (Fig. 35.1).

Developing this model under a short time frame required close and open engagement between the project managers and woodland ecologists and woodland managers. As a consequence, the S&TM for box gum grassy woodland has been effective in:

- helping project field officers convey complex ecological responses to changing land use and management to land managers;
- helping land managers visualise their landscapes and what woodland conservation is trying to achieve;
- assisting field officers to identify the most suitable management actions for a particular woodland state;
- providing decision makers with guidance on what management actions might achieve measurable improvements within Program budget and time frame; and,
- communicating eligibility criteria and purchasing preferences that align with the Program's objectives.

Eligibility for the project has so far been restricted to sites assessed as states 1, 2 or 3 (Figure 35.1). States 4 and 5 are presently ineligible for funding on two grounds: the expectation that all funded sites can be restored to a condition consistent with EPBC Act listing criteria within a 15 year contract and the likely cost inefficiency of improving habitat in the poorest condition. We recognise that the decision to exclude the most degraded woodland sites is not supported by all woodland ecologists, as it can be argued that scarce public funds should also be directed at achieving the largest improvement in ecological condition. We agree and are exploring more sophisticated decision-support tools to take better account of this opportunity (see Lesson #5).

4. The use of a purpose-built, conservation value metric has effectively guided investments and has helped land managers appreciate evidence-based decision-making

All competitive tenders require a robust method for scoring and ranking proposals, and there is a substantial theoretical and empirical literature on metric design and application (O'Connor 2008). When used to support investment decision-making, an environmental metric needs to incorporate three distinct but related considerations. First, it needs to reliably and quantitatively assess the condition of the ecological asset being targeted. Second, it needs to assess the relative contribution of proposed management actions in improving ecological condition. Third, it needs to consider the duration and security provided by the proponent to ensure the long-term persistence of improved condition.

We used the woodland S&TM to develop a conservation metric that generated a score for each bid according to a numerical function that incorporated: i) habitat condition of the woodland patch in relation to condition benchmarks and likely improvement in condition due to conservation management; ii) the regional 'rarity' of the vegetation type; iii) the landscape context of each patch; iv) the area of the patch; v) the duration of the management agreement offered; and vi) additional security as reflected by optional inclusion of a covenant on the land (Gibbons and Ryan 2008; Whitten *et al.* 2009).

The metric score for the woodland offered in each proposal is divided by the proponent's bid price to produce a conservation value index that allows comparisons from all proposals in the tender round. A marginal cost curve (Figure 35.2) illustrates the relative value of each proposal as a function of the total conservation value of the tender round. Such 'hockey-stick'

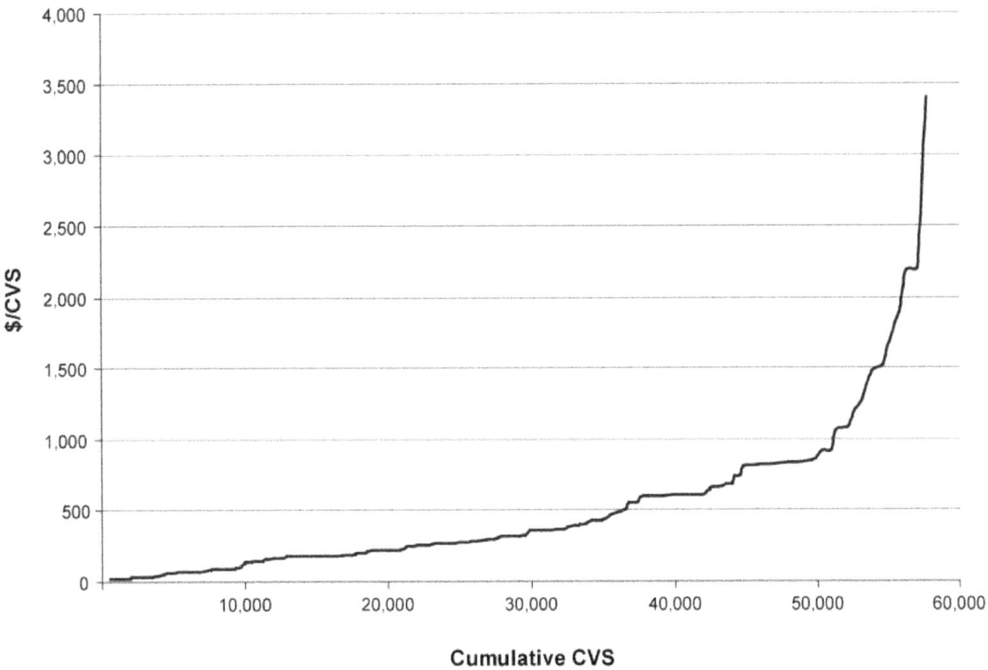

Figure 35.2 A marginal cost curve of cumulative Conservation Value Score (CVS) plotted against the cost per unit score.

curves are typical of conservation tenders and they provide a sound basis for selecting proposals against their relative contribution to the total conservation value generated by the tender round (Stoneham *et al.* 2003).

Our experience with the use of metrics in several market-based conservation programs confirms that private land managers are keen to understand the underlying logic of metric design and its relationship to ecological field measurements. Consequently, we are progressively providing more guidance materials and are introducing information sessions that incorporate 'tender simulation exercises' to allow participating land managers an opportunity to gain some hands-on experience with competitive tenders.

5. Market-based approaches to conservation provide significant opportunities for innovation across the policy-research interface

The four lessons we have drawn from our recent experiences in the use of markets for woodland conservation on private land have revealed significant opportunities for innovation where research and policy intersect. We identify three broad domains of intersection.

First, are technical ecological research issues that will improve the quality of the particular conservation outcome being addressed? For instance, the current woodland metric operates on the assumption that all management actions have an equal probability of success, which in reality is unlikely. The metric is being revised to contain assessments of the likelihood of success of each management action in a given situation. In addition, although the current metric makes some assessment of landscape considerations, the metric and the broader decision-making processes do not explicitly take account of the spatial distribution of bids in a round, or indeed the locations of similar woodland habitats, including those secured in previous bids. It is clear that landscape-scale conservation benefits, including opportunities for connectivity (Crooks and Sanjayan 2006), are not yet optimised in these markets.

Second, are there opportunities for expanding the mix of conservation and other environmental outcomes that the market could deliver? We are mindful that focussing solely on patches of remnant woodland within the agricultural matrix of a property is suboptimal (Attwood *et al.* 2009). There are likely to be additional management actions in the matrix that would benefit the ecological condition or extent of the woodland remnants; for example, incorporating isolated paddock trees into the scope of investments (Manning and Lindenmayer 2009). Moreover, the program will soon expand tenders to include groups of nationally listed ecological communities that co-occur in a region. There is also enormous scope within the emerging markets for biodiversity, water and carbon, for researchers, policy makers and resource managers to work collaboratively on more integrated and sophisticated tools and approaches (TEEB 2009). For example, working in Victoria, Eigenraam *et al.* (2006) have developed a tender process, with supporting metrics, for securing multiple environmental outcomes that include improvements in native vegetation, water quality and carbon.

Third, are there opportunities associated with the broader governance and administration of conservation investments? For example, there is evidence of the success of outcomes-based programs (Kapos 2009), evidence for specific conservation markets being aligned with emerging offsets markets (Galatowitsch 2009) and for further developing biodiversity banking and trading systems (Carroll *et al.* 2008; Ribaudo *et al.* 2008).

We are mindful that the conservation merits of some of these policy-research intersections will be contested for a variety of reasons; however, we welcome this as part of the challenge of innovation.

Conclusions

The use of market-based approaches for conserving temperate woodlands on private land is relatively recent. Our early experiences have been positive, with strong partnerships emerging between policy makers, researchers and private land managers. There is also an unmet community demand for incentive-based approaches like the Environmental Stewardship Program that provide long-term funding. Indications are that land managers understand and welcome the competitive tender process.

As conservation planners and program designers, we are always looking to better harness the policy tools we have – legislation, incentives/markets and information – to deliver the best outcome possible from the investment of limited public funds. Our sense is that conservation agendas will increasingly look to market-based opportunities because they require a clear specification of what is wanted, demand accountable and transparent mechanisms for managing investments, are cost-efficient and create real scope for innovation.

Innovation across the policy–research interface will inevitably challenge some traditional approaches, values and priorities for biodiversity conservation (Gibbons *et al.* 2008). However, many disciplines including ecology, climatology, environmental economics and law, community psychology and public administration are critical contributors in designing more effective policies and programs for the conservation of Australia's temperate woodlands.

Endnote

1 Area of temperate eucalypt woodlands was estimated from data provided in Commonwealth of Australia (2007), Major Vegetation Group 5, with rangelands, tropics

and alpine areas excluded. Analysis was undertaken by Environmental Resources Information Unit in DEWHA.

Acknowledgements

We thank Karl Bossard and Lisa Winsemius from the Department's Environmental Resources Information Network for analysis relating to woodland cover and threatened species records. We also thank Pete Lyon and Fiona Dickson for comments and advice.

Bios

Charlie Zammit has a Ph.D. in plant ecology and has spread his professional career between academia and government positions. He is currently Assistant Secretary, Biodiversity Conservation Branch where he is responsible for national biodiversity, vegetation and forest policy issues and for developing market-based approaches to biodiversity conservation on private land. He has held a variety of senior Commonwealth policy positions in the Department of The Prime Minister & Cabinet, the Department of the Environment and the Department of Foreign Affairs & Trade. He has also held academic positions at the Australian National University, San Diego State University, the University of Southern Queensland and Macquarie University.

Simon Attwood is a Program Officer in the Environmental Stewardship Program. His Ph.D. examined arthropod assemblage responses to agricultural land-use change. He is responsible for the ecological benchmarking and monitoring aspects of Environmental Stewardship. He has been involved in a wide range of research and extension work in the UK and Australia relating to biodiversity conservation in production landscapes, ecosystem service values of natural, semi-natural and anthropogenic habitats, and mechanisms for delivering on-farm incentive schemes.

Emma Burns is an Assistant Director in the Environmental Stewardship Program. She has a Ph.D. in frog conservation genetics and phylogeography in Australia. She is responsible for scientific management issues to support Environmental Stewardship, including survey, tender and metric designs for optimal conservation outcomes. She has worked in various roles with a focus on conservation ecology for a number of years, including positions in research, consultancy and government.

References

Attwood, S.J., Park, S.E., Maron, M., Collard, S.J., Robinson, D., Reardon-Smith, K. and Cockfield, G. (2009). Declining birds in Australian agricultural landscapes may benefit from aspects of the European agri-environment model. *Biological Conservation* **142**, 1981–1991.

Bestlemeyer, B.T., Tugel, A.J., Peacock Jr., G.L., Robinett, D.G., Shaver, P.L., Brown, J.R., Herrick, J.E., Sanchez, H. and Havstad, K.M. (2009). State-and-Transition Models for heterogeneous landscapes: a strategy for development and application. *Rangeland Ecology and Management* **62**, 1–15.

Bishop, J., Kapila, S., Hicks, F., Mitchell, P. and Vorhies, F. (2008). *Building Biodiversity Business.* Shell International Limited and the World Conservation Union: London, UK and Gland, Switzerland.

Carroll, N., Fox, J. and Bayon, R. (2008). *Conservation and Biodiversity Banking.* Earthscan: USA.

Commonwealth of Australia (2007). Australia's Native Vegetation: A Summary of Australia's Major Vegetation Groups. Department of the Environment and Water Resources: Canberra, ACT.

Crooks, K. and Sanjayan, M. (2006). *Connectivity Conservation*. Cambridge University Press: UK.

Donald, P.F. and Evans, E.A. (2006). Habitat connectivity and matrix restoration: the wider implications of agri-environment schemes. *Journal of Applied Ecology* **43**, 209–218.

Eigenraam, M., Strappazzon, L., Lansdell, N., Ha, A., Beverly, C. and Todd, J. (2006). 'EcoTender: Auction for Multiple Environmental Outcomes'. Department of Primary Industry: Victoria.

Galatowitsch, S. (2009). Carbon offsets as ecological restorations. *Restoration Ecology* **17**, 563–570.

Gibbons, P. and Ryan, P. (2008). 'A conservation value index for Box Gum Grassy Woodland'. Unpublished report and metric for the Australian Government.

Gibbons P., Zammit C., Youngentob K., Possingham, H.P., Lindenmayer, D.B., Bekessy, S., Burgman, M., Considine, M., Felton, A., Hobbs, R.J., Hurley, K., McAlpine, C., McCarthy, M.A., Moore, J., Robinson, D., Salt, D. and Wintle, B. (2008). Some practical suggestions for improving engagement between researchers and policy-makers in natural resource management. *Ecological Management & Restoration* **9**, 182–186.

Herzog, F. (2005). Agri-environment schemes as landscape experiments. *Agriculture, Ecosystems & Environment* **108**, 175–177.

Industry Commission (1997). Role of economic instruments in managing the environment. Industry Commission: Melbourne.

Kapos, V., Balmford, A., Aveling, R., Bubb, P., Carey, P., Entwistle, A., Hopkins, J., Mulliken, T., Safford, R., Stattersfield, A., Walpole, M. and Manica, A. (2009). Outcomes, not implementation, predict conservation success. *Oryx* **43**, 336–342.

Manning, A.M. and Lindenmayer, D.B. (2009). Paddock trees, parrots and agricultural production: An urgent need for large-scale, long-term restoration in south-eastern Australia. *Ecological Management and Restoration* **10**, 126–135.

McIntyre, S. and Lavorel, S. (2007). A conceptual model of land use effects on the structure and function of herbaceous vegetation. *Agriculture, Ecosystems and Environment* **119**, 11–21.

McIntyre, S., McIvor, J.G. and Heard, K.M. (2002). *Managing and Conserving Grassy Woodlands*. CSIRO Publishing: Melbourne.

O'Connor, P. (2008). Metrics for market-based instruments: a summary of the literature. Market-Based Instruments Pilot Program, NAPSWQ. http://www.marketbasedinstruments.gov.au/Publicationsproducts/tabid/133/Default.aspx

Productivity Commission (2001). Constraints on private conservation of biodiversity. Commission research paper. AusInfo: Canberra.

Ribaudo, M., Hansen, L., Hellerstein, D. and Greene, C. (2008). 'The use of markets to increase private investment in environmental stewardship'. USDA Economic Research report Number 64.

Stoneham, G., Chaudhri, V., Ha, A. and Strappazzon, L. (2003). Auctions for conservation contracts: an empirical examination of Victoria's BushTender trial. *The Australian Journal of Agricultural and Resource Economics* **47**, 477–500.

TEEB (2009). The Economics of Ecosystems and Biodiversity for National and International Policy Makers. UNEP.

Westoby, M., Walker, B. and Noy-Meir, I. (1989). Opportunistic management for rangelands not at equilibrium. *Journal of Range Management* **42**, 266–274.
Whitten, S., Carter, M. and Stoneham, G. (2004). Market based tools for environmental management. *Proceedings of the Sixth AARES National Symposium*. RIRDC Publication No 04/142.
Whitten, S.M., Gorddard, R., Langston, A. and Reeson, A. (2009). 'A Review of the Box Gum Grassy Woodlands Environmental Stewardship Project Metric'. Unpublished report for the Australian Government.

PART D

Policy: by state

36

PUTTING WOODLANDS IN THEIR PLACE

Keith Bradby

1. It is not woodland, or mallee or heathland that most effectively motivates, inspires and informs; it is their combination into a place that is all those things, and much more.
2. Nothing is relevant to just woodlands: they exist in a social and ecological fabric for which the precise boundaries drawn on vegetation maps have no counterpart.
3. We need to be cleverer, or at least more rational, about where we focus our conservation and science efforts.
4. Woodlands in Western Australia probably were a more liveable landscape for the original Australians, and we are still learning how to respect that.
5. While we would like to achieve improved protection, the more urgent need is often to achieve improved management.
6. Remote and centralised management systems can never manage the range of nuances and interactions involved in ecological systems, particularly when social and organisational complexity is added.
7. Western Australian woodlands have been so badly recognised and supported that in the 21st century it is still possible to 'discover' vast, relatively unknown areas, which have similar global importance to places like the Great Barrier Reef.

Introduction

For me, woodlands are just one facet of the places in which I have lived and for which I have cared. I grew up amidst the forests and woodlands of central Victoria, and learnt about trees when I needed to cut the right species for posts. In the mid-1970s, I headed to Western Australia, reportedly the land of opportunity – particularly for an aspiring beekeeper. I never got as far as the forests: instead I was captured by those wonderful woodlands I first met at Balladonia, and then by the mosaic of mallee, heathlands and Banksia country joining them to the south. I settled near Ravensthorpe, on the edge of mallee and heath, but comfortably tucked into a fold of woodland – tall Flat-topped Yates (*Eucalyptus occidentalis*) meandering along one of the many small creeks.

I made my living directly from the bush for many years – first as a beekeeper, then as a collector and wholesaler of native seeds. The community in which I lived was a pioneering one:

much of the region had been opened up for agriculture only in the previous 20 years. In 1979, the Western Australian Government, with strong local backing, announced an intention to 'open up' further areas of land, until 'up to' a further 3 million hectares of woodland, mallee and heath had been turned into marginal farmland (Bradby *et al.* 1984). Working against this proposal propelled me into advocacy work, and to ongoing involvement with efforts to develop more sensible futures for the region – the landcare movement, through local enterprise development, and through community based programs of biodiversity research and protection. Later, I worked at a policy level in government, to tighten Western Australia's land-clearing controls, and in trying to help farmers adjust to their changed circumstances when permission to clear was refused.

It was during the same period that a number of us started to see both the opportunity and the need to work more cohesively across the larger landscape, and to achieve a much greater increase in environmental restoration. The Gondwana Link collaboration, where I presently work, was born in 2002; partly on the basis that the dry woodlands, the wet forests and the country in between all have some needs in common that are best met through a combined conservation effort (see map above).

Lessons

1. It is not woodland, or mallee, or heathland that most effectively motivates, inspires and informs; it is their combination into a place that is all those things, and much more

I celebrate, cherish and care for 'place' – for 'country', if I can borrow that phrase from the first people – and worry that reductionist phrases like 'woodland' mark the beginning of that slippery slope leading to 'country' being broken down into scientific disciplines and emotionally sterile 'study areas' that can never truly hold the spirit and heart of country.

2. Nothing is relevant to just woodlands: they exist in a social and ecological fabric for which the precise boundaries drawn on vegetation maps have no counterpart

It may be possible to research some aspects of a patch of woodland in isolation, but it is not possible to effectively protect, manage or restore woodlands without considering and working with other vegetation types, people, interests, organisations, processes, and histories. As obvious as that may be, individually and organisationally we are still getting used to the implications of this simple fact. It remains more difficult than it should be for organisations and individuals to work together beyond their specific specialties, interests or tribal groupings.

Woodlands, and other important environments, often are saved by using knowledge and forces beyond the limitations of ecological science. We need to work through broad-based efforts that appeal to the emotions, the bank-book and the ballot box.

Places with tall trees, such as woodlands, do tug at the heart-strings and increasingly are recognised for both their biological richness and their innate beauty. We now rarely refer to woodlands as just 'scrub', but we are still a long way from having woodlands on the public or political radar to the same extent as rainforests and coral reefs. We have to be cleverer about how we engage people in their conservation. We need to stop bickering amongst ourselves about the ecological fine points, and to start talking with the broader community about the fundamental values involved, just as also seems useful in the broader political world (Lakoff 2004). Additionally, there is a significant body of experience on the value of telling stories, of humanising our words, and placing our words in the context of place (e.g. Robertson *et al.* 2000), rather than following the dry and often turgid formulae of a scientific paper.

Allied to this is a need for the scientific community to be more respectful of other forms of knowledge, insight and communication. Most of us learn about a place by our working interactions with it. There is a considerable store of information in the heads and hearts of beekeepers, seed pickers, fence-post cutters, miners, farmers and the like. Yet there is still a scarcity of scientific or land management frameworks where this knowledge can sit alongside the more conventionally gained scientific knowledge. In Gondwana Link, I have found the Conservation Action Planning approach (Low 2003) a good tool for not only blending the different forms of knowledge, but also for drawing forth the less quantifiable insights held by many scientists. We use this as a 'place-based' approach, which appeals to how most people see the world but which also transcends the limitations of a single focus on any particular vegetation community, reserve or other aspect.

In recent years, our work in the Great Western Woodlands, an area of intense activity by the mining and exploration sectors, has shown that support for better management can come from a variety of sources, as long as you are prepared to listen for what motivates others, rather than assuming your own ecological focus can be everyone's driver. And along the way you may discover that there is a broad constituency of people out there who love woodlands, even if they don't call them that.

3. We need to be cleverer, or at least more rational, about where we focus our conservation and science efforts

It seems that most woodland research is done in eastern Australia, where there is the least woodland. The flow of public funds for restoration and management follows a similar pattern, and reflects the fact that most people live in eastern Australia. But I hope no-one is pretending that there is any sort of national scientific or ecological basis to how national priorities are currently set.

4. Woodlands in Western Australia probably were a more liveable landscape for the original Australians, and we are still learning how to respect that

I believe that human occupation and use is part of the culture and ecology of woodland systems, with the rights of aboriginal people being a priority. Near Fitzgerald River National Park, amongst country that was originally woodland, mallee and heath, the woodland areas probably were the most heavily used part of the mosaic and supported more people for more of the year than the other vegetation types. While Europeans cleared most of the woodlands early in the development of agriculture because of the better soils, much of the cultural importance remains even though the original population has been totally disenfranchised.

We have a particular opportunity in Western Australia to achieve meaningful conservation of the Great Western Woodlands through a process that also advances the employment opportunities and land rights of the traditional owners. We don't yet know how to achieve that and need to draw from experiences elsewhere in Australia, including from the savanna woodlands of northern Australia.

5. While we would like to achieve improved protection the more urgent need is often to achieve improved management

Inland Australia has many areas which nominally have the status of a conservation reserve, but where land use and management remain largely unchanged, except perhaps for the provision of signs and basic facilities (e.g. compost toilets) at a few select areas. Major long-term ecological issues, like increased wildfire and the spread of feral animals, remain both critically urgent and largely unchecked. Devising and applying successful ecological management is complex. I worry that decades of conservation campaigning has led both the key support groups and the wider public to expect environmental needs to be met through simple solutions, such as a change in tenure.

This concern is particularly relevant to the woodland patches that remain amongst, and often on, farmland. With Gondwana Link, our first serious work began in the section between Stirling Range and Fitzgerald River National Parks, and the development of a Conservation Action Plan soon identified the Flat-topped Yate (*Eucalyptus occidentalis*) woodlands as a key conservation target for that area (Deegan and Sanders 2008). It is not just that they grew on the more fertile soils and so had been largely cleared, but also that every remaining patch is experiencing ongoing degradation. Over the last 20–30 years, there have been outbursts of leaf-eating lerps, hydrological changes, a drying climate, herbicide drift and no doubt other assaults of which we are not even aware. The trees now all appear in poor condition, with many seeming to be on their 'last legs'. This includes both the areas 'protected' in nature reserves and those areas fenced off and protected by farmers. We are trying to discover what can be done to break this cycle of decline. One approach is to plant large areas of young Yates and hope they can cope better with their agricultural surroundings than the original trees have. But that can only be a partial solution, and further complex management changes are required on the ground.

Similar urgent and complex needs, which do not respect the declaration of National Park and nature reserve boundaries, exist in the Great Western Woodlands where a recent increase in large, intense fires is, in all likelihood, changing the vegetation structure and composition across large areas (Berry *et al.* 2009).

6. Remote and centralised management systems can never manage the range of nuances and interactions involved in ecological systems, particularly when social and organisational complexity is added

I fervently believe that the 'glue' which holds together large-scale landscape management is the country and the people who care deeply for that country. Therefore, I argue that the closer the structures and staff involved are to the place – the geography, the country – then the better the program will work, the more adaptable and nimble it will be, and the more ecologically focused and useful the outcomes will be.

Time and time again I have seen geographically remote and centralised organisations struggle to impose homogenous approaches on heterogeneous landscapes. It doesn't work. Time and time again I have seen seemingly insoluble organisational issues dissolve when the people on the ground come together with the authority to find and implement a solution. It does work.

7. Western Australian woodlands have been so badly recognised and supported that in the 21st century it is still possible to 'discover' vast, relatively unknown areas, which have similar global importance to places like the Great Barrier Reef

Projecting the vision for the larger Gondwana Link and seeking new funding sources generated fresh thinking on the great patch of woodland east of the Rabbit Proof Fence in south-western Australia. A donor, sourced through The Nature Conservancy, was keen to support work there and another key group, The Wilderness Society, was keen to undertake the work. So, with backing from the Nature Conservancy, The Wilderness Society started scoping values and opportunities in the area, and then were able to encourage other donors also to support that work. The Wild Country Hub at the Australian National University secured additional research funding.

One of the early products was the publication in June 2008 of the first cohesive scientific report on the 16 million hectare area now blessed with the name 'The Great Western Woodlands' (Watson *et al.* 2009). And great indeed they are, now achieving local and international recognition as the largest remaining temperate woodland and Mediterranean shrubland on Earth.

Within months of the report being launched, and following astute political campaigning by The Wilderness Society, bipartisan political support was secured for improved conservation management. After the 2008 State election this translated into a trebling of the management budget for the area. It was very heartening to see this initial response. It was in part due to political pressure, but we also feel the campaign tapped a community vein of wonder and excitement about an area almost largely ignored until now.

While the State Government is starting to fund a community dialogue on the area's future, the bulk of the funding to support public discussion on the woodlands is coming from the Wild Australia program, a joint effort of two global organisations, the Pew Environment Group and The Nature Conservancy. While we are not a poor country, even now Australians still need global support to fund the most basic of measures for one of our most significant natural areas.

Conclusion

I will finish by quoting a friend, who summed up far better than I can the disjunction between protecting reserves (woodland or otherwise) and improving ecological health over whole landscapes. In the words of Nathan McQuoid, 'Our nature reserve system is like a rare stamp collection – beautiful and to be treasured, but bloody useless for running a postal service with'. It is good to now be working on a conservation 'postal service', that has potential to extend and link with others across the whole country, and that values woodlands for themselves and for their place in the matrix of ecological systems and human activities in which they are located.

Acknowledgements

I thank the many good people and organisations who, through Gondwana Link, are helping accelerate and expand environmental protection, management and restoration across south-western Australia. First and foremost are the many private and corporate donors who have been both generous and adventurous enough to support our work. Achieving our collective vision relies on the combined efforts of many groups: Bush Heritage Australia, Capes to Capes Catchment Group, Fitzgerald Biosphere Group, Friends of Fitzgerald River National Park, Greening Australia, Green Skills, Oyster Harbour Catchment Group, The Nature Conservancy, Pew Environment Group and The Wilderness Society. For scientific leadership and insightful thinking, I particularly thank Paula Deegan from The Ecology Centre, University of Queensland.

Bio

Keith Bradby lives in south-western Australia, with his wife Margaret and youngest son Jack. He works for the environment through Gondwana Link, where he is the founding Director.

References

Bradby, K., Jasper, R., Pearce, H. and Richards, R. (1984). *Diversity or Dust a review of the impact of agricultural land clearance programs in south west Australia*. Australian Conservation Foundation: Melbourne.

Berry, S., Keith, H., Mackey, B., Brookhouse, M. and Jonson, J. (2009). 'Biomass carbon stocks in the Great Western Woodlands'. Research Report, ANU Enterprise Pty Ltd.

Deegan, P. and Sanders, A. (2008). 'Functional Landscape Plan (FLP). Fitzgerald to Stirling Operational Area. Version 2 April 2008'. Unpublished report for Gondwana Link (summary available on www.gondwanalink.org).

Lakoff, G. (2004). *Don't Think of an Elephant! Know Your Values and Frame the Debate*. Chelsea Green Publishing: Vermont.

Low, G. (2003). *Landscape Scale Conservation: A Practitioner's Guide*. The Nature Conservancy. See also The Nature Conservancy's 'Conservation by Design' website www.nature.org/aboutus/howwework/cbd/

Robertson, M., Nichols, P., Horwitz, P., Bradby, K. and MacKintosh, D. (2000). Environmental narratives and the need for multiple perspectives to restore degraded landscapes in Australia. *Ecosystem Health* **6**(2).

Watson, A., Judd, S., Watson, J. *et al.* (2008). *The extraordinary nature of the Great Western Woodlands*. The Wilderness Society of WA Inc.: Perth.

37

CONSERVATION MANAGEMENT PRINCIPLES FOR TASMANIA'S LOWLAND WOODLANDS

Louise Gilfedder

1. Treasure your woodlands.
2. Value and respect the management knowledge of the woodland landholders.
3. Conservation needs to become a viable business enterprise in agricultural landscapes.
4. Healthy woodlands require appropriate disturbance regimes and landscape-scale management diversity.
5. Manage for woodlands of the future - regeneration of the dominant trees during each generation of ownership is essential.
6. Woodland conservation will be achieved through working in partnership with landholders.

Introduction

Tasmania's temperate woodlands extend from the lowlands of eastern Tasmania and the Bass Strait Islands to the Snow Gum (*Eucalyptus pauciflora*) woodlands of the high country (Duncan and Brown 1985; Kirkpatrick *et al.* 1988; Kirkpatrick and Gilfedder 2000). Much of my working life has been in the Tasmanian Midlands where the vegetation forms a continuum from the lowland temperate grasslands through woodlands to open forests, responding variously to elevation, rainfall and soil types (see map on next page). The woodlands are largely in private ownership, with the exception of coastal woodlands. Woodland types range from grassy understorey through to heathy woodland on less fertile soils (Gilfedder *et al.* 2003), have high levels of species richness, and are a stronghold of globally and regionally rare and threatened species (Kirkpatrick and Gilfedder 1995; Gilfedder *et al.* 2003), and were habitat to the now extinct Tasmanian Emu (*Dromaius novaehollandiae diemenensis)* and Thylacine (*Thylacinus cynocephalus).* Tasmanian woodlands are severely depleted in extent, and the remaining woodlands are small, fragmented and in poor condition (Kirkpatrick and Gilfedder 1995; Gilfedder and Kirkpatrick 1998a).

My early research at the University of Tasmania focused on the conservation biology of threatened woodland and grassland plant species (e.g. Gilfedder and Kirkpatrick 1994, 1995,

1998b), extending to landscape-scale approaches including an understanding of factors influencing the woodland remnant condition (Gilfedder and Kirkpatrick 1996, 1998a; Kirkpatrick and Gilfedder 1995, 1998; Kirkpatrick *et al.* 2007), and an examination of landholder attitudes and intentions (Kirkpatrick and Gilfedder 1997), gaining insight into social and economic drivers. This led to a shift into policy and program development and implementation in the State conservation agency, working with municipal government, natural resource management bodies, local Landcare groups, private land NGOs, and most importantly, individual landholders. A Churchill Fellowship to northern America in 2006 highlighted the potential role of landholder-led conservation approaches (http://www.churchilltrust.com.au).

Lessons

1. Treasure your woodlands

Many Tasmanian landholders, the custodians of our woodlands, are largely unaware that woodlands are an urgent conservation priority. Vital knowledge of how precious their woodlands are, how much has been lost since European settlement, and the importance of the ecosystem services they provide, remains elusive. Conservationists and researchers must articulate compelling cases to woodland owners to protect our woodlands from further loss and degradation.

It is vital that scientists, researchers and policy-makers inform, educate and inspire woodland managers and owners, through provision of targeted technical materials to key landowners. *Managing and Conserving Grassy Woodlands* (McIntyre *et al.* 2002) and *Woodlands: A Disappearing Landscape* (Lindenmayer *et al.* 2005), a coffee-table style woodlands publication, have been influential in Tasmania through conveying information in an attractive accessible format. Popular works in Tasmania, such as the *Tasmanian Bushcare Toolkit* (Kirkpatrick and Gilfedder 1999), the *Nature of the Midlands* (Gilfedder *et al.* 2003), *Managing Tasmanian Native Pastures* (Mokany *et al.* 2006) and *People, Sheep and Nature Conservation* (Kirkpatrick and Bridle 2007), all aimed to educate and inspire landholders about the values and importance of their bush, converting years of research from Tasmania and south-eastern Australia into accessible conservation management advice with a local feel. It is also important that information about conservation values and significance is conveyed to landholders at the property-level scale. Farm-scale property conservation plans, natural values reports and farm poster/maps have all been successfully used in Tasmania to convey conservation management information, presented in formats that don't facilitate disposal such as the usual stapled A4 pages.

2. Value and respect the management knowledge of the woodland landholders

Most landholders do not manage with a single simple approach – they understand the complexities and subtleties of their management and respond accordingly. Researchers need to record and respect this knowledge, sometimes gained over generations.

In the Tasmanian Midlands, as in many other parts of Australia, there is currently a generation of farming elders who have valuable management and natural history information that is rarely recorded, is certainly not in the literature, and is at risk of being lost. This wisdom has been gained from both lifetimes and generations of farming experience undertaken in a variable and ever-changing environment. These knowledgeable farmers are ageing and we urgently need to record their oral history of past woodland management practices. Woodland management by successive generations of landowners on some properties has been informed by knowledge and observations including, in the case of fire management, knowledge transfer from the Tasmanian indigenous people to the European colonisers (Kirkpatrick and Bridle 2007). Interviews with landholders on past management, particularly grazing and fire practices (Kirkpatrick and Gilfedder 1997) and on past biodiversity patterns (Gilfedder *et al.* 2003), have been incorporated into Tasmanian management guides (e.g. Mokany *et al.* 2006).

3. Conservation needs to become a viable business enterprise in agricultural landscapes

New policies and incentive programs are required in agricultural regions where productivity is high and biodiversity loss is greatest, as is the case in woodland regions throughout southern Australia. If we are serious about the protection of woodlands and their biodiversity and ecosystem services, government programs need to ensure that conservation delivers a comparable income stream.

Many landholders in woodland regions see themselves as custodians of nationally important biodiversity and providers of fundamental environmental services. Increasingly, they consider that if they are to protect these natural resources and values on behalf of the broader community, then conservation needs to be a viable business enterprise. The economic returns of conservation stewardship do not currently approach the financial rewards of alternative land uses. Landowners are often frustrated at the reduced scope of government conservation planning that focuses on a single value (e.g. forests, threatened species, grasslands) rather than broader landscape-scale approaches. Government programs need to broaden their scope and the community must be willing to pay the real costs of conservation. Let's get conservation in agricultural landscapes onto the balance sheet.

4. Healthy woodlands require appropriate disturbance regimes and landscape-scale management diversity

Maintaining a variety of management practices in woodland landscapes is essential, along with the avoidance of management homogeneity. Active management must include the reinstatement of historical disturbance regimes or the maintenance of surrogates such as the judicial use of grazing to reduce biomass in the absence of active fire management.

Appropriate ecological disturbance regimes are vital to the protection and active conservation management of woodland patches, with grazing and fire the critical disturbance factors in Tasmania. In addition, the negative impacts of nutrient enrichment on woodland condition is well documented in the literature (e.g. Gilfedder and Kirkpatrick 1998a; Dorrough *et al.* 2005) and cannot be overemphasised. Intermediate disturbance regimes, where disturbance is neither rare nor too frequent, have been shown to be the most compatible with the maintenance of natural values in Tasmanian woodlands (Gilfedder and Kirkpatrick 1998a). This is most likely due to closely matching management with the historical disturbance regimes of biomass reduction by native herbivores and indigenous fire management.

An excellent conceptual framework to predict the effects of livestock grazing and grazing exclusion on the conservation values of native vegetation across natural ecosystems in Australia has been developed by Lunt *et al.* (2007), one of the most useful reviews on woodland management in the past few decades. The decision tree to assess the likely impacts of livestock grazing is recommended for ecologists, managers and landowners.

The decreasing use of fire as a management tool in woodlands is of considerable concern, and landowners need clear management advice and implementation support. The restoration of fire regimes of varying intensity, frequency and timing is advocated to favour a range of biodiversity – we need woodlands that are regularly burned, never burned, burned in all seasons, hot fires, cool burns, and so on.

The Tasmanian Bushcare Toolkit (Kirkpatrick and Gilfedder 1999) encourages management by need, not by formulae - simple 'recipes' may not provide for the range of desired disturbance regimes across a landscape. Finally, a reminder that the set of disturbance conditions that favour the persistence of threatened plant species can be very different to the regimes that favour good condition and minimise exotic species invasion (Gilfedder and Kirkpatrick 1998a).

5. Manage for woodlands of the future – regeneration of the dominant trees during each generation of ownership is essential

Each generation of woodland owners needs to allow for a regeneration event during their watch, particularly of the dominant tree species, to ensure the presence of woodlands of the future.

Managing for the replacement of woodlands is not straightforward. In recent decades, Tasmania's woodlands have been beset by two major ecological issues – the loss of dominant tree species through severe dieback, and a lack of regeneration. Appropriate disturbance regimes have not had the desired results. Regeneration planning also needs to consider more than simply the dominant tree species. Landholders need to capitalise on favourable seasons, resting woodlands from grazing in good times to allow herbaceous understorey species, including grasses, to flower and set seed. Similarly, protection from grazing after fire is vital, as is the early protection of young tree saplings and seedlings until they are tall enough to escape damage and loss to grazers. A simple trick, learned from a Midlands farmer, is to drop a 'cage' of dead branches over young saplings or special plants such as palatable orchids and peas, protecting them in the short-term from grazing until they set seed or are tall enough to escape browsing.

In addition to the loss of woodlands per se, the Tasmanian Midlands now faces the loss of important old-growth woodland trees. Fencing off small clumps of isolated old-growth trees (perhaps three to five per patch) to encourage regeneration is essential, with experimentation on suitable techniques including soil scarification, and the use of fire or crash-grazing to stimulate germination of the dominant eucalypt species. Supplementary planting of appropriately provenanced tree species is also to be encouraged, including species more suited to predicted future climates.

6. Woodlands conservation will be achieved through working in partnership with key woodland landholders

Woodland conservation on privately owned land will best be achieved through active partnerships with landowners, preferably initiated and led by the landowners themselves. The journey from initial engagement, knowledge sharing and the establishment of trust and mutual respect can be a slow process that can lead, in time, to the formalising of long-term commitments, often in the form of conservation covenants that may be perpetual. Patience is required.

Sustainable management of natural resources (including woodlands) requires a partnership approach that gives recognition and respect for farming traditions and addresses the need to have viable business enterprises. Research scientists and policy makers need to be strategic and work where they will get the best return for their efforts. Focusing on the largest and most intact woodland remnants is a good start, but by no means the only approach. Strategically targeting critically endangered natural values is also vital. The most effective conservation efforts in agricultural landscapes will be those initiated and led by the landowners themselves. The Malpai Borderlands Group provides an inspirational model from North America that aims to 'unite scientists, landholders, agencies and conservationists on a community-based effort to understand, protect and enhance a large complex landscape' (Sayre 2005). The book is a great read. The Malpai Group have identified that it may take a generation to achieve their vision. Building partnerships takes time and patience, and landholders are frustrated by the short-term nature of many incentive programs, and the lack of continuity of extension staff.

Conclusions

Much of the research and observations reported here were largely undertaken in the 1990s and there has been considerable ecological and social change in the Tasmanian Midlands since that time. We need to understand the policy implications of generational handover, challenging economic times and demographic change. Questions such as the long-term effects of stock grazing, the impacts of changes to grazing management systems, and the interactive effects of livestock and native herbivores are knowledge gaps. The impact of a predicted climate of lower rainfall, warmer temperatures and reduced frost is a key challenge. In Tasmania, the establishment of a Long-term Ecological Research (LTER) program in the dry forest ecosystems could help address these issues, complementing the Warra LTER site in the wet forests of southern Tasmania (http://www.forestrytas.com.au) by building on research on ecological and agricultural factors. Another key challenge is to communicate the results of research to landholders, and more broadly to natural resource management extension staff – a major issue given the reduction in government and industry extension officers. The employment of farmers who are trusted and respected by a local rural community as extension and program delivery officers can be an important step. In the Tasmanian Midlands, new agricultural opportunities, particularly for irrigated cropping, heralds dramatic land-use change with associated conservation challenges and opportunities. Ensuring the protection of important woodland remnants, along with other important biodiversity, through innovative partnerships and approaches is critical. The woodland landscape is disappearing and we risk the loss of a landscape that many of us define as quintessentially Australian.

Acknowledgements

Thanks to Professor Jamie Kirkpatrick at the University of Tasmania for steering me into the challenging work of undertaking conservation science and policy in agricultural landscapes.

Thanks also to Kerry Bridle, Matthew Appleby, Daniel Sprod and others in Tasmania who have provide stimulating discussions over the years on these themes. But most of all, my grateful acknowledgement goes to the landowners and farming families of Tasmania from whom I have learned so much practical knowledge over the years. And thanks also to the Winston Churchill Memorial Trust for awarding me a Fellowship in 2006 that enabled me to broaden my horizons.

Bio

Louise Gilfedder is an ecologist and conservation scientist with the Department of Primary Industries, Parks, Water and Environment in Hobart, Tasmania. Louise served several terms on the board of Bush Heritage Australia and still contributes through membership of the Conservation Science Committee. Much of her working life has been involved with achieving conservation in agricultural landscapes and integrating biodiversity protection into production systems, largely focusing on grassland and woodland ecosystems, and in particular working with sectors such as the wool industry. Her current work, developing adaptation approaches to build resilience against climate change into Tasmania's natural values, highlights the fragility of these important ecosystems and the urgency of the need for woodland conservation actions.

References

Dorrough, J., Moxham, C., Turner V. and Sutter, G. (2006). Soil phosphorus and tree cover modify the effects of livestock grazing on plant species richness in Australian grassy woodland. *Biological Conservation* **130**, 394–405.

Duncan, F. and Brown, M.J. (1985). 'Dry sclerophyll vegetation in Tasmania'. Wildlife Division Tas. N.P.W.S. Tech. Report 85/1.

Gilfedder, L. and Kirkpatrick, J.B. (1994). Climate, grazing and disturbance and the population dynamics of *Leucochrysum albicans* at Ross, Tasmania. *Australian Journal of Botany* **42**, 417–430.

Gilfedder, L. and Kirkpatrick, J.B. (1996). Improving the conservation of vegetation remnants in Tasmania. In *Nature Conservation 4: The Role of Networks*. (Eds D. Saunders, J.L. Craig and E.M. Mattiske.) pp. 613–621. Surrey Beatty & Sons: Chipping Norton, NSW.

Gilfedder, L. and Kirkpatrick, J.B. (1998a). Factors influencing the integrity of remnant bushland in subhumid Tasmania. *Biological Conservation* **84**, 89–96.

Gilfedder, L. and Kirkpatrick, J.B. (1998b). Distribution, disturbance tolerance and conservation of *Stackhousia gunnii* in Tasmania. *Australian Journal of Botany* **46**, 1–13.

Gilfedder, L., Kirkpatrick J.B., Wapstra, A. and Wapstra, H. (2003). 'The Nature of the Midlands.' Midlands Bushweb, Northern Midlands Council: Longford, Tasmania.

Kirkpatrick, J.B. and Bridle, K. (Eds). (2007). *People, Sheep and Nature Conservation: The Tasmanian Experience*. CSIRO Publishing: Melbourne.

Kirkpatrick, J.B. and Gilfedder, L. (1995). Maintaining integrity compared with maintaining rare and threatened taxa in remnants in subhumid Tasmania. *Biological Conservation* **74**, 1–8.

Kirkpatrick, J.B and Gilfedder, L. (1997). The influence of owner's intentions on remnant bush conservation in Tasmania. In *Conservation Outside Nature Reserves*. (Eds P. Hale, and D. Lamb.) pp. 217–225. Centre for Conservation Biology, University of Queensland: Brisbane.

Kirkpatrick, J.B and Gilfedder, L. (1998). Conserving woody natives: two Tasmanian endangered herbs in the Brassicaceae. *Australian Journal of Ecology* **23**, 466–473.

Kirkpatrick, J.B. and Gilfedder, L. (1999). 'Tasmanian Bushcare Toolkit.' Department of Primary Industries, Water and Environment: Hobart.

Kirkpatrick, J.B. and Gilfedder, L. (2000). Distribution, ecology and conservation status of lowland and subalpine woodlands in Tasmania. In *Temperate Eucalypt Woodlands in Australia: Biology, Conservation Management and Restoration*. (Eds R.J. Hobbs and C.J. Yates) pp. 45–46. Surrey Beatty and Sons: Chipping Norton.

Kirkpatrick, J.B., Gilfedder, L. and Fensham, R.J. (1988). 'City parks and cemeteries: Tasmania's remnant grasslands and grassy woodlands.' Tasmanian Conservation Trust: Hobart.

Kirkpatrick, J.B., Gilfedder, L., Mendel, L. and Jenkin, E. (2007). Run country on the run. In *People, Sheep and Nature Conservation: The Tasmanian Experience*. (Eds J.B. Kirkpatrick and K. Bridle) pp. 161–181. CSIRO Publishing: Melbourne.

Lindenmayer, D., Crane, M. and Damian, M. (2005). *Woodlands: A Disappearing Landscape*. CSIRO Publishing: Canberra.

Lunt, I.D., Eldridge, D.J., Morgan, J.W. and Witt, G.B. (2007). Turner Review. A framework to predict the effects of livestock grazing and grazing exclusion on conservation values in natural ecosystems in Australia. *Australian Journal of Botany* **54**, 401–415.

McIntyre, S., McIvor, J.G. and Heard, K.M. (2002). *Managing and Conserving Grassy Woodlands*. CSIRO Publishing: Melbourne.

Mokany, K., Friend, D., Kirkpatrick, J. and Gilfedder, L. (2006). *Managing Tasmanian Native Pastures: A Technical Guide for Graziers*. Tasmanian Institute of Agricultural Research: Hobart.

Sayre, N. (2005). *Working Wilderness: The Malpai Borderlands Group and the Future of the Western Range*. Rio Nuevo Publishers: Tucson, Arizona.

38

EXPERIENCES IN PLANNING AND IMPLEMENTING WOODLAND CONSERVATION PROJECTS IN NORTHERN VICTORIA

Geoff Park

1. Effective woodland conservation must be supported by high quality ecological data, collected and applied at an appropriate scale.
2. Achieving tangible improvements in the extent and condition of woodland ecosystems requires substantial and sustained resource allocation.
3. Increasing demand for funding and a requirement to demonstrate the effectiveness and value-for-money of past investment has led to greater sophistication in program design.
4. Ecological assessment, analysis and planning at a relevant ecological scale, the bioregion, has supported improved targeting of investments.
5. Sustained and respectful involvement of local people including farmers, 'lifestyle' land managers, and Landcare and community environmental groups, is fundamental to woodland conservation.
6. It is important to recognise, value and maintain landscape memory from a diversity of people with an interest in woodland conservation.
7. Moving towards spatially explicit approaches to engagement, planning and implementing is required to underpin future woodland conservation efforts.
8. Understanding and harnessing scientific and social networks is crucial to effective sharing of knowledge and information.
9. The key ingredients of successful programs are:
   - a grassroots approach to planning and governance;
   - stating a clear vision for the creation of healthy, resilient and productive landscapes;
   - a focus on community engagement;
   - building habitat connectivity; and
   - the use of planning, good science and monitoring.

Introduction

Since 1998 I have been involved in a range of initiatives aimed at achieving improved conservation outcomes for woodlands in south-eastern Australia, with a focus on northern Victoria (see map above). My involvement has spanned the dimensions of policy development (at regional and state-wide scales), native vegetation and biodiversity planning (property, landscape and bioregional scales), project and program development, implementation, monitoring and evaluation. My experiences have principally been at the interface between institutions such as Catchment Management Authorities and the community.

The woodland ecosystems of northern Victoria have undergone dynamic change since European settlement. Research currently being undertaken as part of the Landscape Logic Commonwealth Environmental Research Facilities (CERF) program (www.landscapelogic.org.au) has identified key drivers of change in extent and condition for woodland and forest ecosystems. The interplay of factors such as rabbits, myxomatosis, tax breaks for clearing, government subsidies for superphosphate, and drought have played out in complex ways. Detailed comparative analysis of historical aerial photography from the late 1940s to the present day reveals fascinating patterns of vegetation change across spatial and temporal dimensions. Influences on the extent of native vegetation can be immediate (e.g. conversion to cropping) or incremental (e.g. native species favoured by decline in use of fertiliser), and direct (e.g. revegetation or seed-sowing) or indirect (e.g. control of rabbits, demographic change resulting in a shift from agriculture to lifestyle 'land management') (Race *et al.* 2009). The story is not a simple one of loss and gain, but of a highly dynamic landscape where initial clearing of woodland may be followed by a series of regeneration and clearing events at intervals between 10 and 50 years. What is clear, however, is the relatively minor impact of government investment in woodland protection in recent years being overwhelmed by passive and active land management actions.

The observations that follow draw on these experiences, combined with previous work on community involvement in natural resource management and concurrent involvement with philanthropic support for community driven landscape restoration.

Lessons

1. Effective woodland conservation must be supported by high quality ecological data, collected and applied at appropriate scale

Regional vegetation planning during the late-1990s was a critical step in developing a more systematic approach to on-ground actions. Detailed vegetation mapping highlighted the extent of depletion and fragmentation of temperate woodlands. The North West Ecological Vegetation Class (EVC) Mapping project (White *et al.* 2003), complemented previous EVC mapping for the Box Ironbark Forests and Woodlands Investigation (ECC 2001) and Regional Forest Agreement process (VicRFASC 1997) to provide complete coverage at a 1:100 000 scale across northern Victoria for native vegetation. The concept of Ecological Vegetation Classes (EVCs) has been a powerful force, not only to support planning and investment, but particularly as a tool for better interpretation and reading of landscape heterogeneity. High quality spatio-temporal information on the influence of habitat elements (for example, remnant patch size, connectivity and landscape position) on woodland birds (Radford *et al.* 2005) has been assembled. This has been invaluable in improving our understanding of the conservation values of woodlands and led to more sophisticated approaches to planning and implementing landscape and site-based actions. It has proven more difficult to compile supporting information on the current state and trajectory of critical attributes such as habitat condition and ecological processes in woodland ecosystems. Without this information, it is difficult to be confident that the nature and location of conservation actions are anywhere near optimal.

2. Achieving tangible improvements in the extent and condition of woodland ecosystems requires substantial and sustained resource allocation

The advent of the Natural Heritage Trust (NHT) and National Action Plan for Salinity and Water Quality (NAPSWQ) saw a significant injection of funding for planning and on-ground works associated with woodland protection and enhancement. The NHT Bushcare program commenced in 1998 with major funding, approximately $2–3 million annually, until 2006. In particular, it supported projects at a range of scales (sub-catchment to regional), with a range of collaborative partnerships (CMAs, NGOs, state agencies, local government and community groups) and enabled the exploration of a diversity of approaches (e.g. direct grants, extension, rate rebates and covenanting programs). The NAPSWQ program began in 2002. While it had a deliberate focus on treatment of salinity and water quality threats, establishment and protection of native vegetation remained a crucial tool. In the North Central CMA region alone, an estimated $40 million has been directed to native vegetation management between 1998 and 2009. As these programs unfolded, a number of key limitations emerged that were addressed to some extent in subsequent initiatives. These were:

- Restrictions on incentive rates for activities such as fencing and revegetation combined with the preference of regional bodies to use flat rate systems for incentive allocation. This led to the evolution of 'sliding scale systems' where higher levels of funding were provided to projects in priority areas or where environmental benefits were seen to be greater.

- A focus on annual delivery programs that rewarded those projects and organisations able to maximise expenditure of available funds. This has led to some perverse outcomes (e.g. funding some projects that were delivering largely private benefits with little public environmental benefit) and some dramatic revegetation failures as a result of unfavourable climatic conditions and ongoing drought.
- A comprehensive failure to monitor the success (or otherwise) of investment, with little systematic collection of baseline data and subsequent site revisitation. This has meant that a wonderful opportunity was lost to learn from the diversity of activities and approaches used to protect and enhance woodland ecosystems. Victoria is fortunate to have a combination of quality spatial data, robust and accepted habitat assessment tools (e.g. Habitat Hectares) (Parkes *et al.* 2003) and powerful spatially enabled project management tools such as the Catchment Activity Management System (CAMS) and eFARMER (Roberts *et al.* 2009). This suite of methods and tools has significant potential to support sophisticated monitoring and evaluation of the impact of on-ground interventions. Unfortunately, the potential benefits have largely gone unrealised due to a combination of factors including rapid staff turnover, inadequate institutional support and training, and human failure to adhere to agreed protocols and processes for managing data and information.

3. Increasing demand for funding and a requirement to demonstrate the effectiveness and value-for-money of past investment has led to greater sophistication in program design

Most woodland conservation programs in northern Victoria have evolved over the past 10 years from flat rate grants to the application of sliding scale incentives based on conservation priorities and the crude application of environmental benefits indices. This was, in part, due to the oversubscription of project funds as landholder interest in woodland conservation has increased, but also recognition that investment should adhere more closely to evidence-based landscape restoration principles. Victoria led the way in the development, trialling and evaluation of market-based approaches (e.g. Bush Tender) aimed at securing cost-effective biodiversity outcomes that met agreed state-wide and regional ecological priorities. The initial trial of Bush Tender in Victoria occurred in the North Central and North East CMA regions with a focus on box-ironbark forests and woodland ecosystems. It is important, however, to realise that government investment is not the only source. The Connecting Country project (www.connectingcountry.org.au) in Central Victoria started in 2007 with the development of a 'Biodiversity Blueprint' that included an audit of existing vegetation management activities across the project area. This information, collected from more than 100 site visits and meetings with private land managers, demonstrated that privately funded management actions exceeded publicly funded actions by nearly 2:1, and showed the need to understand and harness private investment in woodland conservation actions.

4. Ecological assessment, analysis and planning at a relevant ecological scale, the bioregion, has supported improved targeting of investment

The Victorian Biodiversity Strategy (NRE 1997) advocated a bioregional approach to native vegetation and ecosystem management. Bioregions were promoted for regional-scale planning 'because they are based on the patterns of ecological characteristics and the underlying environmental features. They therefore reflect natural boundaries and relationships between biodiversity assets and natural resource based activities' (NRE 1997). The adoption of this important policy position led to the development of bioregional and landscape-scale biodiversity action planning across Victoria. This planning approach included:

- The strategic and planning framework provided by the State biodiversity strategy, native vegetation management framework, and the Flora and Fauna Guarantee Act.
- Tools, including: 1) Bioregional Network Analysis (BNA) for identifying priority species for action and threatening processes affecting biodiversity; 2) the habitat-hectare approach for assessing native vegetation quality; and 3) the focal species concept whereby management targeted at particular species can be used as a cost-effective approach for broader biodiversity conservation.
- The scientific basis underpinning biodiversity conservation management, particularly the direct and indirect effects of habitat loss and fragmentation on local extinction, and some of the critical thresholds.
- Guidelines for compiling landscape-scale biodiversity action plans, which direct on-ground actions by the community and other stakeholders.

Bioregional planning had important implications for woodland ecosystems in northern Victoria. It highlighted the dramatic loss of woodland communities across the Victorian Riverina, Goldfields and Murray Mallee bioregions, identified key threatening processes in a systematic way, and made a cogent case for increased investment in the protection and enhancement of woodland remnants, especially on private land (NRE 1997). Part of the underlying conceptual framework for bioregional planning was the need to consider the importance of scale. Implementation of the approach required consideration of regional, catchment, landscape and local scales in identifying priorities, selecting policy and planning tools, targeting of on-ground conservation actions and monitoring and evaluating outcomes (Wilson and Lowe 2003).

A key but often overlooked component of bioregional planning is Biodiversity Action Planning (BAP). BAP was conceived by policy makers, planners and practitioners in the late-1990s as a tool for action planning at multiple scales. It is a structured approach to identifying priorities and mapping significant areas for native biodiversity conservation at bioregional, landscape and local scales. The stated aims of Biodiversity Action Planning (Platt and Lowe 2002) were to:

- conserve native biodiversity *in situ* by maintaining viable examples of the range of ecosystems occurring naturally in Victoria;
- encourage a more strategic approach and a shift in public expenditure toward the protection, restoration and ongoing management of priority sites for biodiversity; and
- achieve community support for landscape planning and the conservation of strategic assets, particularly in rural landscapes.

The approach has contributed to improved strategic planning and priority setting by state agencies and CMAs, but the key plank of community involvement remains disappointingly unrealised.

5. Sustained and respectful involvement of local people, including farmers, 'lifestyle' land managers, Landcare and community environmental groups, is fundamental to woodland conservation

Many woodland remnants are located on private land and managing them for biodiversity generally relies on collaboration with private landholders. Without their cooperation, the best science-based planning will count for little. Biodiversity Action Planning set out to integrate information and decision making at multiple scales, including sub-catchment, neighbourhood and property levels. At these finer scales, innovative work was done in developing community-owned, spatially explicit plans (DSE 2003). This was the result of the dedicated work of local extension officers who employed participatory processes to gain community support and involvement in developing plans and subsequently implementing management actions. This type of support for community processes (such as inclusion of local knowledge and learning from local experimentation), has proven difficult to maintain in times of diminishing extension budgets, rapid staff turnover, and an often misguided desire for 'new' approaches.

A key ingredient provided by local extension officers is 'trust', an overlooked and underrated factor in conservation programs. It is built through sustained collaboration between landholders, groups and agencies, with an understanding of their respective knowledge and aspirations. Taking time and expending resources to establish trustful relationships is a key element of recent initiatives, such as Connecting Country. A feature of Connecting Country is its governance and management structure where project planning is the responsibility of an elected Management Committee supported by a Reference Group open to all consortium members including agencies, authorities and NGOs. This approach is based on learning from people in the area who have had long-standing involvement in previous government-driven programs that failed to realise their promised potential.

6. It is important to recognise, value and maintain landscape memory from a diversity of people with an interest in woodland conservation

The design and delivery of effective conservation programs is challenging. In my experience, too little emphasis has been placed on carefully examining site characteristics and interpreting landscape history in project planning. Significant benefits can be gained from reconstructing landscape and site histories as a precursor to intervention, through incentives and extension. Recognising local knowledge and tapping into the wisdom and experience of local networks is a key factor underpinning successful woodland conservation. A distinguishing feature of such projects has been the use of innovative techniques to assist vegetation establishment (e.g. direct seeding) or enhance natural regeneration processes (e.g. soil disturbance by ripping and scalping). These techniques have generally been employed by experienced extension officers in collaboration with farmers. In tandem, these people have a long-standing knowledge of local landscape history and joint interest in the integration of environmental and production matters. The relationships that exist are characterised by a high degree of trust and a willingness to 'give new ideas a go'. It is perhaps counterintuitive that even over the past 10 years of below-average rainfall in northern Victoria, farmers have been even more willing to embark on woodland conservation work. As one highly regarded extension officer said to me, 'farmers like to grow things and if it's too dry to grow a crop then why not grow native vegetation' (Lindsay Ezard, personal communication).

A key challenge for government agencies and non-government organisations is to support better documentation and monitoring of management actions, so that both scientific and tacit knowledge is available to improve future activities. There are some exciting uses of technology that support the recording and valuing of community landscape knowledge. For example, the 'Community Web Mapping Portal' (http://cwmp.spatialvision.com.au) is designed to enable users to view, create and share information about their local landscape including flora and fauna observations, weed and pest infestations, significant sites and the location of on-ground works. The application is built using Google Maps technology with expandable capability allowing integration of new tools and datasets as the project develops.

7. Future success in woodland conservation will require a move towards more spatially explicit approaches

A move toward spatially explicit approaches is needed to specify which conservation outcomes are sought by the governments and the wider community, and to underpin more critical examination of the cost and feasibility of projects to reach desired goals. There are, however, significant implications associated with moving to spatially explicit approaches that in practice are often ignored.

It has been interesting (and somewhat alarming) to witness the recent proliferation of ambitious ecoregional and continental 'biolink' projects. The visions they articulate are inherently appealing but generally lack detailed spatial planning or acknowledgment of the need for local community input and empowerment. There is a danger that a 'top–down approach' will fail to properly examine feasibility issues associated with landscape change on the scale proposed, and in a way that truly involves local communities.

The cost, and technical and socio-economic feasibility of large-scale conservation projects are seldom examined seriously. Woodland ecosystems are often associated with more fertile and agriculturally productive parts of the landscape. This means that priority remnants are frequently on private land, where restoration and changed management will require farmers to move from primary production to conservation on a significant scale. The ability and willingness of landholders to adopt these new practices is largely ignored (Pannell *et al.* 2006) when landscape-scale projects such as biolinks are promoted. Along with changes on private land, complementary work on public land such as pest control and weed management is essential, at a time when funding for public land management is seriously lacking.

We need to adopt more spatially explicit approaches to woodland conservation if we are to achieve cost-effective biodiversity outcomes. A key step in the evolution of this approach is being clear about which conservation outcomes are sought. This may be expressed in terms of which assets or conservation targets should be the focus of attention: for example, particular threatened species, ecological communities or landscape elements identified through systematic conservation planning. A range of tools such as Conservation Action Planning (CAP) (Groves 2003) and the Investment Framework for Environmental Resources (INFFER, www.inffer.org) have been developed that assist with prioritisation and resource allocation.

8. Understanding and harnessing scientific and social networks is crucial to effective sharing of knowledge and information

Knowledge brokering is a means of providing access to the underpinning science of woodland management and restoration and translating this to landholders, communities, extension staff and NRM planners. Fundamentally, this is a 'people-centred' process that requires understanding the power of social networks, acknowledging the value of tacit and formal knowledge, fostering collaborative and participatory research partnerships, and the ability to integrate information from a range of disciplines such as ecology, social science and economics.

Over the past five years, I have worked as a Biodiversity Knowledge Broker with a focus on priority ecological communities across northern Victoria. Knowledge brokering facilitates the process of turning information into knowledge, by enabling research (or information) to be accessed by those who need it, in a form that is relevant to their situation and needs. The ultimate aim is for end-users to adopt behaviour for sustainable natural resource management (Land & Water Australia 2005). Australian NRM organisations such as the CRC for Freshwater Ecology, the Australian Biosecurity CRC, Greening Australia and, more recently, CSIRO (Sustainable Ecosystems) use knowledge brokers to ensure their strategic commitments to knowledge exchange are met.

9. There are some key elements in planning and implementing successful woodland conservation programs

From my experience the following elements are associated with successful initiatives.

1. A grassroots approach – projects led by the community and based on existing Landcare, Friends and other concerned groups. This means that project governance is community rather

than agency-focused. Participants in these projects help shape what they want the landscape to look like in the future and how to achieve this. There is no single view of what is best, but robust discussion of options. Participants can contribute at any scale: from the backyard to bush block, paddock to farm, neighbourhood to catchment. These projects are not static – there is room to develop, change course and evolve.

2. A vision for the creation of healthy, resilient and productive landscapes. These projects aim to build an understanding of the importance of biodiversity in supporting agriculture by improving land health. They envision productive landscapes which are able to maintain ecological processes in a time of changing climate, land use and settlement patterns.

3. Community engagement. This is done by encouraging the active involvement of local people: farmers, small landholders, town and country folk, public land managers, and schools in gathering, sharing and recording information about the local landscape. Participation in a range of activities (e.g. mapping, on-ground works, information sharing, monitoring) enables a diversity of individuals and groups to contribute.

4. Building habitat connectivity and promoting on-ground works such as fencing, planting, natural regeneration, weed and pest control, to improve the extent and quality of habitat.

5. Use of planning, good science and monitoring through the application of high quality ecological information and community knowledge. Such projects seek to establish baseline data and undertake long-term ecological monitoring to enable the impact of management to be measured and to inform future planning. These projects use knowledge of ecological principles for improving resilience and connectivity (Radford *et al.* 2007) through developing collaborations with leading landscape ecologists and restoration professionals.

Conclusions

Restoring the health of woodland ecosystems is a monumental task that will require action across entire landscapes. The goals and targets that have been set as part of the development of Regional Catchment Strategies show the magnitude of this challenge (e.g. NCCMA 2003). A key challenge in delivering integrated catchment management is to deal with the issue of scale. Most individuals, groups and communities operate within a frame of reference that is meaningful to their needs and aspirations. There is an emerging consensus on the scale, urgency and nature of the challenges, and the interconnectedness of biodiversity, vegetation, ecosystem function and landscape health (MDBC 2001). There is less clarity about what the options are for achieving the desired outcomes, either in terms of how to fund landscape change, how to use various tools for prioritisation and targeting, the appropriate choice of policy instruments or how to mobilise community support and resources. Hopefully some of the observations outlined in this contribution shed some light on the critical ingredients required to support woodland conservation outcomes into the future.

Acknowledgements

I would like to thank the landholders, researchers, extension staff and community groups that have so willingly shared with me their knowledge and passion for woodland conservation over the past 20 years.

Bio

Geoff Park works as a Knowledge Broker with the North Central Catchment Management Authority. In this role, he is responsible for the development of collaborative partnerships between researchers, policy makers, extension staff and landholders that lead to improved knowledge exchange and on-ground biodiversity outcomes. His background is in landscape ecology, teaching and community education. He has a long-standing interest and involvement with communities working to improve biodiversity conservation in agricultural landscapes. Geoff Park is also currently working on a range of projects that improve community and landholder access to NRM knowledge and spatial tools that aim to improve on-ground environmental outcomes. As a trustee of the Norman Wettenhall Foundation, an environmental philanthropic organisation, he is supporting local scale community participation in biodiversity restoration projects across south-east Australia.

References

Department of Sustainability & Environment (DSE), North Central Catchment Management Authority and Sandy Creek Landcare Group (2003). Sandy Creek Local Area Biodiversity Plan. DSE: Melbourne. Online at http://www.dse.vic.gov.au/conservation and environment/biodiversity action planning/landscape planning/muckleford.

ECC. (2001). 'Box-Ironbark – Forests and Woodlands Investigation – Final Report'. Environment Conservation Council: Melbourne.

Groves, C.R. (2003). *Drafting a Conservation Blueprint: A Practitioner's Guide to Planning for Biodiversity.* Island Press: Washington, D.C.

Land & Water Australia (2005). Knowledge and adoption strategy – managing information and knowledge for adoption outcomes. Land & Water Australia: Canberra.

MDBMC (2001). Integrated Catchment Management in the Murray-Darling Basin 2001–2010. Delivering a Sustainable Future. Murray Darling Basin Ministerial Council: Canberra.

NCCMA (2003). North Central Regional Catchment Strategy 2003–2007. North Central Catchment Management Authority: Huntly, Victoria.

NCCMA (2006). North Central Native Vegetation Plan. North Central Catchment Management Authority: Huntly, Victoria.

NRE (1997). Victoria's Biodiversity – Our living Wealth – Sustaining Our Living Wealth and – Directions in Management. Department of Natural Resources and Environment: East Melbourne, Victoria.

Pannell, D.J., Marshall, G.R., Barr, N., Curtis, A., Vanclay, F. and Wilkinson, R. (2006). Understanding and promoting adoption of conservation practices by rural landholders. *Australian Journal of Experimental Agriculture* **46**, 1407–1424.

Parkes, D., Newell, G. and Cheal, D. (2003). Assessing the quality of native vegetation: the 'habitat hectares' approach. *Ecological Management & Restoration* **4**, Supplement: S29–S37.

Platt, S.J. and Lowe, K.W. (2002). Biodiversity Action Planning: action planning for native biodiversity at multiple scales – catchment, bioregional, landscape, local. Department of Natural Resources and Environment, Melbourne.

Race, D., Duncan, D., Kyle, G., Merritt, W. and Park, G. (2009). Landscape history and vegetation change in the Muckleford region of Victoria. Online at www.landscapelogic.org.au.

Radford, J.Q., Bennett, A.F. and Cheers, G.J. (2005). Landscape-level thresholds of habitat cover for woodland-dependent birds. *Biological Conservation* **124**, 317–337.
Radford, J.Q., Williams, J.E. and Park, G. (2007). Effective Landscape Restoration for Native Biodiversity in Northern Victoria. North Central Catchment Management Authority: Huntly, Victoria.
Roberts, A.M., Park, G., Melland, A.R. and Miller, I. (2009). Trialling a web-based spatial information management tool with land managers in Victoria. *Australia Journal of Environmental Management* **91**, 523–531.
VicRFASC (1997). Biodiversity Technical Report – Central Highlands Comprehensive Regional Assessment. Commonwealth and Victorian Regional Forest Agreement Steering Committee: Canberra.
White, M., Oates, A., Barlow, T., Pelikan, M., Brown, J., Rosengren, N., Cheal, D., Sinclair, S. and Sutter, G. (2003). 'The Vegetation of North-West Victoria – A report to the Wimmera, North Central and Mallee Catchment Management Authorities'. Victorian Department of Sustainability and Environment: Melbourne.
Wilson, J.A. and Lowe, K.W. (2003). Planning for the restoration of native biodiversity within the Goulburn Broken Catchment, Victoria, using spatial modelling. *Ecological Management & Restoration* **4**, 212–219.

39

WORKING WITH LANDHOLDERS TO PROTECT WOODLAND BIRDS: AN 18-YEAR LESSON FROM NORTHERN VICTORIA

Doug Robinson

1. Landholders are pivotal.
2. Conservation actions must match social expectations.
3. Scientific knowledge and dissemination of that knowledge is crucial.
4. Conservation objectives need to be defined specifically but implemented broadly.
5. Conservation actions need to be developed broadly but implemented locally.
6. Focal species are a valuable part of extension folklore.
7. There is no single solution to woodland conservation.

Introduction

In the late 1980s and early 1990s, the perilous state of Australia's temperate woodlands began to capture the attention of biologists and conservationists as the next major environmental issue to be addressed, following a previous focus on forests and rainforests during the 1970s and 1980s. The increased awareness of the plight of Australia's temperate woodlands was also driven by concern from within the agricultural sector, as dryland salinity emerged as a major threat to farm sustainability through much of southern Australia.

In 1991, with Birds Australia, I applied for funding to investigate the ecology and causes of decline of one of Victoria's well-known woodland bird species – the Grey-crowned Babbler *Pomatostomus temporalis* – as an initial means of tackling woodland conservation.

As a newish biologist, recently returned to Victoria in 1990 after a year spent working in Arnhem Land and other parts of the Northern Territory, the contrast between the fecundity and diversity of Northern Territory's natural environment and the comparative paucity of Victoria's natural environments was stark. The contrast also heightened an awareness of how much natural diversity had disappeared in Victoria and the urgent need to focus conservation attention on the protection and restoration of Victoria's endangered woodland and grassland ecosystems.

The following lessons are based principally on the conservation process we have followed for the Grey-crowned Babbler in northern Victoria's box-ironbark forests and plains grassy woodlands over the last 18 years (see map above), from initial research and problem identification, through to implementation and extension activity, and focusing now on ongoing monitoring and adaptive management.

Lessons

1. Landholders are pivotal

More than 95% of the woodland estate in Victoria (about 7 million ha, Lunt and Bennett 1999) is privately owned by thousands of landholders (Barr 2005; Mansergh *et al.* 2006). Any conservation actions proposed for woodlands in Victoria must align with these landholders' wishes. Most landholders care about saving threatened species but will not save those species at a substantial cost to farm productivity.

Most of these landholders are knowledgeable about their land and the wider environment. They currently farm the land as a major means of income and food/fibre production, and they have clear views about the future use of their land when they decide to sell it or transfer the title. Conservation planning therefore needs to parallel farm planning, as far as possible, providing resources and labour to support environmental works. The Grey-crowned Babbler project has largely focused on widening roadside habitat into adjacent properties – landholders receive a new boundary fence; babblers are slowly being provided twice as much habitat as they could use previously.

2. Conservation actions must match social expectations

Conservation actions taken by individual landholders, Landcare groups and other community organisations represent donations of land, labour and goodwill towards the public good; in this case, the conservation of a bird. Conservation actions being delivered by the community on their own land must therefore match the community's expectations for the land and district.

When the Grey-crowned Babbler conservation project started in the early 1990s, it was at the beginning of the National Landcare Program and Landcare groups were a vibrant part of local rural communities. Because of community interest in natural resource and biodiversity issues at the time, it was possible to encourage local groups to adopt babbler conservation as a part of their landcare activities. When the project began, however, sheep farming was the dominant land use in our project area and was still profitable. Farmers were consequently unwilling to donate too much land towards the babbler project even when they were supportive of its aims. Initially, we fenced roadside buffers into paddocks that were only 12 m wide – that being the maximum amount of land landholders were prepared to donate and the minimum amount that we thought would achieve some conservation benefit. Fifteen years later, our standard width is 25 m for buffers and 40 m for corridors.

Land ownership and community expectations of appropriate land use also have changed in the 15 years since the project began: many landholders have received government grants to undertake natural resource management works; some landholders have obtained annual payments through a range of conservation tender programs to manage larger portions of their property for conservation (Mansergh *et al.* 2006; Robinson 2006; Miles 2008); some landholders have permanently protected parts of their property under a conservation covenant with Trust for Nature (Miles 2008); and organisations including Trust for Nature and the State Government have purchased land for conservation (Mansergh *et al.* 2006).

3. Scientific knowledge and dissemination of that knowledge is crucial

Landholders typically seek evidence-based land-management advice and apply that advice on their land. One of the strengths of the babbler project has been the scientific underpinning of the management recommendations made to landholders and other land managers, followed by an ongoing monitoring and review program (Robinson 1994; Robinson 2006; Smallbone *et al.* unpublished data). These results are circulated annually through a newsletter to all landholders, land managers and supporters involved in the project.

Providing evidence to landholders that their conservation efforts have made a real difference to the survival of the Grey-crowned Babbler has been a powerful motivator of the local community (Robinson 2006). One community's response, for example, has been to publish postcards which feature the babbler's conservation story for their local town. Basing land-management investment decisions on scientific criteria has also helped to make decisions about where to spend conservation funding much more objective. As part of the Longwood Plains Biodiversity Project, a community-based conservation project on the riverine plains of north-eastern Victoria, the steering committee has set minimum standards for the types of works it will subsidise with public funding (Robinson *et al.* 2003; DSE 2007). Where proposals for environmental works do not fit the project's guidelines, the explanation given to landholders is based on the ecological requirements of the species or ecosystem we are targeting, so that landholders understand why their project is not being funded. As part of the broader extension program, we follow up by providing the landholder with a wildlife booklet (Robinson *et al.* 2003) to help them appreciate the ecological needs of the target species.

4. Conservation objectives need to be defined specifically but implemented broadly

The ecological requirements of different groups of biota do not generally correspond closely and there is a poor correlation between the distribution of different groups in relation to ecological features (Robinson 1998; Mac Nally *et al.* 2002). Even within groups of biota such as woodland birds, there are many factors influencing the distribution of different species (Radford and Bennett 2007).

It is consequently critical that conservation objectives developed for any given landscape or property recognise the different needs of the various groups of biota and individual taxa. It is also important that those objectives be defined as explicitly as possible to maximise the potential conservation gains we can achieve at a site. Once developed, however, those conservation objectives should be implemented as broadly as possible to achieve conservation gains at the landscape or population scale, as these are the scales at which we will make a real difference in terms of nature conservation.

5. Conservation actions need to be developed broadly but implemented locally

The conservation of species in agricultural landscapes is dependent on the coordination of individual conservation actions at multiple sites at a scale sufficient to achieve the desired population change (e.g. Peach *et al.* 2001. Conservation actions therefore need to be designed at the population or landscape-scale, but implemented at the site scale.

One of the early mistakes of the babbler conservation project was a failure to develop a comprehensive conservation plan based on population models which established the level of change needed to make populations viable. Instead, actions were developed at the site scale within the key populations, focusing on three main actions identified to help babbler conservation. These were: improving habitat quality at the site scale to increase group size and breeding success; improving habitat connectivity at the population scale to improve opportunities for dispersal and reduce the risk of extinction to the population; and increasing the total amount of available habitat to facilitate population increases (Robinson *et al.* unpublished data).

Although we have since done population modelling for the Grey-crowned Babbler, in hindsight it would have been more efficient to have collected enough ecological and demographic data to estimate conservation needs for the population and then undertake site-based conservation planning based on those population models.

6. Focal species are a valuable part of extension folklore

'Biodiversity' is an elusive concept – it is difficult to describe well, difficult to describe simply and difficult to describe emotively. By contrast, descriptions about the particular life-history details and conservation needs of individual species, especially fauna, do motivate landholders, land managers and other citizens to support conservation work in the woodlands. Regardless of the debate about the effectiveness of focal species as a conservation-planning tool (Lindenmayer and Fischer 2003; Roberge and Angelstam 2004), focal species (*sensu* Lambeck 1997) are a powerful means of engaging with landholders and encouraging conservation works at the site scale.

To maximise that engagement, it is important that landholders are provided with feedback: are there more of the focal species now than there were previously? Has the focal species colonised their site? What other interesting species of animals or plants are now using the site?

To maximise the conservation gain achieved, it is also important to have multiple focal-species stories to tell which encompass as many elements of biodiversity as possible. Landholder motivation relies strongly on this ability to engage with landholders and provide the necessary support so that they want to initiate and continue with conservation activities on their land.

7. There is no single solution to woodland conservation

Woodlands occur across different tenures but primarily on private land which is also being farmed. Conservation actions thus need to align with the landholders' wishes. But the actions also need to align with the specific management issues facing the ecosystem or species that is being targeted. The actions furthermore need to align with the scale of response required to make a difference. The actions need to be the most effective responses possible to make a difference. And the actions need to be integrated, wherever possible, to maximise the conservation gain being achieved at a particular site. Accordingly, detailed, holistic conservation planning from the landscape scale to the site scale is crucial if we want to change the trajectory of temperate woodlands and their biota.

Conclusions

There has been substantial progress in our knowledge of temperate woodlands and their biota in the last 20 years. The level of conservation-focused research and conservation delivery also has increased rapidly. Nevertheless, some important gaps in our knowledge still remain. One of these gaps is identifying which conservation action or combination of actions will make the greatest difference for conservation of particular species or particular landscapes. Conservation planning of this type has been done in terms of habitat reconstruction (Thomson *et al.* 2009) but we still need additional planning-decision tools and research to help us decide what allocation of resources should be assigned to different management issues when we are considering conservation of a particular ecosystem, landscape or taxon. For example, we have shown recently (Wilson *et al.* 2009) that habitat-enhancement works have made a positive difference to babbler survival. But we also know that road-traffic volume, landscape characteristics, population density and adjoining land use affect babbler populations in various ways (Robinson *et al.* unpublished data). It would be invaluable to be able to decide which of these management issues was the most significant for each babbler population in terms of impacts on survival and the scale of change we could achieve. Future conservation planning for the species could then assign the amount of investment we should allocate to different conservation actions.

We also need to trial solutions for woodland fauna recovery, particularly in terms of the provision of particular resources. Such studies are well established in Europe as part of agri-environment schemes (Grice *et al.* 2004) and focus on increasing the availability of breeding-season food resources, winter food resources and nesting habitat. The current collapse of our woodland avifauna (Mac Nally *et al.* 2009) has also been linked to declines in food resources as a consequence of climate change. Research is needed to trial options for restoring critical resources for woodland birds in order to improve survival and breeding success.

The next major challenge for woodlands management accordingly revolves around how to manage woodland patches to maximise their biological productivity, hence the survival and fecundity of their associated wildlife. Five management themes have emerged as critical issues to be addressed in order to improve woodland productivity. These are: development of long-term data on demographic responses by target species of woodland wildlife in order to evaluate responses to climate change and management interventions; management of dense regrowth stands of trees to improve overall plant diversity and habitat structure; maintenance of habitat patchiness and diversity at the landscape scale; management of native herbivore numbers to restore diversity and structural complexity of understorey; and control of the Noisy Miner. Ironically, therefore, some of our major challenges involve the control of native vegetation and native fauna.

Acknowledgements

This chapter has benefited from many discussions over the years with landholders, colleagues in the natural resource management field and volunteers from the Friends of the Grey-crowned

Babbler group. I am especially grateful to Ian Davidson, Peter Adam, Andrew Bennett and Sally Mann for many discussions on natural resource management over the years. Special thanks go to the Davidson and Adam families who helped embed us in the local community and encouraged us to stay.

Bio

Doug Robinson is a conservation biologist currently working for Trust for Nature as a Conservation Science Coordinator. For the past 15 years he has primarily worked in an extension capacity, encouraging landholders to implement conservation actions on private land throughout the riverine woodlands, grassy woodlands and box-ironbark forests of northern Victoria. Doug Robinson began working on the Grey-crowned Babbler with Birds Australia in 1991 and has focused on their conservation ever since. He initially moved to northern Victoria on a one-year grant to investigate their ecology and conservation needs. Eighteen years later, he and his family are still living there, embedded in the local landscape and community.

References

Barr, N. (2005). The changing social landscape of rural Victoria. Department of Primary Industries: Melbourne.

Bennett, A.F. (1993). Fauna conservation in box and ironbark forests: a landscape approach. *Victorian Naturalist* **110**, 15–23.

Davidson, I. and Robinson, D. (1992). Grey-crowned Babbler (*Pomatostomus temporalis*) FFG Action Statement No. 34. Department of Conservation and Natural Resources: Melbourne.

DSE (2007). Conservation Plan for the Longwood Landscape Zone. Department of Sustainability and Environment: Benalla.

Grice, P., Evans, A., Osmond, J. and Brand-Hardy, R. (2004). Science into policy: the role of research in the development of a recovery plan for farmland birds in England. *Ibis* **146** (Suppl. 2), 239–249.

Lambeck, R.J. (1997). Focal species: a multi-species umbrella for nature conservation. *Conservation Biology* **11**, 849–856.

Lindenmayer, D.B. and Fischer, J. (2003). Sound science or social hook: a response to Brooker's application of the focal species approach. *Landscape and Urban Planning* **62**, 149–158.

Lunt, I. and Bennett, A.F. (1999). Temperate woodlands in Victoria: distribution, composition and conservation. In *Temperate Eucalypt Woodlands in Australia: Biology, Conservation, Management and Restoration.* (Eds R.J. Hobbs and C.J. Yates) pp. 17–31. Surrey Beatty & Sons: Chipping Norton, NSW.

Mac Nally, R., Bennett, A.F., Brown, G.W., Lumsden, L.F., Yen. A., Hinkley, S., Lillywhite, P. and Ward, D. (2002). How well do ecosystem-based planning units represent different components of biodiversity? *Ecological Applications* **12**, 900–912.

Mac Nally, R., Bennett, A.F., Thomson, J.R., Radford, J.Q., Unmack, G., Horrocks, G. and Vesk, P.A. (2009). Collapse of an avifauna: climate change appears to exacerbate habitat loss and degradation. *Diversity and Distributions* **15**, 720–730.

Mansergh, I., Anderson, H. and Amos, N. (2006). Victoria's living capital – decline and replenishment 1800–2050. Part 2. The new millennium; replenishment. *Victorian Naturalist* **123**, 288–313.

Maron, M. (2007). Threshold effect of eucalypt density on an aggressive avian competitor. *Biological Conservation* **136**, 100–107.
Miles, C. (2008) Testing market-based instruments for conservation in northern Victoria. In *Biodiversity: Integrating Conservation and Production: Case Studies from Australian Farms, Forests and Fisheries.* (Eds T. Lefroy, K. Bailey, G. Unwin, and T. Norton, T) pp. 133–146. CSIRO Publishing: Melbourne.
Peach, W. J., Lovett, L. J., Wotton, S. R. and Jeffs, C. (2001). Countryside stewardship delivers cirl buntings (*Emberiza cirlus*), in Devon, UK. *Biological Conservation* **101**, 361–373.
Radford, J.Q and Bennett, A.F. (2007). The relative importance of landscape properties for woodland birds in agricultural environments. *Journal of Applied Ecology* **44**, 737–747.
Roberge, J-M and Angelstam, P. (2004). Usefulness of the umbrella species concept as a conservation tool. *Conservation Biology* **18**, 76–85.
Robinson, D. (1994). 'Research Plan for threatened woodland birds in southeastern Australia. ARI Technical Report No. 133'. Department of Conservation and Natural Resources: Melbourne.
Robinson, D. (1998). Lowering the goals? Habitat requirements of vertebrates and other wildlife groups in Victoria's Northern Plains and how well they match up. In *Down to Grass Roots: Proceedings of a Conference on Management of Grassy Ecosystems.* (Eds V. Craigie and C. Hocking) pp. 49–59. Victoria University: Melbourne.
Robinson, D. (2006). Is revegetation in the Sheep Pen Creek Area, Victoria, improving Grey-crowned Babbler habitat? *Ecological Management and Restoration* 7, 93–104.
Robinson, D., Davidson, I. and Tzaros, I. (in prep.). 'Management plan for the conservation of the Grey-crowned Babbler in Victoria, Flora and Fauna Technical Report Series No. 148'. Department of Sustainability and Environment: Melbourne.
Robinson, D., Stothers, K. and Howell, M. (2003). 'A Wildlife Guide for Landholders in the plains and box-ironbark regions of the Goulburn-Broken Catchment'. Goulburn Broken Catchment Management Authority: Shepparton.
Smallbone, L. Robinson, D. and Sleigh, S. (in prep.). The biodiversity benefits of an agri-environment scheme in the wheat and sheep belt of Victoria, Australia.
Thomson, J.R., Moilanen, A.J., Vesk, P.A., Bennett, A.F. and Mac Nally, R. (2009). Where and when to revegetate: a quantitative method for scheduling landscape reconstruction. *Ecological Applications* **19**, 817–828.
Wilson, C. W., Robinson, D., van der Ree, R., McCarthy, M., Vesk, P. and Saywell, S. (2009). The effectiveness of habitat works for the survival and population status of the Grey-crowned Babbler *Pomatostomus temporalis.* A report to The Norman Wettenhall Foundation and the Goulburn Broken Catchment Management Authority. University of Melbourne: Melbourne.

40

KNOWLEDGE, REGULATION AND INCENTIVES IN PROTECTING TEMPERATE GRASSY WOODLANDS: SIX LESSONS FROM NEW SOUTH WALES

John Benson

1. Document the resource through biological survey, classification and mapping.
2. Complete targeted research on threatening processes such as fragmentation, weed invasion and native species recruitment.
3. Enact laws and regulations for establishing public reserves and secure property agreements as a conservation core.
4. Implement assessment tools that allow scientifically credible decisions on land use or incentive payments using criteria such as habitat rarity/threat/condition.
5. Protect travelling stock routes, roadside remnants and other public land.
6. Support local community information networks and learn from adaptive management.

Introduction

More than three decades of ecological research and dealing with environmental policy in the State of New South Wales (NSW) in south-eastern Australia have provided me with insight into the information, regulations and education programs that are effective in protecting biodiversity in a variety of landscapes. This understanding inspired my main current research to classify and assess the threat and protected area status of plant communities over the 80 million hectares of NSW into the NSW Vegetation Classification and Assessment database (Benson 2006; Benson *et al.* 2006; Benson 2008; see summary of the project at: http://www.rbgsyd.nsw.gov.au/science/research/vegetation_of_nsw). It is anticipated that about 1400 plant communities will eventually be described across NSW, of which at least 100 will be grassy woodland types. My work in grassy woodlands and native grasslands has mainly involved botanical surveys on the NSW southern tablelands, north coast, northern tablelands, western slopes and the eastern section of the western plains (see map on next page). Additionally, I have collated hundreds of botanical surveys and mapping projects to populate the 90 information fields in the NSWVCA database.

While many ecological communities and species are now well represented in Australia's protected area system that comprises over 10% of the landmass, some vegetation types, plant species and wide-ranging fauna are difficult to adequately represent in protected areas. Temperate grassy woodlands, which occupy relatively high-nutrient soils on flat to undulating terrain across southern Australia, are an example. These landscapes are now intensively developed for various forms of agriculture. Prior to European settlement, grassy woodlands covered large swaths of countryside, with the explorer Thomas Mitchell labelling the western Victoria grassy woodland region as *Australia Felix* (Mitchell 1838). Such glowing commentary encouraged rapid agricultural transformation of grassy woodlands through clearing, grazing and cropping. What remains are highly fragmented patches that occur mainly on private land, with limited areas on public lands such as travelling stock routes, roadsides reserves and public conservation reserves. A minor proportion of grassy woodland remnants are in 'good' condition (*sensu* Parkes *et al.* 2003); that is, containing a diverse composition of native species and a complex habitat structure (multiple vegetation strata, logs, tree hollows).

Temperate grassy woodlands are unquestionably among Australia's most threatened ecological communities. Many are listed as endangered ecological communities under state and national laws. The lessons outlined below are based on work in NSW, with its array of laws and regulations that could be considered as some of the best in the world – at least on paper.

Lessons

1. Document the resource through: biological survey, classification and mapping

Classifying or mapping the different types of grassy woodland is a prerequisite to protecting them through land-use decisions or management. Without this work it is difficult to undertake conservation programs, ranging from projects to enhance landscape connectivity to

Table 40.1 Broad categories of grassy woodlands in New South Wales.

Black Box (*Eucalyptus largiflorens*) grassy open woodlands on outer inland floodplains
River Red Gum (*Eucalyptus camaldulensis*) grassy woodlands on inland floodplains
Weeping Myall (*Acacia pendula*) grassy low woodlands of central NSW alluvial plains
Western Grey Box (*Eucalyptus microcarpa*) and Poplar Box (*Eucalyptus populnea*) grassy woodlands of the inland alluvial plains and lower Western Slopes
White Box (*Eucalyptus albens*), Blakely's Red Gum (*Eucalyptus blakelyi*), Yellow Box (*Eucalyptus melliodora*), Red Box (*Eucalyptus polyanthemos*), Long-leaved Box (*Eucalyptus goniocalyx*) grassy woodlands of the Western Slopes and Tablelands
Apple Box (*Eucalyptus bridgesiana*) grassy woodlands on valley flats and terraces of the NSW Western Slopes and Tablelands;
Snow Gum (*Eucalyptus pauciflora*), Black Sallee (*Eucalyptus stellulata*), Ribbon Gum (*Eucalyptus viminalis*), Candlebark (*Eucalyptus rubida*), Mountain Gum (*Eucalyptus dalrympleana*) tall grassy woodlands of the Tablelands
New England Peppermint (*Eucalyptus nova-anglica*) grassy woodland in the New England Bioregion
Forest Red Gum (*Eucalyptus tereticornis*), Cabbage Gum (*Eucalyptus amplifolia*), *Eucalyptus seeana*, Grey Box (*Eucalyptus moluccana*), Woollybutt (*Eucalyptus longifolia*), ironbark grassy woodlands on clay to loam soils in coastal valleys

research and management of threatened species. Classification requires documenting biotic and abiotic features through field survey and remote sensing to deliver descriptions and maps of woodland types. It typically involves subjecting sample data to exploratory statistics, such as ordination and cluster analysis (Belbin 1991), to reduce the natural complexity of species assemblages. Other analyses test the relationships of defined types of woodlands with environmental variables such as climate, soils, substrate or hydrology. If sample data are not available, expert description can be used to produce classifications. A broad grouping of NSW temperate grassy woodlands is outlined in Table 40.1. These broad groups would be split into many types in regional or local-scale vegetation classification or mapping.

The advantage of developing a jurisdiction-wide woodland classification is that it can be incorporated into state-wide or national information systems, thus providing a 'language' on grassy woodland type and threat status for use in assessments and setting priorities in conservation and management programs. In NSW, the NSWVCA project (cited above) fulfils this requirement. It cross-links defined vegetation types to other landscape classifications, assesses threats, applies threat criteria and audits areas in all protected areas. This delivers an overall threat and protected area assessment of grassy woodland types and this information can be incorporated into the main NSW decision-support tool (the Biometric tool) for land-use change and incentive payments in property planning in the NSW Native Vegetation Act 2003 (Gibbons *et al.* 2005). The data are also used in general environmental assessments under other laws. Additionally, well labelled photographs of woodland types are useful as a visual educational guide to local community groups or regional management authorities.

While regional-scale mapping provides context on range and extent of vegetation, it often fails to map small remnants vital in corridors. If the aim is to re-establish woodlands, more accurate and finer-scale mapping is required that accurately (perhaps >90%) depicts woodland types in small remnants (e.g. Davidson 2008; Benson and Ashby 2000). Such mapping can also show where native-derived grasslands remain where trees have been cleared. These sites often form suitable core planting areas for restoration of trees (Figure 40.1). Advances in resolution of remote sensing imagery are delivering a paradigm shift for mapping vegetation. In NSW,

Figure 40.1 While the tree cover of this hill near Albury has largely been removed, a groundcover of native species provides a basis for re-establishment of a woodland tree structure to enhance woodland links across the landscape. (Photo: John Benson)

Figure 40.2 Isolated trees of Red Box (*Eucalyptus polyanthemos*) with an exotic groundcover may not persist over decades or produce seedlings without substantial intervention. (Photo: Jaime Plaza)

high-resolution ADS40 digital airborne aerial photography can be interpreted at a 1:2000 scale and be digitised directly into geographic information systems. Complementary use of regularly flown satellite imagery can be used to document changes in mapped extent of temperate grassy woodland, both of regrowth and clearing. Reporting these changes helps local and regional authorities to set priorities for future expenditure.

2. Complete targeted research on threatening processes such as fragmentation, weed invasion and native species recruitment

If we do not understand why species go extinct in woodland remnants and conversely what they need to persist, woodland conservation programs will fail. Research aimed at understanding the functioning of grassy woodlands is fundamental to guiding management and ecological restoration programs. One of the most informative avenues of research has been investigating the spatial size and connectivity of woodland patches for the survival of particular species. For example, Reid (1999) delivers guidelines on woodland patch size and spatial connectivity required for the persistence of woodland bird species. Such research on landscape links is emphasised by Bennett (1998). These research results can guide public land acquisition or strategies for off-reserve protection with private landholders. Other types of research important for managing grassy woodlands are those focused on controlling weed species, reducing artificially high nutrient levels and understanding the population dynamics of common or rare species. Research into genetic diversity (Prober 2000) can guide the size and arrangement of woodland remnants to be protected. The clearing and thinning of grassy woodlands has left isolated paddock trees over large areas (Figure 40.2). Trees in these thinned areas are less able to outcross as part of reproduction, thus lowering seed production (Burrows 2000). The loss of trees through senescence, with little compensating recruitment from seedlings, can substantially reduce woody native vegetation cover over large areas (Gibbons and Boak 2002) with a consequent loss of tree hollows important to some fauna (Gibbons and Lindenmayer 2002).

Of considerable urgency is research on controlling weed species that threaten large areas of grassy woodlands. For example, two exotic grasses are rapidly invading NSW temperate grassy woodlands. Coolatai Grass (*Hyparrenia hirta*) is invading grassy woodlands on the North Western Slopes of NSW and African Love Grass (*Eragrostis curvula*) is spreading into woodlands of the Southern Tablelands. Unfortunately, most of Australia's grassy woodlands coincide with regions experiencing falling rainfall and possibly more frequent droughts. Implications of changed climate for the recruitment and persistence of a range of native species in grassy woodlands requires long-term research.

3. Enact laws and regulations for establishing public reserves and secure property agreements as a conservation core.

Laws and regulations are fundamental to protecting ecological communities and species. The introduction of the NSW Threatened Species Conservation Act in 1995 and the NSW Native Vegetation Act in 1998/2003 substantially increased leverage to protect natural habitat across all tenures of land in NSW. These laws were pre-dated by wildflower protection legislation in 1928, the establishment of the Fauna Protection Panel in 1948 and Australia's first National Parks and Wildlife Act in 1967. In 1987, two pieces of contrasting environmental legislation were passed by the New South Wales Parliament. The first Wilderness Act in Australia was enacted to protect the State's wildest lands. The other law, pertinent to grassy woodland protection, involved amending the NSW National Parks and Wildlife Act to allow for voluntary property agreements between landholders and the NSW National Parks and Wildlife Service. Caveats were always available under Section 88 of the NSW Conveyancing Act 1919 but had rarely been used for biodiversity protection. Victoria's 1970s Land Conservation Trust was a pathfinder for non-reserve conservation and it influenced the content and introduction of the NSW Conservation Trust Act that came into force in 2001.

The occurrence at a site of threatened species or Endangered Ecological Communities (EECs) listed under the NSW TSC Act or Federal EPBC Act significantly influences the assessment of importance of that site. Since many temperate grassy woodlands are listed as EECs or are recorded as over-cleared vegetation types (>70% cleared) in regulations and biodiversity planning tools, they are largely protected from legal clearing with provisions for offsets in cases of sites of very low condition. Illegal clearing, however, is another issue and monitoring of it and prosecuting cases remains a substantial challenge for regulatory authorities.

4. Implement assessment tools that allow scientifically credible decisions on land use or incentive payments using criteria such as habitat rarity/threat/condition

Publicly accountable decision-support tools have changed the way that business is done in natural resource management on private land. They are now the most important vehicle for the protection and recovery of endangered grassy woodlands across much of southern Australia. In NSW, the gazettal of the Native Vegetation Act Regulations in 2005 introduced the NSW Biometric tool (Gibbons *et al.* 2005) as a jurisdictional-wide method for environmental assessment at the property scale on proposals for land-use change or incentive payments. Earlier in Victoria, the Habitat-Hectares rule-set (Parkes *et al.* 2003) provided a method for auctioning incentive funding to maximise return in biodiversity benefits. Both these approaches require condition benchmarks for vegetation types that are developed from field data and expert input. Benchmark assessment provides a means of establishing incentive payments to landholders to restore condition in a woodland remnant or provides a basis for deliberating on land-use change proposals. Importantly, it is beneficial to link sound vegetation classifications and endangered species databases to these tools to ensure jurisdictional-wide perspectives are available when making decisions about land-use change of incentive payments to landholders. A summary of actions and use of NSW regulations in relation to site attributes is provided in Table 40.2.

5. Protect travelling stock routes roadside remnants and other public land

Some of the best patches of grassy woodlands and native grasslands in NSW are located on travelling stock routes (TSRs) or reserves. The prevention of naturally vegetated TSRs being converted from public to private land along with improved management is paramount to protecting grassy woodlands over the tablelands, western slopes and western plains sections of NSW. TSRs cover about 2% of NSW and in some regions they house much of the remaining native vegetation (Davidson *et al.* 2005). Many TSRs have not been cleared, are generally less intensively grazed than adjoining private lands and have rarely been cropped or sown to exotic pasture. Regularly, proposals are made to sell-off TSRs to private landholders to ease financial burdens on TSR management authorities. Petitions to governments have helped raise the profile of TSRs but sell-offs could still happen. A single decision to increase funding to the authorities managing TSRs, would achieve as much as any other action in protecting the temperate grassy woodlands of NSW.

6. Support local community information networks and learn from adaptive management

Without local community support, most conservation programs are doomed to failure – especially programs focused on private land. An increased awareness by landholders of the importance of native vegetation for ecological services has helped argue the case for protecting vegetation remnants on private land. The 1980s Landcare movement tapped local community

Table 40.2 Protection actions in relation to grassy woodland site attributes on type, condition and connectivity.

Site attributes (and/or)	Actions	Key NSW Regulations
Irreplaceable habitat in terms of type representation and condition; Endangered Ecological Community; very poorly represented (for example, < 3% of original extent) in protected areas; medium to large patch; or contains populations of threatened species.	Purchase and protect in public reserve or secure (99 year) property agreement (equivalent to IUCN protected areas).	National Parks & Wildlife Act; Native Vegetation Act; NSW Conservation Trust; Action Targets under NSW Catchment Management Authorities; Threatened Species Conservation Act.
Habitat in good condition; >3% of original extent represented in protected areas; medium or small patch; may contain populations of threatened species.	Protect in secure (99 year) property agreement or short term PA (15 years) with management such as fencing, weeding, feral animal control.	Native Vegetation Act; NSW Conservation Trust; Action Targets under NSW Catchment Management Authorities; Zoning under State, Regional & Local Plans under Environmental Planning & Assessment Act.
Habitat in poor condition but important to landscape connectivity and should persist with management; not containing key populations of threatened species.	Incentive payments for short or secure property agreements and management actions such as fencing, weeding, feral animal control.	Native Vegetation Act; Action Targets under NSW Catchment Management Authorities; NSW Conservation Trust.
Habitat in very poor condition including some areas of senescing isolated trees; small patches; not likely to persist or recover composition; not important to landscape connectivity;	Possible tradeoff for improvements in similar vegetation type in the same vicinity, depending on assessment of importance to key species (e.g. tree hollows).	Native Vegetation Act.

Note: in NSW most grassy woodland types are >70% cleared and therefore clearing is discouraged under the Biometric tool in the NSW Native Vegetation Act)

concern about salinity, soil erosion and loss of native vegetation, including grassy woodlands. In NSW, initiatives such as the Grassy Box Woodlands Conservation Management Network produced on-ground protection actions including conservation agreements and a newsletter titled 'Woodland Wanderings' that contained informative articles by farmers, scientists and others. Continued government support of local networks is important for assembling people with different backgrounds and views on land management. Improved auditing of the impacts of expending funds on actions for grassy woodland recovery (including documenting successes and failures) would contribute to improving adaptive management across planning regions.

Conclusions

The future of temperate grassy woodlands will largely depend on a range of measures including classifying and mapping the different types of woodlands, targeted research on woodland species persistence or re-establishment, instigation and wise use of laws and regulations, re-establishment of woodland cover to form larger links, protecting remaining public lands such as travelling stock reserves and cooperation between landholders, scientists and government agencies.

About 25 years ago a mentor once said to me: 'You can have the best laws in the world but if people don't support them they are next to useless'. In NSW we now have some of the best environmental regulations anywhere in the world but they will not work to protect grassy woodlands that primarily occur on private land without gaining local support. Local networks are critical for relaying scientific results to landholders and for feedback by landholders to scientists and planners. The Habitat-Hectares (Parkes *et al.* 2003) and Biometric (Gibbons *et al.* 2005) decision-support tools have improved the objectivity and methods used in allocating incentive payments to landholders to manage or improve private land remnants. Improving the measurements used in the parameters of these schemes will improve decision outcomes. Judicious auditing of local and regional expenditures allocated by regional authorities such as CMAs could refine how limited funds should be spent.

Even if broadscale clearing of grassy woodlands is stopped through regulations and incentives, insidious threats remain. Native plant and animal populations continue to be lost from patches due to the impacts of previous clearing and fragmentation. Some weeds are increasing in abundance. Higher temperatures and lower rainfall across southern Australia (Hennessy *et al.* 2004), along with more frequent drought, are placing stress on remaining grassy woodlands causing dieback of trees. These threats highlight the urgency of a coordinated effort across jurisdictions and spatial scales to better protect Australia's remaining temperate grassy woodlands.

Bio

John Benson (B.Sc. Macquarie, Ph.D. University of Sydney) is Senior Plant Ecologist at the Royal Botanic Gardens, Sydney. Over 34 years, his projects have included survey and protection of NSW rainforests; establishing conservation reserves; survey/research on threatened plant species; development of rare species databases and threatened species laws; survey/mapping vegetation in a number of regions including of native grasslands and woodlands; media articles on biodiversity and land management; and he has acted as an adviser to government to end broadscale land clearing. Since 1999, his key project has been developing a comprehensive database system classifying NSW plant communities and assessing their threat and protected area status for use in environmental assessment and land use regulations (see http://www.rbgsyd.nsw.gov.au/science/current_research/vegetation_of_nsw).

References

Belbin, L. (1991). The analysis of pattern in bio-survey data. In *Nature Conservation: Cost-effective Biological Surveys and Data Analysis.* (Eds C. R. Margules and M. P. Austin) pp. 176–190. CSIRO: Melbourne.

Bennett, A.F. (1998). 'Linkages in the Landscape: The Role of Corridors and Connectivity in Wildlife Conservation.' IUCN: Gland, Switzerland and Cambridge, UK.

Benson, J.S. (2006). New South Wales Vegetation Classification and Assessment: Introduction – the classification, database, assessment of protected areas and threat status of plant communities. *Cunninghamia* **9**, 331–382.

Benson, J.S. (2008). New South Wales Vegetation Classification and Assessment: Part 2 Plant communities in the NSW South-western Slopes Bioregion and update of NSW Western

Plains plant communities. Version 2 of the NSWVCA database. *Cunninghamia* **10**, 599–673.
Benson, J.S., Allen, C., Togher, C. and Lemmon, J. (2006). New South Wales Vegetation Classification and Assessment: Part 1 Plant communities of the NSW Western Plains. *Cunninghamia* **9**, 383–451.
Benson, J.S. and Ashby, E.M. (2000). Vegetation of the Guyra 1:100 000 map sheet New England Bioregion, New South Wales. *Cunninghamia* **6**, 747–872.
Burrows, G.E. (2000). Seed production in woodland and isolated trees of *Eucalyptus melliodora* (yellow box, Myrtaceae) in the South Western Slopes of New South Wales. *Australian Journal of Botany* **48**, 681–685.
Davidson, I. (2007). Developing a local landscape biodiversity plan in the Woomargama & Jindera local landscapes (Greater Hume Shire). Regeneration Solutions Ltd: Glenrowan.
Davidson, I., Scammell, A., O'Shannassy, P., Mullins, M. and Learmonth, S. (2005). Travelling stock reserves: refuges for stock and biodiversity? *Ecological Management & Restoration* **6**, 5–15.
Gibbons, P., Ayers, D., Seddon, J., Doyle, S. and Briggs, S. (2005). BIOMETRIC Version 1.8. 'A Terrestrial Biodiversity Assessment Tool for the NSW Property Vegetation Plan Developer.' CSIRO Sustainable Ecosystems: Canberra.
Gibbons, P. and Boak, M. (2002). The value of paddock trees for regional conservation in an agricultural landscape. *Ecological Management & Restoration* **3**, 205–210.
Gibbons, P. and Lindenmayer, D. (2002). *Tree Hollows and Wildlife Conservation in Australia.* CSIRO Publishing: Melbourne.
Hennessy, K., McInnes, R., Abbs, D., Jones, R., Bathols, D., Suppiah, R., Ricketts, J., Rafter, T., Collins, D. and Jones, D. (2004) 'Climate change in New South Wales. Part 2: Projected changes in climate extremes'. Consultancy Report to New South Wales Greenhouse Office by Climate Impact Group CSIRO Atmospheric Research and National Climate Centre. Australian Government Bureau of Meteorology: Melbourne.
Mitchell, T.L. (1938). *Three Expeditions into the Interior of Eastern Australia with Description of the Recently Explored Region of Australia Felix and of the Present Colony of New South Wales.* T. & W. Boone: London.
Parkes, D., Newell, G. and Cheal, D. (2003). Assessing the quality of native vegetation: the 'habitat hectares' approach. *Ecological Management and Restoration* **4**, S29–S38.
Prober, S. M. and Brown, A. H. D. (1994). Conservation of the grassy White Box woodlands: population genetics and fragmentation of *Eucalyptus albens. Conservation Biology* **8**, 1003–1013.
Reid, J.R.W. (1999). Threatened and declining birds in the New South Wales sheep-wheat belt. I. Diagnosis, characteristics and management. CSIRO Wildlife and Ecology: Canberra.

41

SEVEN LESSONS FROM SOUTHERN QUEENSLAND WOODLANDS

Teresa J. Eyre

1. Provide effective biodiversity science at the science–policy interface.
2. Maintain keystone habitat features throughout the landscape.
3. Acknowledge that intact landscapes are a mosaic.
4. Manage the land mosaic with clear goals.
5. 'Biodiversity' is not just about numbers of species.
6. Develop and use appropriate concepts of woodland condition.
7. Establish an effective monitoring program and commit to it.

Introduction

This chapter outlines seven key lessons for improving the practicability and effectiveness of conservation and management of Queensland's southern woodlands, both on and off-reserve (see map on next page). The first lesson is a goal for ecological scientists, and the remaining six are primarily principles or prescriptions for decision makers and land managers.

In southern Queensland, clearing woodland for agricultural development has been extensive and indeed once endorsed. Broadscale clearing is now illegal under the *Vegetation Management Act 1999*, however, and sustainable grazing land management is promoted through lease renewal under the *Delbessie Agreement* (2007). Consequently, interest in biodiversity conservation management in Queensland's rural landscapes has increased. In response, we have focused research on biodiversity values in the pastoral landscapes of the fragmented Poplar Box (*Eucalyptus populnea*) and Brigalow (*Acacia harpophylla*) woodlands, and the 'intact' semi-arid Mulga (*A. aneura*) woodlands. Vascular plants, vertebrates and habitat features have been the major foci for two reasons. First, their responses correspond with the scale of management (landscape and property or site scale); and second, policy and land managers relate best to them. In southern Queensland, the interactions between variable rainfall, total grazing pressure, fire, historic clearing, exotic pasture and regenerative capacity creates an interesting and complex landscape mosaic of varying vegetative states. My experiences in these landscapes form the basis of my seven lessons.

Lessons

1. Provide effective biodiversity science at the science-policy interface

As a government scientist, I have been involved in the development of policies that have had a profound impact upon the status of Queensland's woodlands and the people who manage them. Getting the science into policy, however, is not easy. The formulation of policy frequently occurs within extremely short time frames, which do not correspond with those of research. Science can get entwined in values, suffer from 'paralysis-by-analysis' and be denigrated by stakeholders. Frustrated policy makers then make decisions without appropriate scientific foundation. As scientists, we know the importance of being objective and avoiding advocacy, '... the active, covert or inadvertent support of a particular policy ...' (Lackey 2007). The reality is that advocacy cannot be totally avoided. Policy makers need to hear a clear message, as they have little time to work out the scientific nuances for themselves. Indeed, a review by Nelson and Vucetich (2009) reveals that advocacy is almost unavoidable, and that environmental scientists should aim to improve their ability to advocate in an honest, justified and transparent manner. As effective scientists, we must keep our language value-neutral (e.g. avoid use of 'poor' and 'good'). We need to constrain our advice to matters informed by science, explicitly quantify uncertainty in our data, be adaptive in our research approach, and maintain close links with policy makers during all stages of research. Policy and land managers get their information from a relatively small pool of scientists. As scientists, we have a responsibility to actively and effectively participate in policy matters, and arbitrate the use of advocacy as an honest pursuit amongst ourselves.

2. Maintain keystone habitat features throughout the landscape

Numerous investigations of habitat use by woodland fauna have identified a range of keystone habitat features (see review by McElhinny *et al.* 2006). These are the structural and floristic attributes of woodlands that provide essential foraging, nesting, breeding and sheltering resources for species at the site scale. Although there has been no formal, national agreement on what constitutes keystone habitat features for fauna of Australian woodlands, suites of features commonly used in vegetation condition protocols include large trees, shrub cover, downed coarse woody debris, trees with hollows, litter cover and grass and forb cover. In Australia, forest management agencies have developed and implemented specific prescriptions to ensure continuance of habitat features (e.g. the retention of up to six hollow-bearing trees per ha and their recruits). In agricultural landscapes, a prescriptive approach may require incentives, such as extensions to lease terms, or tax benefits. This is because we need to encourage maintenance of features such as large trees and logs across all vegetative states, including the productive paddocks as well as the remnant patches.

3. Acknowledge that intact landscapes are a mosaic

It cannot be assumed that large patches of retained woodland comprise continuous tracts of functional habitat, as they would be defined in dichotomous 'habitat' versus 'non-habitat' landscape models. Essential to effective conservation management is the recognition that intact woodlands are constructed of patches reflecting fine-scale disturbance that determines local habitat condition. For example, complexity of habitat features in the lower vegetation layers in intact landscapes in the Brigalow Bioregion is quite variable. This is a consequence of patchily applied grazing and fire management and interactions with spatially variable rainfall. In response, the composition of bird assemblages varies markedly, with small, insectivorous birds notably absent from patches with little mid and lower layer complexity (Eyre *et al.* 2009a). The boundaries between patches within the mosaic of intact landscapes are gradual rather than abrupt when compared with variegated and fragmented landscapes (*sensu* McIntyre and Hobbs 1999). This often renders patches difficult to identify and map. Furthermore, there is the problem of what scale and elements to use when describing and quantifying habitat patches within a mosaic (Lindenmayer *et al.* 2008). For example, in south-east Queensland, we were able to map the mosaic of an intact landscape based on the distribution of large, mature trees using high-resolution remote sensing products. In this study, mosaic configuration and composition significantly influenced the tree-hollow dependent, wide-ranging Yellow-bellied Glider (*Petaurus australis*), but not the ground-foraging Eastern Yellow Robin (*Eopsaltria australis*) (McAlpine and Eyre 2002). A clearly defined management objective will guide the setting of the appropriate scale and mosaic classification.

4. Manage the land mosaic with clear goals

This lesson relates to Lessons #2 and #3, and represents a scaling up of management from the site scale through to the property and landscape scales. Consideration of the mosaic acknowledges that ecological interactions exist between and among elements in the landscape (Lindenmayer *et al.* 2008), relevant to an overall conservation goal (Bennett *et al.* 2006). The benefit of managing the mosaic is that it allows consideration of both grazing land production and conservation goals. It also goes beyond simple reservation models and moves towards closing the 'sustainability gap' (Fischer *et al.* 2007). Properties of the mosaic include composition, habitat extent, spatial configuration and environmental variation (Bennett *et al.* 2006). This is illustrated by the spread of the transformer pastoral species Buffel Grass (*Pennisetum ciliare*) into non-target Poplar Box woodlands. The likelihood of Buffel Grass occurring in a remnant woodland patch depends on the configuration and extent of cleared vegetation in the surrounding landscape, as well as the variation in rainfall within the mosaic (Eyre *et al.* 2009b). Of course, different species will respond to different properties of the mosaic, and management action taken will depend on the overall goal. If the goal is to increase Buffel Grass cover in the landscape, then the action will be the opposite of that taken to conserve the ecological integrity of retained woodland by keeping Buffel Grass out. A clear definition of the management goal for both production and conservation is imperative. In my experience, grazing land managers do set a general goal of sustainable production for the continued benefit of the next generation. Specific sub-goals are then set (e.g. increase evenness of grazing by reducing distance to water). The same specific sub-goals are required from a conservation perspective (e.g. protect regrowth and increase mid and ground-layer habitat in remnant vegetation for small woodland birds through fire and grazing management).

5. 'Biodiversity' is not just about numbers of species

The term 'biodiversity' needs to be effectively marketed to increase accessibility and use by land managers. The broad adoption of 'biodiversity' in policy and management is positive, but can be vague. For example, a key concept underpinning Queensland's *Vegetation Management Act 1999* is the management of vegetation 'to maintain or increase biodiversity'. There is a tendency to oversimplify the use of 'biodiversity' so that it basically represents number of species (species richness).

In southern Queensland woodlands, we are seeing major shifts in bird and reptile species assemblages in response to habitat modification at the site and landscape scales. Yet, there is no discernible change in species richness. Does this constitute maintenance of biodiversity? Not to ecologists familiar with the terminology and who routinely deal with the complexities of 'biodiversity'. But what about the rest of the community? The imprecise use of 'biodiversity' has been identified as an impediment to the development and use of effective policy and research, and alternative concepts have been endorsed such as 'biological integrity' (Angermeier and Karr 1994). Regardless of this, the term 'biodiversity' is ensconced in legislation and many strategies and management manuals. The key is to use the term precisely and simply, while acknowledging that it is more complicated than just species diversity. Habitat structure, ecosystem function and interactions between species at multiple scales are also features. Embracing the complexity of 'biodiversity' will help formulate practical on-ground management objectives. When land managers ask what I mean by biodiversity, I explain that it is the 'the variety of life, its composition, structure and function, at a range of scales' (Freudenberger and Harvey 2003).

6. Develop and use appropriate concepts of woodland condition

The recent preoccupation with assessing vegetation condition along a gradient of 'poor' to 'good' has led to the development of a number of conceptual frameworks based on the successional theory of Clements (1928) and the edaphic polyclimax concept of Tansley (1935). This means that 'condition' is based on quantifying and comparing attributes to those of a 'climax' or 'potential natural' state of the same community type. The use of such models in ecosystems driven by climatic events has been much criticised, however, because they do not adequately incorporate multiple pathways of change, complex vegetation dynamics and natural fluctuations in some attributes (Friedel 1991). Furthermore, describing vegetation as 'poor' suggests that it is of little service to biodiversity, which is not always the case. For example, regrowth, thickened vegetation, and even low cover of exotic grass all represent transitional vegetative states that provide some service to fauna in the landscape (Bowen *et al.* 2007; Eyre *et al.* 2009a,b). For some attributes, we still do not really understand the complexities of their response to disturbance events. What constitutes a 'natural state' remains questionable. A dense midstorey (vegetation thickening) is often described as symptomatic of inappropriate land management, but instead may be reflecting natural ecological dynamism (Fensham

2008). Whichever the case, we do know that a dense midstorey provides an important refuge for a number of small woodland bird species (Eyre *et al.* 2009a).

An alternative assessment and management framework includes the use of state-and-transition models. These have been tested and applied most widely in rangelands (Westoby *et al.* 1989) and more recently to endangered woodland (Spooner and Allcock 2006). These models therefore provide an effective framework for future research, and improved management and restoration activities. To date, adoption of state-and-transition models appears hampered by limited broadscale acceptance of conceptual definitions, and there is little guidance available for the development and application of models. Bayesian belief networks (McCann *et al.* 2006) could provide a useful, formalised way forward.

7. Establish an effective monitoring program and commit to it

Ecological monitoring need not be a waste of time and money. Unfortunately, many agencies perceive site-based monitoring as ineffective. A primary reason is because political commitment tends to be relatively short-term; a consequence of electoral cycles and major policy change. For example, two long-term, site-based woodland monitoring programs are no longer supported, in response to major change in forest and vegetation policy in Queensland. A robust, adaptive institutional setting that transcends short-term and short-sighted political priorities is required to acknowledge that, although timber harvesting will phase out and vegetation clearing has stopped, these landscapes are still managed and manipulated. Collaborative arrangements with non-bureaucratic institutions that help develop, drive and validate monitoring programs would be ideal. In the meantime, we need to justify why long-term, site-based monitoring is required. Southern Queensland woodlands are ecologically dynamic, in response to highly variable climatic conditions and diffuse management regimes. Thus long-term data are required to adequately detect change, and increase our understanding of triggers of change. These factors are often not well detected from short-term space-for-time approaches. Long-term ecological monitoring in Queensland's woodland landscapes is required to:

- determine if an extinction debt is yet to manifest since broadscale clearing in Queensland has only recently ceased;
- document the rate and extent of encroachment by Buffel Grass throughout the landscape, and how this affects native species;
- clarify the relative roles of management, seasonal variability and climate change on woodland structural and floristic change, and how faunal species respond; and
- determine if sustainable management approaches are working (e.g. under the *Delbessie Agreement*).

Conclusion

The seven lessons learned from southern Queensland woodlands are for conservation on both public and privately owned land. Cessation of broadscale clearing of remnant vegetation in Queensland has been important for conservation of woodland biodiversity. It is vital, however, to recognise the existence of mosaics within and between patches of remnant vegetation, and the need for appropriate management to effectively enhance biodiversity values. Setting consistent conservation management goals and implementing them should be relatively achievable on public lands, which are largely managed by a single, and theoretically suitably resourced, agency.

Biodiversity conservation activities on privately owned land are disparate, and tend to rely on short-term government-funded schemes such as Envirofund. Current incentive programs in Queensland focus on portions of properties that either merit protection, (e.g. fencing sensitive riparian areas or covenants on high conservation areas) or need restorative action (e.g. weed eradication). These programs are of course important, but do not encourage the 'manage-the-mosaic' approach. The concept of mosaics, and acknowledgment that grazing land properties can be managed with dual goals of production and conservation, must be integrated into the science–policy–management interface. Effective marketing to simply present ecological concepts to the general community would be useful. Reinstating informed regionally based extension officers, who assist in building capacity among land managers, would contribute to effective knowledge transfer between biodiversity science, policy and management.

Acknowledgements

Thanks to the land owners, policy makers, Land & Water Australia and Meat & Livestock Australia who have contributed to expanding ecological knowledge of Queensland's woodlands through their support and funding of research. The editors and J. Rylance greatly improved this chapter. Special thanks to the team for their dedication and professionalism.

Bio

Teresa Eyre is a research field ecologist who manages a small team of zoologists, botanists and ecologists within the Biodiversity Sciences Unit of the Queensland Department of Environment and Resource Management. She has spent the last 20 years researching the responses of fauna in managed forest and woodland ecosystems of Queensland, with the primary objective of informing management and government policy. Teresa Eyre and the team are currently spending most of their time on family-owned properties in the Brigalow and Mulgalands Bioregions. Her favourite scar was made by a wild caught-and-released Mahogany Glider in 1992.

References

Angermeier, P.L. and Karr, J.R. (1994). Biological integrity versus biological diversity as policy directives. *BioScience* **44**, 690–697.

Bennett, A.F., Radford, J.Q. and Haslem, A. (2006). Properties of land mosaics: implications for nature conservation in agricultural environments. *Biological Conservation* **133**, 250–264.

Bowen, M.E., McAlpine, C.A., House, A.P.N. and Smith, G.C. (2007). Regrowth forests on abandoned agricultural land: a review of their habitat values for recovering forest fauna. *Biological Conservation* **140**, 273–296.

Clements, F.E. (1928). Plant succession and indicators: a definitive edition of plant succession and plant indicators. Hafner Press: New York.

Eyre, T.J., Maron, M., Mathieson, M.T. and Haseler, M. (2009a). Impacts of grazing, selective logging and hyper-aggressors on diurnal bird fauna in intact forest landscapes of the Brigalow Belt, Queensland. *Austral Ecology* **34**, 705–716.

Eyre, T.J., Wang, J., Venz, M.F., Chilcott, C. and Whish, G. (2009b). Buffel grass in Queensland's semiarid woodlands: response to local and landscape scale variables, and relationship with grass, forb and reptile species. *The Rangeland Journal* **31**, 293–305.

Fensham, R. (2008). Leichardt's maps: 100 years of change in vegetation structure in inland Queensland. *Journal of Biogeography* **35**, 141–156.

Fischer, J., Manning, A.D., Steffen, W. *et al.* (2007). Mind the sustainability gap. *TRENDS in Ecology and Evolution* **22**, 321-624.
Freudenberger, D. and Harvey, J. (2003). 'Assessing the Benefits of Vegetation Enhancement for Biodiversity: A Framework. Report for Environment Australia'. CSIRO Sustainable Ecosystems: Canberra.
Friedel, M.H. (1991). Range condition assessment and the concept of thresholds: a viewpoint. *Journal of Range Management* **44**, 422–426.
Lackey, R.T. (2007). Science, scientists, and policy advocacy. *Conservation Biology* **21**, 12–17.
Lindenmayer, D., Hobbs, R.J., Montague-Drake, R. *et al.* (2008). A checklist for ecological management of landscapes for conservation. *Ecology Letters* **11**, 78–91.
McAlpine, C.A. and Eyre, T.J. (2002). Testing landscape metrics as indicators of habitat loss and fragmentation in continuous eucalypt forests (Queensland, Australia). *Landscape Ecology* **17**, 711–728.
McCann, R.K., Marcot, B.G. and Ellis, R. (2006). Bayesian belief networks: applications in ecology and natural resource management. *Canadian Journal of Forest Research* **36**, 3053–3062.
McElhinny, C., Gibbons, P., Brack, C. and Bauhus, J. (2006). Fauna-habitat relationships: a basis for identifying key stand structural attributes in temperate Australian eucalypt forests and woodlands. *Pacific Conservation Biology* **12**, 89–110.
McIntyre, S. and Hobbs, R. (1999). A framework for conceptualising human effects on landscapes and its relevance to management and research models. *Conservation Biology* **13**, 1282–1292.
Nelson, M. P. and Vucetich, J. A. (2009). On advocacy by environmental scientists: what, whether, why, and how. *Conservation Biology* **23**, 1090–1101.
Spooner, P.G. and Allcock, K.G. (2006). Using a state-and-transition approach to manage endangered *Eucalyptus albens* (White Box) woodlands. *Environmental Management* **38**, 771–783.
Tansley, A.J. (1935). The use and abuse of vegetational concepts and terms. *Ecology* **16**, 248–307.
Westoby, M., Walker, B. and Noy-Meir, I. (1989). Opportunistic management for rangelands not at equilibrium. *Journal of Range Management* **42**, 266–274.

PART E

Conclusion

42

HOW FAR HAVE WE COME? PERSPECTIVES ON ECOLOGY, MANAGEMENT AND CONSERVATION OF AUSTRALIA'S TEMPERATE WOODLANDS

David Lindenmayer, Andrew F. Bennett and Richard Hobbs

Introduction

Australia's temperate woodlands are environments deeply embedded in the nation's history, heritage and culture. They were occupied by indigenous people for tens of thousands of years and are culturally important for Australians of European descent. They are the places where gold was first discovered, were home to some of the nation's most notorious bushrangers, provided the backdrop for classic paintings by masters like Tom Roberts and were the inspiration for iconic story-tellers such as Banjo Patterson and Henry Lawson. Temperate woodlands are also critically important repositories of Australian and global biodiversity. There are, therefore, compelling historical, cultural and ecological reasons for the informed management and conservation of these environments. These compelling reasons, however, are mismatched with the reality of the environmental status of the nation's temperate woodland environments.

For the most part, Australia's temperate woodlands are heavily cleared, much of the remaining vegetation is in poor condition, and large numbers of the plants and animals found in these environments are threatened or endangered. This situation, coupled with the fact that the vast majority of remaining temperate woodland is on private land that has primary management objectives other than conservation, means that the management of temperate woodlands is characterised by a raft of ecological, social and economic challenges. Many of the challenges appear prominently in the chapters in this book. In this final chapter, we provide a brief overview of some of these challenges through discussing what we consider to be 10 emergent themes and issues arising from the work presented by the various authors. We give a short commentary on each of these issues. The first seven themes are largely ecological ones: they are followed by two topics that are either cross-disciplinary or have direct socio-ecological or direct science policy implications. The tenth and final theme explores some future perspectives in woodland conservation and management. In keeping with the general spirit of this book, we have kept this chapter short; it is far from an exhaustive treatise of all issues nor a detailed exploration of any single issue in great depth.

1. The importance of spatial scale

Spatial scale is a theme common to many of the chapters in this book. For example, Bennett and Radford (Chapter 13) discussed the cumulative positive effects of different patch, paddock, farm and landscape elements on biodiversity as well as the incremental and cumulative negative effects on biota of losing these various elements. Similarly, Lindenmayer *et al.* (Chapter 21) highlighted how the interaction of various vegetation assets on a farm (e.g. paddock trees, woodland remnants, native pastures) had a cumulative effect on biodiversity at the farm level. Spooner (Chapter 22) also canvassed scale issues, but in a different way in that he expressed concern that a sole focus on those remnants on a farm in good condition may lead to a degradation of the remainder of a farm. This is a key issue because temperate woodlands are easy to degrade but very difficult to restore (Lunt *et al.*, Chapter 20).

Watson (Chapter 19) and Paton *et al.* (Chapter 10) discussed the potential risks of mismatches in the scale of key ecological processes driving biotic responses (e.g. those at a landscape or regional scale, such as hydrological processes or dispersal movements of animals) and the scale of management actions (e.g. vegetation restoration at a patch scale). They (and other authors) highlighted the need to think about matching processes and actions for the latter to be most effective in achieving 'whole-of-landscape change' (Lefroy, Chapter 6). A research question remains, however, about how much incremental action is needed (e.g. at a patch scale) to produce effective landscape-scale and regional-scale solutions to environmental problems in temperate woodlands.

Scale effects in biodiversity survey, in biotic responses (Watson, Chapter 19), and for restoration actions (Paton *et al.*, Chapter 10; Streatfield, Chapter 4) also have implications for woodland vegetation mapping. In particular, the scale of resolution of mapped entities (Thackway, Chapter 31) that is useful for guiding management actions and for informed conservation practice. Notably, Major (Chapter 24) argued that key attributes of woodland vegetation for fauna (such as shrubs, logs and dead trees) are not well captured in existing mapping (Major, Chapter 24; Maron, Chapter 29). This is important as it highlights the need for those involved in detailed and extensive vegetation mapping programs to: (1) engage more widely with potential users of mapped information; (2) carefully articulate the objectives of their mapping work and what it can and can not be effectively used for (e.g. Benson, Chapter 40); and (3) explicitly highlight where other mapping tools or approaches may be needed to used for particular applications or to solve particular problems.

The importance of spatial scale is also important in considerations of the viability of farming enterprises and hence the opportunities (or otherwise) for other management activities on a farm, like conservation actions. For example, Lefroy (Chapter 6) noted that farming enterprises focus on improving productivity at the paddock scale and economic viability at the farm scale. Yet it is usually at the catchment and larger scales that goals of conserving soils, water and biodiversity become meaningful. This is an important insight that drives to the heart of issues that are discussed further in subsequent sections of this chapter like: (1) the appropriate scale or scales to best integrate woodland conservation and agricultural production; (2) the ecological and economic viability of schemes like offsetting programs on a farm; and (3) ways to collectively manage the different parts of a portfolio of native vegetation assets on a farm (see Cunningham *et al.* 2008).

2. The importance of a temporal perspective

Changes to temperate woodland environments have been extraordinarily rapid in ecological time: vast changes in many parts of Australia described in this book have occurred in less than one to two centuries as demonstrated by the continued presence of trees of pre-European settlement origin. Many authors note that the woodlands and biota of today are a legacy of historical patterns of land use and land management practices. This has a number of implications. First, there has been disproportionate loss of the most productive parts of woodland environments (like those in valley floors on more fertile soils) and, of the remaining areas, those subject to less intensive and extensive levels of modification often are located on land less suitable for agriculture such as steep slopes, skeletal soils and rocky terrain (e.g. Dorrough, Chapter 17). This means that differential restoration efforts are required to recover some types of woodland vegetation.

Second, past historical practices are responsible for some of the best examples of woodlands today, like the travelling stock reserves set aside for the movement of sheep and cattle in an era before mechanised transport. Many authors have canvassed the need for effective protection of these wide (up to 1.6 km or more) reserves, some of the most valuable grassy woodlands for plant conservation and faunal habitat (e.g. chapters by Benson, Spooner, Lunt *et al.*, Michael and Lindenmayer). Third, the rapid rate of change means that many populations, communities and ecological processes are still in a state of flux (Watson, Chapter 17), still responding to aspects of human landscape modification (like land clearing) that took place years or even many decades ago. As a consequence, we are still to pay the 'extinction debt', with future losses of species (e.g. Benson, Chapter 40; Driscoll, Chapter 23; Ford, Chapter 25; Martin, Chapter 30) and changes in ecosystem function likely to occur in many regions over coming decades.

A temporal perspective is also essential in management and restoration. For example, Mac Nally (Chapter 12) highlighted the time lags associated with the development of key structural elements of woodland stands, like trees with hollows and large fallen logs. These attributes of woodlands are critical habitat components for many elements of the biota, but take a prolonged time to develop. Restoration of ground-layer plant communities also may take many years, particularly where the legacy of past land use is elevated levels of soil nutrients (such as phosphorous and nitrates), which place exotic plants at an advantage (Lunt *et al.*, Chapter 20; Dorrough, Chapter 17). Thus, a long-term approach is required for restoration and associated monitoring and research (Williams, Chapter 33), recognising that inherent natural variation in climate and extreme events may obscure ecological patterns and responses in the short term (Watson, Chapter 19).

Temporal changes in the human and social dimensions of woodland management have potentially profound implications for conservation. Lefroy (Chapter 6) pointed out the 'changing human face' of rural Australia, including a marked decline in recent decades in the proportions of landholders who are active farmers. This trend is most prominent within commuting distance of regional centres, where increasing numbers of properties are now purchased for lifestyle purposes. The implications of such ongoing changes in these areas, which have been termed 'transitional landscapes', are complex: changes in land ownership offer new opportunities in land management, but non-farmer owners are not necessarily more likely to undertake conservation activities (Lefroy, Chapter 6). Gilfedder (Chapter 37) and Kirkpatrick (Chapter 11) highlight the need to recognise the insights of landholders who have gained decades of management experience and knowledge of their land and district. This often contrasts with the short-term tenure of agency staff whose role is to liaise with landholders in natural resource management programs. Rapid turnover of agency staff and the lack of long-term continuity of many government programs (Williams, Chapter 33), make it difficult to build long-lasting relationships and the essential ingredient, trust (Park, Chapter 38), that are vital for bringing about lasting change in woodland management on private land.

3. A complementary understanding of pattern and process

As in most areas of ecology and ecological research, the vast majority of woodland research studies have documented the pattern of occurrence of particular organisms rather than the key ecological processes giving rise to those patterns. This is particularly true for woodland fauna, although work on species such as the Noisy Miner (*Manorina melanotis*) (Clarke and Grey, Chapter 14) and Red-capped Robin (*Petroica goodenovii*) (Major, Chapter 24) are valuable exceptions. Woodland plant ecology is more advanced in studying and understanding underlying ecological processes (e.g. Lunt *et al.*, Chapter 20; McIntyre, Chapter 27). Much more work needs to focus on the ecological processes giving rise to spatial or temporal patterns of both woodland plants and animals. This includes, for example, better understanding of the traits of species most vulnerable to decline and the mechanisms of such decline in modified landscapes (Driscoll, Chapter 23; Maron, Chapter 29); and the mechanistic relationships between essential resources (e.g. food, shelter, breeding sites) and the abundance of species in different woodland types and configurations (Watson, Chapter 19).

Such process-based work is important for many well known reasons, but three key ones are that: (1) it increases understanding of the generality of findings; (2) it improves the ability to make accurate predictions from one landscape or region to another; and (3) it creates an improved capacity to forecast biotic responses to management interventions and conservation actions. These advantages are particularly pertinent in temperate woodlands. Improvements in woodland vegetation condition (as well as corresponding responses of some elements of the biota) will benefit from a better understanding of ecological processes, to then guide appropriate management interventions such as the strategic application of fire (Kirkpatrick, Chapter 11, Lunt *et al.*, Chapter 20) or reducing excessive levels of soil nutrients (Dorrough, Chapter 17).

4. Woodland patches, linear networks and paddock trees are critical elements in woodland mosaics

Woodlands occur naturally as mosaics: the composition and structure of woodland vegetation is subtly influenced by topography, geology, soil nutrient status, climate, edaphic patterns, and disturbance regimes. Clearing and landscape change over the last 200 years has superimposed further, and visibly obvious, complexity on this environmental mosaic. Remnants of woodland in the form of patches (large and small), linear strips and scattered trees, along with remnant grasslands, rocky outcrops and wetlands, make up a suite of natural components amidst anthropogenic ones like croplands, exotic pastures, irrigated lands and replanted vegetation.

Many chapters discuss the importance of remnant vegetation and the relative value for biota of different kinds of remnants in the landscape (e.g. chapters by Eyre, Ford, Kirkpatrick, Lindenmayer *et al.*, MacAlpine, Majer *et al.*, Martin, Spooner, Watson). The conclusions of these authors sometimes differ, but a common theme is the critical need to retain what is left (and to increase it), given the extent that has been lost. Some of these differences relate to the taxa being studied. For example, Kirkpatrick (Chapter 11) noted that size, shape and duration of isolation had little influence on the flora of woodland remnants, with plant species composition more influenced by environmental conditions and management. Alternatively, while small and isolated woodland remnants do have value for vertebrate fauna (e.g. Ford, Chapter 25; Lumsden and Bennett, Chapter 16), greater emphasis is given to the spatial context of patches in the landscape and the amount of habitat they provide (e.g. Driscoll, Chapter 23; Lindenmayer *et al.*, Chapter 21; McAlpine *et al.*, Chapter 28; Martin, Chapter 30).

In many regions, strips of remnant vegetation along roadsides and streams form linear networks through otherwise cleared landscapes. While many of these strips are narrow and essentially 'edge', others may be hundreds of metres wide (e.g. travelling stock routes and some streamside strips). Many authors described the value of linear networks in supporting a wide range of taxa, including ants, beetles and other invertebrates (Majer *et al.*, Chapter 9; Major, Chapter 24); reptiles (Brown, Chapter 15; Driscoll, Chapter 23); birds (MacAlpine *et al.*, Chapter 28) and mammals (van der Ree, Chapter 18). Their ongoing value as habitat, however, is strongly influenced by the management regime they experience, and the consequent capacity to provide important components such as ground cover refuge for reptiles (Brown, Chapter 15) and large old trees for the Squirrel Glider, *Petaurus norfolcensis* (van der Ree, Chapter 18). Major (Chapter 24) cautions that we need to understand the population dynamics of species in such areas and to be sure they are not acting as demographic 'sinks'.

Several authors highlighted the range of ecological roles played by scattered trees (Manning and Fischer, Chapter 5; Paton *et al.*, Chapter 10). Yet it is also clear that these trees are widely targeted for clearing as part of agricultural development such as the establishment of centre-point pivot irrigation systems (Maron, Chapter 29). The loss of these trees as part of agricultural development is understandable but it clearly also has significant negative ecological consequences. Ways to resolve this 'wicked environmental problem' (*sensu* Ludwig *et al.* 2001) have yet to be adequately resolved and, again, this remains a topic for (urgent) further research. One proposed solution has been the use of offsets, but this approach needs to be applied in ways that truly leads to a no net loss of habitat for wildlife (Gibbons and Lindenmayer 2007).

5. The pervasive effects of the Noisy Miner, and other 'problem' native species

Landscape change has favoured some native species to the point that they become a management 'problem' due to their effects on other species. Examples noted by authors in this book include macropods which may over-browse ground-layer plants (Robinson, Chapter 39), the browsing impacts of the Common Brushtail Possum (*Trichosurus vulpecula*) (Kirkpatrick, Chapter 11) and, especially, the competitive effects of the hyper-aggressive Noisy Miner on other woodland birds. The problems created by the Noisy Miner were prominent in many chapters – both in terms of their effects on other bird species (Clarke and Grey, Chapter 14; Maron, Chapter 29; Major, Chapter 24), as well as their impacts on woodland vegetation, such as tree dieback (Clarke and Grey, Chapter 14; Kirkpatrick, Chapter 11). Notably, the Noisy Miner does not have a continent-wide distribution, but Recher *et al.* (Chapter 8) discussed the potential for equivalent problems created by the Yellow-plumed Honeyeater (*Lichenostomus ornatus*) in the woodlands of south-western Australia.

Clarke and Grey (Chapter 14) recommended that shooting the Noisy Miner was the most cost-effective and humane way to significantly reduce populations of this over-abundant native species. The scientific basis for doing this will need to be argued articulately by ecologists, resource managers and policy makers to ensure that the case for culling is not thwarted by ill-informed animal rights activists. In temperate woodlands of Tasmania, problems associated with tree dieback arise from the over-abundance of both the Noisy Miner and the Common Brushtail Possum (Kirkpatrick, Chapter 11). Kirkpatrick does not discuss the prospects of culling the Noisy Miner, but does recommend the need to either exclude or kill the Common Brushtail Possum to prevent tree death – although he cautions this should be done only with appropriate permits and where populations are large.

This example strongly emphasises the need for high quality scientific information to guide evidence-based conservation policy and on-ground management actions in what can be an emotionally charged social and political context. Like so many others throughout this book, it also underscores the fundamental importance of links between science, policy and management actions.

6. The challenge of active management and restoration

The need for active measures to manage and restore woodland ecosystems is a theme canvassed by many authors (e.g. see Hobbs, Chapter 7; Kirkpatrick, Chapter 11; Lunt *et al.*, Chapter 20; Manning and Fischer, Chapter 5; Martin, Chapter 30; Paton *et al.*, Chapter 10; Spooner, Chapter 22). Many woodland patches are only a skeleton of the original system as a consequence of decades of grazing, fertiliser use, tree clearing, loss of fallen timber, weed invasion, altered fire regimes and other impacts. Other woodlands may retain significant components of the flora or fauna, but their condition is deteriorating; while others may lack specific habitat components required by particular species (like shrub cover or fallen timber). In all of these situations, protection by changing the tenure or reservation status is not enough; active measures are needed.

Maintaining or restoring suitable disturbance regimes in woodlands, such as by the use of fire and herbivore grazing is advocated by several authors (Gilfedder, Chapter 37; Spooner, Chapter 220), but there remains considerable uncertainty about what a 'suitable' regime might be, how widely applicable it might be, and how it may affect different components of the biota. Kirkpatrick (Chapter 11) suggests there is a high degree of contingency in how species respond to variation in management regimes at different times in different places. For these reasons a number of authors argue for the need for 'management diversity', using different management approaches and actions in different locations (e.g. Dorrough, Chapter 17; Ford, Chapter 25). If we are to learn from these diverse approaches, it is essential they are coupled with effective monitoring programs (see Section 7 below). Further, there is a critical need for carefully designed research to test different management and restoration techniques. Manipulative experimental studies are a powerful approach; such as work on the relationship between soil nutrients and plant species composition in grassy woodlands reported by Lunt *et al.* (Chapter 20), or the effect on woodland fauna of manipulating levels of woody debris (Mac Nally, Chapter 12). Other research approaches include observational studies that stratify by different management treatments, such as Martin's (Chapter 30) work isolating the relative effects of stock grazing and tree clearing on woodland birds.

Which woodlands should receive the highest priority for investment in management and restoration? This question, of immediate relevance to investors and management agencies, is not straightforward (Hobbs, Chapter 7). Two perspectives are given here by authors, largely representing different ways of framing the debate. The first (e.g. Spooner, Chapter 22) argues that we must protect the best patches of woodland, those of highest quality where plant communities are most intact – as such woodlands are now scarce. An alternative view (e.g. Gibbons, Chapter 3) argues that we should invest where scarce conservation dollars will achieve the greatest gain, and suggests that this is more likely to be in remnants currently in moderate to poor condition. As identified by Gibbons (Chapter 3), a key knowledge gap that will help resolve these differing perspectives is a better understanding of the trajectory of woodland systems under the *status quo*. Other issues are also relevant. Vegetation types disproportionately lost by clearing, such as those on fertile soils, are a high priority (Mac Nally, Chapter 12). Likewise, social factors, such as the willingness and commitment of individuals or communities to be involved will influence priorities for restoration.

In addition to restoration of *existing* woodlands, active measures are needed to 'put back' a substantial proportion of that which has been lost – either by natural regeneration or replanting. Much effort has gone into improved technical knowledge of how best to successfully re-establish vegetation (Carr *et al.*, Chapter 2), and there is increasing recognition of guidelines such as the use of indigenous species, matching species to landform, enhancing structural complexity, and so on. Knowledge of the value of revegetation at the landscape scale, and its complementary role to that of remnant woodland, also is increasing (Lindenmayer *et al.*,

Chapter 21). Here too, however, we need to employ a diversity of approaches (including monitoring and experimental management) to test the relative benefits of different spatial patterns of revegetation. Streatfield *et al.* (Chapter 4) discuss the benefits of 'whole paddock rehabilitation' as compared to revegetation in small blocks or strips. Driscoll (Chapter 23) recommends the re-establishment of large reserves (e.g. of at least 200–300 ha and more) at regular intervals across rural landscapes to account for regional species turnover and to conserve intraspecific genetic variation.

7. The essential need to do effective monitoring and gather long-term data

Several authors lamented the dearth of long-term ecological monitoring in woodland conservation and management (e.g. Hobbs, Chapter 7; Williams, Chapter 33; Spooner, Chapter 22; Park, Chapter 38; Eyre, Chapter 41). The paucity of monitoring and resulting long-term data have made it impossible to gauge the effectiveness of billions of dollars of investment through major government programs like the Natural Heritage Trust (Halkowicz 2009). We add to the chorus of concern from many ecologists and resource managers that high quality, long-term monitoring data for woodland ecosystems are extremely sparse. Indeed, it would be fair to say that this fundamental problem has featured as a sadly recurring theme at what must be hundreds of workshops around Australia on woodland conservation and management over the past 10–15 years. This problem will not be addressed until scientists, policy makers, regional NRM bodies and State and Federal Governments take monitoring and long-term research seriously – and make serious investments to gather the data over prolonged periods that are needed to make informed decisions and informed policy.

Long-term monitoring data, such as that based on well designed longitudinal studies, experiments and natural experiments, are particularly important in woodland environments for two key reasons. First, woodland environments are notorious for considerable inter-annual variability in rainfall and to a lesser extent, temperature. Such variation can, in turn, have profound effects on many elements of the biota (Watson, Chapter 19; Recher *et al.* Chapter 8). Long-term data are essential to tease apart seasonal variations from other longitudinal trends, such as increases or decreases of animal and plant populations. Second, as outlined above, management interventions such as those involving the application of fire, the control of grazing pressure, the manipulation of soil nutrients, and ecological thinning of stands will often be critical for improved conservation outcomes in woodlands (Kirkpatrick, Chapter 11; Lunt *et al.*, Chapter 20), including those made in major programs like woodland stewardship schemes (Dorrough, Chapter 17). Long-term data are essential to quantify the effectiveness of these (and other) management interventions, and to demonstrate such outcomes to those who have undertaken the work. For example, Robinson (Chapter 39) noted that the regular monitoring of the Grey-crowned Babbler (*Pomatostomus temporalis*) combined with annual feedback to landholders that their actions have made a real difference to the species, has been a powerful motivator in the community. Data from long-term monitoring programs are also critical for evaluating whether particular management priorities are appropriate ones or whether they need to be altered (Gibbons, Chapter 3; Martin, Chapter 30).

8. Beyond reserves – woodland conservation and management on private land

A traditional realm of conservation biology has been the dedicated conservation management of large intact areas of native vegetation and large ecological reserves, and the development of methods to efficiently select complementary new reserves (Driscoll, Chapter 23; Recher *et al.*, Chapter 8). These activities remain important in all environments, including those dominated by temperate woodland. Other approaches, however, particularly those associated with off-reserve management, are needed in temperate woodlands. The vast majority of land is in private ownership and the remaining native vegetation primarily exists as small patches, but these nevertheless have considerable value for the conservation of plants (Lunt *et al.*, Chapter 20; Kirkpatrick, Chapter 11) and animals (Lindenmayer *et al.*, Chapter 21; Major, Chapter 24; Bennett and Radford, Chapter 13; Ford, Chapter 25). The success or failure of conservation in temperate woodlands will largely depend on the extent to which conservation measures can be integrated with the management of land for agricultural production.

The authors of chapters in this book have highlighted several broad issues which are central to the challenge of effective nature conservation in landscapes predominantly held in private ownership. A key question from an ecological perspective is: 'What is the best way to do wildlife-friendly farming' (Fischer *et al.* 2008)? This, in turn, leads to questions about the most appropriate scale/s at which to integrate conservation and agricultural production – within paddocks, between paddocks, across whole farms, or even at landscape scales. The answer to this critical problem is not known: it is an outstanding research topic that needs to be addressed urgently. The solution will involve a balance between different kinds of conservation measures that complement each other across different scales (Bennett and Radford, Chapter 13). Experience in forest environments has demonstrated the need for conservation at a range of spatial scales and that management actions need to be guided by well articulated questions and context-specific considerations like the kinds of biota, landscape characteristics and key ecological processes, such as the scale, frequency and severity of disturbance regimes (Lindenmayer and Franklin 2002).

A second issue for conservation on private land revolves around the kinds of institutional mechanisms that will be most effective in bringing about change at the scale required to make a difference. Many approaches and programs have been tried, variously based on information and education, incentive schemes, stewardship approaches, covenants and property agreements, legislative controls, land purchase and, more recently, market-based incentive approaches (see Park, Chapter 38; Pearson and Donaldson, Chapter 32; Streatfield *et al.*, Chapter 4; Zammit *et al.*, Chapter 35). While these programs are admirable and have achieved some gains, Lefroy (Chapter 6) argues that awareness, moral persuasion and public programs will never be sufficient at the scale required. He advocates the need for conservation farming practices that deliver private as well as public good; or, as Gilfedder (Chapter 37) expresses it, conservation needs to become a viable business enterprise and deliver a comparable income stream to that from alternative uses of the land. The challenge is to move from small-scale actions measured in hectares or tens of hectares, to approaches that enhance conservation outcomes over thousands and tens of thousands of hectares. A mix of approaches and programs will be needed, as well as novel mechanisms such as an ecosystem services framework and payment scheme as suggested by Williams (Chapter 33).

Several authors (e.g. Gilfedder, Chapter 37; Robinson, Chapter 39; Park, Chapter 38; Streatfield, Chapter 4) emphasise the importance of community engagement, and the necessity for conservation programs to align with the interests of landholders and local communities if gains are to be made in conservation of woodlands on private land. This leads to the broader question of links across disciplines, institutions and community, which we discuss below.

9. The criticality of cross-disciplinary linkages

A consistent and recurring theme is the need for cross-disciplinary linkages and dialogue between scientists, woodland managers and policy-makers. Indeed, it was the primary topic in contributions from several authors (e.g. Williams, Chapter 33; Person and Donaldson, Chapter 32; Gilfedder, Chapter 37; Paton *et al.*, Chapter 10). This critical need was expressed by those who are primarily researchers, as well as those who are primarily resource managers and policy makers.

A number of reasons were given for why cross-disciplinary linkages must be strengthened. First, many authors argued that it is fundamental for scientific knowledge and research findings to being more effectively incorporated into on-ground management of woodlands and policy development (Williams, Chapter 33; Pearson and Donaldson, Chapter 32). Lefroy (Chapter 6) considered that cross-disciplinary links were essential to advance conservation and management efforts quickly and more effectively. Second, some authors argued that because woodland conservation efforts require targeted management interventions, greater dialogue between practitioners and scientists is essential to guide these interventions. For example, success in several woodland projects was attributed to strong connections between researchers and landholders (e.g. Kirkpatrick, Chapter 11; Robinson, Chapter 39; Major, Chapter 24).

A third reason for strong links between scientists, policy-makers and resource managers is the need to support stakeholders in dealing with the rapid shifts in how Australians value woodland environments. Indeed, it has been only a matter of decades between the time when government policies offered tax incentives to undertake widespread land clearing, to now, when landholders are paid to replant or regenerate the trees they cleared and to undertake management to improve woodland condition. Cross-disciplinary links are critical so that the scientific and policy reasons for such rapid and radical changes are clearly and repeatedly communicated.

The importance of cross-disciplinary links demands more, and better, communication between practitioners and scientists to improve mutual understanding of the different work and management cultures, demands and approaches of different groups of people. Developing partnerships is essential because the different groups speak different 'languages', have different work cultures, different reward systems, and different sets of skills (Lindenmayer *et al.* 2008). Trust and mutual respect is essential; in part because various kinds of knowledge – scientific, policy and political knowledge - together influence decision-making in natural resource management (Pielke 2007). Hence, for example, good science alone is insufficient as often it will need to be translated into a form that policy makers and resource managers can understand and use (King 2004). Increasingly, there is a role for people with specialist skills to foster such cross-disciplinary linkages; for example, 'knowledge brokers' who foster partnerships and integrate and translate knowledge from a range of disciplines (Park, Chapter 38).

Other aspects of cross-disciplinary work also appear to be important for improved woodland conservation and management. For example, there is scope for greater learning from ecological research and management in other ecosystems. Williams (Chapter 30) points out the value of the long-term monitoring programs in Australia's rangeland environments (e.g. Bastin 2008). Forests provide another example: there has been a much longer and more extensive history of research in forests than in woodlands. Part of that research effort has highlighted the critical role of large trees for biodiversity conservation (Gibbons and Lindenmayer 2002). The role of large trees in conserving woodland biota, such as scattered paddock trees or trees along roadsides, is now being widely recognised (Manning and Fischer, Chapter 5; Lumsden and Bennett, Chapter 16; van der Ree, Chapter 18). We believe there is value in examining environments other than woodlands to identify insights which benefit the management of woodland environments.

10. Emerging issues

The final theme in this chapter focuses on what we perceive to be emerging and potential future issues in woodland conservation and management. First, it appears to us that woodland conservation will require continued (and likely increasing) focus on non-traditional areas of conservation biology, particularly off-reserve management. This will need to include the conservation (and sometimes highly targeted management) of small patches of remnant native vegetation and sometimes even the maintenance of individual structures (e.g. paddock trees). We have alluded to this earlier in our commentary on the integration of conservation and agricultural production and we reiterate that a future area of work must be to identify appropriate scale/s for effective integration. There will, however, be no 'magic bullets' for effective integration as a range of conservation, management and policy approaches will be needed (McAlpine *et al.*, Chapter 28) – reserve and off-reserve conservation (including targeted management of old growth woodland, regrowth, plantings) and facilitated through volunteer agreements, stewardship programs and other market-based instruments.

The general perspectives of temperate woodlands are ones of small, heavily modified and highly degraded patches with a greatly impoverished biota. Such kinds of woodlands are undoubtedly depressingly widespread. Clearly then, change is needed in the way remaining areas of temperate woodland are managed. Indeed, degraded temperate woodland remnants may well be those where management interventions and conservation investments lead to the greatest improvements in vegetation condition and the greatest conservation gains (see Gibbons, Chapter 3). We agree with the incisive comments of Lefroy (Chapter 6) that there is an urgent need to document the characteristics of approaches and programs that lead to change. This underpins the basis for knowledge adoption, altered management practices, enhanced policy making, and increasing the pace and effectiveness of all of these (Pearson and Donaldson, Chapter 32). There is also a critical need to highlight conservation success stories, like the value of planted areas for biodiversity conservation (Cunningham *et al.* 2008) including such high profile species as the Grey-crowned Babbler (Robinson, Chapter 39). These stories are essential to inspire much needed additional management and restoration efforts and increased levels of investment and community support for such efforts.

Although small, heavily modified and highly degraded woodland patches may be the norm, there are important exceptions. Part of an approach to improving woodland management outcomes is for researchers and managers to see and study these places to help guide conservation and management efforts. In an example from an entirely different ecosystem, forest ecologists in Sweden took forest managers from that country to unmodified forests in Russia to show them the kinds of structures that needed to be restored in Swedish forests as part of recovering threatened biota (Angelstam, Majewski and Bondrup-Neilsen 1995).

Intact or relatively intact Australian temperate woodland are valuable places to: (1) provide reference conditions against which to benchmark the extent and effectiveness of changes elicited by management intervention; (2) help generate and further refine conceptual models of how woodland ecosystems function; (3) help document the magnitude of change that has occurred as a result of human activities in temperate woodland ecosystems; and (4) provide ecological guideposts and goals for woodland restoration efforts as well as re-vegetation programs. For example, the massive areas of woodlands in south-western Western Australia, including large areas with limited grazing history (Bradby, Chapter 36; Recher *et al.*, Chapter 8) provide extraordinary opportunities for ecological and management learning (Recher *et al.*, Chapter 8). They are worthy of comprehensive conservation for these reasons alone.

Almost all informed discussions of environmental management and biodiversity conservation now encompass concerns about rapid future climate change. Discussions about woodland conservation and management are no different in this regard (Hobbs, Chapter 7; Lunt *et al.*, Chapter 20). Recher *et al.* (Chapter 8), for example, discussed how even relatively small changes in the moisture levels of the litter layer could have major consequences for invertebrates and, in turn, those taxa associated with them, like insectivorous birds. Of course, there may well be many other kinds of changes in woodland ecosystems resulting directly or indirectly from rapid climate change (see Steffen *et al.* 2009, for a recent Australian review) and it is beyond the scope of this book to examine them. However, the prospects of rapid climate change create not only problems but also opportunities. For example, carbon sequestration is widely proposed as a key part of reducing levels of greenhouse gases in the terrestrial biosphere. Freudenberger (Chapter 34) discussed the enormous opportunities for extensive and environmentally informed revegetation programs that could be created by global and national carbon markets. These kinds of changes stimulated by carbon markets also have the potential to lead to changes in the socio-economic fabric of rural communities, such as generating additional and alternative income streams. Current massive levels of under-investment in woodland environments (Pearson and Donaldson, Chapter 32; Mac Nally, Chapter 12) may be, in part, addressed through the financial opportunities associated with carbon markets, with large benefits for improved biological conservation and significant rural social renewal (Bekessy and Wintle 2008). Large-scale carbon sequestration programs, if they eventuate, may usher changes in the kinds of outputs generated from farms (Freudenberger, Chapter 34), much like environmental stewardship programs aim to make improve conservation outcomes a new kind of farm output (Zammit *et al.*, Chapter 35).

It is important not to ignore the ultimate driver of almost all environmental and conservation management problems – the size and levels of resource consumption of human populations. This is surprisingly often overlooked by researchers, policy makers and resource managers alike, although it was explicitly recognised in several chapters in this book (Gibbons, Chapter 3; Martin, Chapter 30; Bennett and Radford, Chapter 13). It is particularly pertinent because it has recently been projected that Australia's population will nearly double to 35 million by 2050. This will exert considerable additional pressure on land and water resources. For example, it may magnify competition for water and compromise environmental flows (e.g. for wetting programs for woodland environments such as those dominated by Black Box *E. largiflorens* and River Red Gum *E. camaldulensis*). There also may be increased pressure to intensify agricultural production with negative impacts on biota (see Maron, Chapter 29; Lunt *et al.*, Chapter 20), particularly given risks of global food shortages in coming decades. Some of the projected increase in population size will lead to urban expansion at the peri-urban margins of major cities and regional centres. This will create further pressure to clear existing woodlands, including stands of threatened or endangered ecological communities. We believe that the environmental consequences of large and rapid expansion of Australia's population need detailed analysis and closer scrutiny (e.g. see Foran and Poldy 2002), particularly given the major problems that have ensued from a current population size that is far smaller than the one likely to be occupying the continent in four decades from now.

Finally, an important message in the chapter by McIntyre (Chapter 27) was the critical need to take the time for recreation and reflection in temperate woodlands, including thinking laterally about their spatial and temporal patterns and dynamics and the underlying processes that give rise to these patterns. Such lateral thinking led to exciting new discoveries such as those of models of landscape cover (McIntyre and Barrett 1992; McIntyre and Hobbs 1999). The ability to generate these insights is in danger of being lost amid the epidemic of busyness that pervades modern research and policy-making life. All of us need to take steps to ensure that the ability and opportunities to make these discoveries is not lost.

References

Angelstam, P., Majewski, P. and Bondrup-Neilsen, S. (1995). West-East Cooperation in Europe for sustainable Boreal forests. *Water, Air and Soil Pollution* **82**, 3–11.

Bastin, G. (2008). Rangelands 2008 – Taking the pulse. In *National Land and Water Resources Audit*. Commonwealth of Australia: Canberra.
Bekessy, S.A. and Wintle, B.A. (2008). Using carbon investment to grow the biodiversity bank. *Conservation Biology* **22**, 510–513.
Cunningham, R.B., Lindenmayer, D.B., McGregor, C., Crane, M. and Michael, D. (2008). The combined effects of remnant vegetation and replanted vegetation on farmland birds. *Conservation Biology* **22**, 742–752.
Fischer, J., Brosi, B., Daily, G., Ehrlich, P., Goldman, R., Goldstein, J., Lindenmayer, D.B., Manning, A., Mooney, H., Pejchar, L., Ranganathan, J. and Tallis, H. (2008). Should agricultural policies encourage land sparing or wildlife-friendly farming? *Frontiers in Ecology and Environment* **6**, 380–385.
Foran, B. and Poldy, F. (2002). Future dilemmas. Options to 2050 for Australia's population, technology, resources and environment. In *Working Paper 02/01*. CSIRO Sustainable Ecosystems: Canberra.
Gibbons, P. and Lindenmayer, D.B. (2002). *Tree Hollows and Wildlife Conservation in Australia*. CSIRO Publishing: Melbourne.
Halkowicz, S. (2009). The evolution of Australia's natural resource management programs: Towards improved targeting and evaluation of investments. *Land Use Policy* **26**, 471–478.
King, L. (2004). Impacting policy through science and education. *Preventative Veterinary Medicine* **62** 185–192.
Lindenmayer, D., Hobbs, R.J., Montague-Drake, R., Alexandra, J., Bennett, A., Burgman, M., Cale, P., Calhoun, A., Cramer, V.A., Cullen, P., Driscoll, D., Fahrig, L., Fischer, J., Franklin, J., Haila, Y., Hunter, M., Gibbons, P., Lake, S., Luck, G., MacGregor, C., McIntyre, S., Mac Nally, R., Manning, A., Miller, J.R., Mooney, H.A., Noss, R., Possingham, H., Saunders, D.A., Schmiegelow, F., Scott, M., Simberloff, D., Sisk, T., Tabor, G., Walker, B.H., Wiens, J., Woinarski, J. and Zavaleta, E. (2008). A checklist for ecological management of landscapes for conservation. *Ecology Letters* **11**, 78–91.
Pielke, R.A. (2007). *The Honest Broker. Making Sense of Science in Policy and Politics* Cambridge University Press: Cambridge, England.
Steffen, W., Burbidge, A., Hughes, L., Lindenmayer, D.B., Musgrave, W., Stafford-Smith, M. and Werner, P. (2009). *Australia's Biodiversity and Climate Change*. CSIRO Publishing: Melbourne.

Index

www.ingramcontent.com/pod-product-compliance
Ingram Content Group UK Ltd.
Pitfield, Milton Keynes, MK11 3LW, UK
UKHW061956041026
15591UKWH00016B/1029

9 780643 100374